BLACK HOLES

NEW HORIZONS

BLACK HOLES
NEW HORIZONS

Editor

Sean Alan Hayward

Shanghai Normal University, China

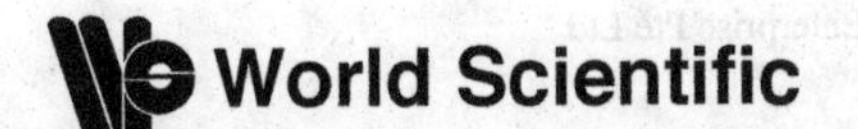

NEW JERSEY • LONDON • SINGAPORE • BEIJING • SHANGHAI • HONG KONG • TAIPEI • CHENNAI

Published by

World Scientific Publishing Co. Pte. Ltd.
5 Toh Tuck Link, Singapore 596224
USA office: 27 Warren Street, Suite 401-402, Hackensack, NJ 07601
UK office: 57 Shelton Street, Covent Garden, London WC2H 9HE

British Library Cataloguing-in-Publication Data
A catalogue record for this book is available from the British Library.

BLACK HOLES
New Horizons

ISBN 978-981-4425-69-8

Printed in Singapore by B & Jo Enterprise Pte Ltd

Preface

This volume collects together summaries of work on local (more correctly quasi-local) approaches to black holes, over fields as diverse as classical and numerical general relativity (GR), differential geometry, quantum field theory (QFT) and quantum gravity. Each chapter is written by an expert or experts on a particular topic. It is intended for anyone interested in black holes, as a reference for the many applications and issues.

Existing textbooks define a black hole by an *event horizon*. However, its teleological nature means that it is epistemologically unsound, empirically unverifiable, even theoretically impractical and not directly related to local physics. The last decade or so has seen increasing evidence that more local notions are more useful and have understandable associated physics. There is heretofore no one volume which covers all the main aspects, hence the timeliness of this attempt.

The key idea at the core of almost all such local approaches to black holes is a *marginal surface*, a.k.a. marginally trapped surface, where a light wavefront is instantaneously trapped by the gravitational field. If such surfaces sweep out a hypersurface in space-time, this is a *trapping horizon* in the author's terminology. The earliest definition is an *apparent horizon*, which however was defined in an indirect way, then sketchily proved to be a marginal surface. Many authors continue to use "apparent horizon" to mean a marginal surface, with or without extra conditions, sometimes not clearly defined.

Some extra conditions are needed to separate black holes from, say, white holes, wormholes or cosmological horizons, and here is where definitions diverge. The author proposed a *future outer* trapping horizon, intended to capture the idea of a just-trapped outgoing wavefront with future trapped surfaces inside. This has the advantage that one may easily prove some expected properties of black holes, namely topology, causal nature and an area-increase law. In numerical relativity, usually one works with a foliation by time, with marginal surfaces located by where an outgoing light-ray expansion changes sign in the right direction within the chosen constant-time hypersurface. This has the advantages of practical locatability, and computability of physical quantities such as mass and angular momentum. Included here are a comprehensive review (chapter 1) and discussion of physical aspects (chapter 2). A precise definition is a *stable MOTS*, for which existence and uniqueness theorems may be proved (chapter 3). Another definition is *dynamical horizons*, which are spatial future trapping horizons, using more intrinsic characterizations of the foliated hypersurface. It should be stressed that while all these definitions are inequivalent in general, in practical situations such as a numerical black-hole evolution, they are likely to agree, except where the horizon is not outer in one sense or another.

Then there are special cases, such as slowly *evolving horizons*, describing black holes near equilibrium (chapter 4). Specializing further, there is a hierarchy of *isolated horizons*, which are null trapping horizons with additional conditions to describe a black hole in equilibrium to some degree (chapter 5). A key advantage here is that they provide a comparatively simple arena for ideas from quantum gravity, including derivations of black-hole entropy from loop quantum gravity. At the most restricted level, there is the much earlier idea of *Killing horizons* of stationary black holes, which allows comparatively simple study of QFT on such backgrounds, relating to thermodynamics (chapter 6).

An introduction to relativistic thermodynamics is also included (chapter 7). This should at least make clear the difference between the first law, expressing energy conservation, and the Gibbs equation (away from equilibrium), and the correct form of the second law as expressing entropy production, not entropy increase (for an open system), both of which are still routinely confused in work on black holes.

Quite recently it has been found that the relation between *trapped surfaces* and marginal surfaces is more complex than previously thought (chapters 8 and 9). This led to yet another definition, the *core* of a black hole.

It should also be mentioned that a great deal of work on alternative theories has been omitted here, due to the current state of observational evidence as consistent with standard GR and QFT, and that the basic ideas mostly generalize straightforwardly.

Twenty years ago, almost everyone accepted what the author has previously called the 1970s paradigm for black holes, applying mostly to Killing horizons, with event horizons as the general definition, leading to practical difficulties, a general lack of physically interpretable quantities and equations, and even a so-called paradox concerning information. By now, the change has been immense. As the contributions here show, the question is no longer whether local approaches to black holes are useful, but which is most useful for a given purpose, and how to understand issues in ways that are geometrically clear, physically interpretable and practically implementable.

Sean Hayward
2012 April 28th, Shanghai

Contents

Chapter 1

An Introduction to Local Black Hole Horizons in the 3+1 Approach to General Relativity*

José Luis Jaramillo
Max-Planck-Institut für Gravitationsphysik, Albert Einstein Institut
Am Mühlenberg 1, Golm D-14476, Germany
Jose-Luis.Jaramillo@aei.mpg.de

We present an introduction to dynamical trapping horizons as quasi-local models for black hole horizons, from the perspective of an Initial Value Problem approach to the construction of generic black hole spacetimes. We focus on the geometric and structural properties of these horizons aiming, as a main application, at the numerical evolution and analysis of black hole spacetimes in astrophysical scenarios. In this setting, we discuss their dual role as an *a priori* ingredient in certain formulations of Einstein equations and as an *a posteriori* tool for the diagnosis of dynamical black hole spacetimes. Complementary to the first-principles discussion of quasi-local horizon physics, we place an emphasis on the *rigidity* properties of these hypersurfaces and their role as privileged geometric probes into near-horizon strong-field spacetime dynamics.

1. Black Holes: Global vs (Quasi-)Local Approaches

1.1. *Establishment's picture of the gravitational collapse*

Our discussion is framed in the problem of gravitational collapse in General Relativity. The current understanding is summarized in what one could call the *establishment's picture* of gravitational collapse,[1] a heuristic chain of results and conjectures:

(1) *Singularity theorems*: If gravity is able to make all light rays locally *converge* (namely, if trapped surfaces exist), then a spacetime singularity forms.[2–5]
(2) (*Weak*) *Cosmic censorship* (Conjecture): In order to preserve predictability, the formed singularity is not visible for a distant observer.[6]
(3) *Black hole spacetimes stability* (Conjecture): General Relativity gravitational dynamics drives eventually the black hole spacetime to a stationary state.
(4) *Black hole uniqueness theorem*: The final state is a Kerr black hole spacetime.[7]

*This chapter was first published in International Journal of Modern Physics D, Vol. 20, No. 11 (2011) 2169–2204

Light bending is a manifestation of spacetime curvature and black holes constitute a dramatic extreme case of this. The standard picture of gravitational collapse above suggests two (complementary) approaches to the characterization of black holes:

(a) *Global approach*: (weak) cosmic censorship suggests black holes as *no-escape* regions not extending to infinity. Its boundary defines the event horizon $\mathscr{E}$.
(b) *Quasi-local approach*: Singularity theorems suggest the characterization of a black hole as a spacetime *trapped region* where all light rays locally converge.

The *establishment's picture* of gravitational collapse depicts an intrinsically dynamical scenario. Hence, a systematic methodology to the study of dynamical spacetimes is needed. We adopt an *Initial (Boundary) Value Problem* approach, which offers a systematic avenue to the *qualitative* and *quantitative* aspects of generic spacetimes.

1.2. *The black hole region and the event horizon*

The traditional[5] approach to black holes involves global spacetime concepts, in particular a good control of the notion of infinity. Given a (strongly asymptotically predictable) spacetime $\mathcal{M}$, the *black hole region* $\mathscr{B}$ is defined as $\mathscr{B} = \mathcal{M} - J^-(\mathscr{I}^+)$, where $J^-(\mathscr{I}^+)$ is the causal past of future null infinity $\mathscr{I}^+$. That is, $\mathscr{B}$ is the spacetime region that cannot communicate with $\mathscr{I}^+$.

We are particularly interested in characterizing a notion of boundary *surface* of black holes. In this global context this is provided by the event horizon $\mathscr{E}$, defined as the boundary of $\mathscr{B}$, that is $\mathscr{E} = \partial J^-(\mathscr{I}^+) \cap \mathcal{M}$. Interesting geometric and physical properties of the event horizon are: (i) $\mathscr{E}$ is a null hypersurface in $\mathcal{M}$; (ii) it satisfies an *Area Theorem*,[8,9] so that the area of spatial sections $\mathcal{S}$ of $\mathscr{E}$ does not decrease in the evolution and, beyond that, (iii) a set of *black hole mechanics* laws are fulfilled.[10]

However, the global aspects of the event horizon also bring difficulties: (a) it is a *teleological* concept, i.e. the knowledge of the full (future) spacetime is needed in order to locate $\mathscr{E}$, and (b) the black hole region and the event horizon can enter into flat spacetime regions. In sum, the notion of event horizon is too global: it does not fit properly into the adopted Initial Value Problem approach.

1.3. *The trapped region and the trapping boundary*

The global approach requires controlling structures that are not accessible during the evolution. In this context, the seminal notion of *trapped surface*[2] plays a crucial role, capturing the idea that all light rays emitted from the surface locally converge. Through the singularity theorems and weak cosmic censorship, it offers a benchmark for the existence of a black hole region: in strongly predictable spacetimes with proper energy conditions, trapped surfaces lie inside the black hole region.[5] Moreover, their location does not involve a whole future spacetime development.

1.3.1. *Trapped and outer trapped surfaces. Apparent horizons*

Given a closed spatial surface $\mathcal{S}$ in the spacetime, we can consider the light emitted from it along outer and inner directions given, respectively, by future null vectors ℓ^a and k^a. Then, light locally converges (in the future) at $\mathcal{S}$ if the area of the emitted light-front spheres decreases in both directions (see though Ref. 11). Denoting the area element of $\mathcal{S}$ as $dA = \sqrt{q}d^2x$, the infinitesimal variations of the area along ℓ^a and k^a define outgoing and ingoing expansions $\theta^{(\ell)}$ and $\theta^{(k)}$ (see Sec. 2.1 for details)

$$\delta_\ell \sqrt{q} = \theta^{(\ell)} \sqrt{q}, \quad \delta_k \sqrt{q} = \theta^{(k)} \sqrt{q}. \tag{1}$$

A trapped surface is characterized by $\theta^{(\ell)}\theta^{(k)} > 0$. In the black hole context, in which the singularity occurs in the future, we refer to $\mathcal{S}$ as a future *trapped surface* (TS) if $\theta^{(\ell)} < 0, \theta^{(k)} < 0$ and as future *marginally trapped surface* (MTS) if one of the expansions, say $\theta^{(\ell)}$, vanishes: $\theta^{(\ell)} = 0, \theta^{(k)} \leq 0$. If a notion of *naturally expanding* direction for the light rays exists (e.g. in isolated systems, the *outer* null direction ℓ^a pointing to infinity), a related notion of *outer trapped surface* is given[5] by $\theta^{(\ell)} < 0$. *Marginally outer trapped surfaces* (MOTS) are characterized by $\theta^{(\ell)} = 0$.

Before proceeding to a characterization of black holes in terms of trapped surfaces, let us consider trapped surfaces from the perspective of a spatial slice of spacetime Σ. The *trapped region in* Σ, $\mathcal{T}_\Sigma \subset \Sigma$, is the set of points $p \in \Sigma$ belonging to some (outer) trapped surface $\mathcal{S} \subset \Sigma$. The Apparent Horizon (AH) is then the outermost boundary of the trapped region $\mathcal{T}_\Sigma$. A crucial result is the following characterization[5,12,13] of AHs: if the trapped region $\mathcal{T}_\Sigma$ in a slice Σ has the structure of a manifold with boundary, the AH is a MOTS, i.e. $\theta^{(\ell)} = 0$.

Given a 3+1 foliation of spacetime $\{\Sigma_t\}$, let us consider the worldtube obtained by piling up the two-dimensional AHs $\mathcal{S}_t \subset \Sigma_t$. Such an AH-worldtube does not need to be a smooth hypersurface (it is not even necessarily continuous, as discussed in Sec. 5.1.1). This is our first encounter with the notion of a spacetime worldtube foliated by MOTS. Though these worldtubes are slicing-dependent, their characterization in terms of MOTSs makes them very useful from an operational perspective.

1.3.2. *The trapped region: Definition and caveats*

From a spacetime perspective, no reference to a slice Σ must enter into the characterization of the trapped region. The spacetime *trapped region* $\mathcal{T}$ is defined as the set of points $p \in \mathcal{M}$ belonging to some trapped surface $\mathcal{S} \subset \mathcal{M}$. Its boundary is referred[14] to as the *trapping boundary.* These concepts offer, in principle, an intrinsically quasi-local avenue to address the notion of black hole region and black hole horizon, with no reference to asymptotic quantities.

In spite of their appealing features, there are also important caveats associated with the *trapped region* and the *trapping boundary.* In particular, we lack an operational characterization of the *trapping boundary* (see also the contribution by

J. M. M. Senovilla[15]). A systematic attempt to address this issue is provided by the notion of *trapping horizon*,[14] namely smooth worldtubes of MOTS (see Sec. 2.2), as a model for the trapping boundary. Trapping horizons, that are nonunique, have led to important insights into the structure of the trapped region, though an operational characterization of the *trapping boundary* is still missing.

The difficulties are illustrated in the discussion of the relation between the trapping boundary and $\mathscr{E}$. In strongly predictable spacetimes with appropriate energy conditions (see, though Ref. 16), the trapped region $\mathcal{T}$ is contained in the black hole region $\mathscr{B}$. In attempts to refine this statement, support was found[17,18] suggesting that the trapping boundary actually coincides with the event horizon, though later work[19] showed that the trapped region not always extends up to $\mathscr{E}$. The question is still open for (*outer*) *trapped regions* constructed on *outer* trapped surfaces, rather than on TSs. Important insight into these issues has been gained in recent works[20,21] demonstrating truly global features of the trapped region $\mathcal{T}$. In particular:

(i) The trapping boundary cannot be generically foliated by MOTS.
(ii) Closed trapped surfaces can enter into the flat region. This is an important issue in this approach to black holes, since it was a main criticism in Sec. 1.2.
(iii) Closed trapped surfaces are *clairvoyant*, that is, they are *aware* of the geometry in noncausally connected spacetime regions. This nonlocal property challenges their applicability in an operational characterization of black holes.

1.4. *A pragmatic approach to quasi-local black hole horizons*

Trapping horizons offer a sound avenue towards the quasi-local understanding of black hole physics. They provide crucial insight in gravitational scenarios where a quasi-local notion of black hole horizon is essential, such as black hole thermodynamics beyond equilibrium, the characterization of physical parameters of strongly dynamical astrophysical black holes (notably in numerical simulations), semi-classical collapse, quantum gravity or mathematical relativity (cf. A. Nielsen's contribution[22]). But, on the other hand, issues like their nonuniqueness or the *clairvoyant* properties of trapped surfaces pose fundamental questions that cannot be ignored.

We do not aim here at addressing first-principles questions about the role of trapping horizons as a characterization of black hole horizons. We rather assume a *pragmatic approach* to the study of gravitational dynamics, which underlines the role of trapping horizons as hypersurfaces of remarkable geometric properties in black hole spacetimes. More specifically, our main interests are:

(i) The construction and diagnosis of black hole spacetimes in an Initial (Boundary) Value Problem approach.
(ii) Identification of a geometric probe into near-horizon spacetime dynamics.

Point (ii) is particularly important in the study of gravity in the strong-field regime, where the lack of *rigid structures* (e.g. symmetries, a *background* spacetime ...) is a generic and essential problem. Given our interests and the adopted pragmatic methodology, we look for a geometric object such that: (a) it represents a footprint of black holes, providing a probe into their geometry; (b) it is adapted, by construction, to an Initial-Boundary Value Problem approach; and (c) although not necessarily unique, it provides a geometric structure with some sort of *rigidity* property. As we shall see in the following, dynamical trapping horizons fulfill these requirements.

1.5. *General scheme*

In Sec. 2 we introduce the basics of the geometry of closed surfaces in a Lorentzian manifold and motivate quasi-local horizons in stationary and dynamical regimes. Section 3 reviews their geometric properties and their special features as *physical* boundaries. Sections 4 and 5 are devoted to applications in a 3+1 description of the spacetime. Section 4 shows the use of quasi-local horizons as inner boundary conditions for elliptic equations in General Relativity, whereas Sec. 5 discusses some applications to the analysis of spacetimes, in particular their role in a *correlation approach* to spacetime dynamics. In Sec. 6 a general overview is presented.

2. Quasi-Local Horizons: Concepts and Definitions

2.1. *Geometry of spacelike closed 2-surfaces $\mathcal{S}$*

2.1.1. *Normal plane: Outgoing and ingoing null vectors*

Let us consider a spacetime $(\mathcal{M}, g_{ab})$ with Levi-Civita connection ∇_a. Given a spacelike closed (compact without boundary) 2-surface $\mathcal{S}$ in $\mathcal{M}$ and a point $p \in \mathcal{S}$, the tangent space splits as $T_p\mathcal{M} = T_p\mathcal{S} \oplus T_p^\perp\mathcal{S}$. We span the normal plane $T_p^\perp\mathcal{S}$ either by (future-oriented) null vectors ℓ^a and k^a (defined by the intersection between $T_p^\perp\mathcal{S}$ and the null cone at p) or by any pair of normal timelike vector n^a and spacelike vector s^a. Let us denote conventionally ℓ^a to be the *outgoing* null normal and k^a the *ingoing* one. We choose normalizations:

$$\ell^a\ell_a = 0, \quad k^a k_a = 0, \quad \ell^a k_a = -1, \quad n^a n_a = -1, \quad s^a s_a = 1, \quad n^a s_a = 0. \tag{2}$$

Directions ℓ^a and k^a are uniquely determined, but a *normalization-boost* freedom

$$\ell'^a = f\ell^a, \quad k'^a = f^{-1}k^a, \tag{3}$$
$$n'^a = \cosh(\sigma)n^a + \sinh(\sigma)s^a, \quad s'^a = \sinh(\sigma)n^a + \cosh(\sigma)s^a,$$

remains for some arbitrary rescaling positive function f on $\mathcal{S}$ (where $\sigma = \ln(f)$ and $\ell^a = \lambda(n^a + s^a)/\sqrt{2}$ and $k^a = \lambda^{-1}(n^a - s^a)/\sqrt{2}$, for some function λ on $\mathcal{S}$).

2.1.2. *Intrinsic geometry of* $\mathcal{S}$

The induced metric on $\mathcal{S}$ is given by

$$q_{ab} = g_{ab} + k_a \ell_b + \ell_a k_b = g_{ab} + n_a n_b - s_a s_b, \tag{4}$$

so that $q^a{}_b$ is the projector onto $\mathcal{S}$

$$q^a{}_b q^b{}_c = q^a{}_c, \quad q^a{}_b v^b = v^a \ (\forall\, v^a \in T\mathcal{S}), \quad q^a{}_b w^b = 0 \ (\forall\, w^a \in T^\perp\mathcal{S}). \tag{5}$$

We denote the Levi-Civita connection associated with q_{ab} as ${}^2D_a$. The volume form on $\mathcal{S}$ will be denoted by ${}^2\epsilon = \sqrt{q}dx^1 \wedge dx^2$, i.e. ${}^2\epsilon_{ab} = n^c s^d\, {}^4\epsilon_{cdab}$, though we will also employ the area measure notation $dA = \sqrt{q}d^2x$.

2.1.3. *Extrinsic geometry of* $\mathcal{S}$ *in* $(\mathcal{M}, g)$

We define the *second fundamental tensor* of $(\mathcal{S}, q_{ab})$ in $(\mathcal{M}, g_{ab})$ (also, *shape tensor* or *extrinsic curvature tensor*) as

$$\mathcal{K}^c_{ab} = q^d{}_a q^e{}_b \nabla_d q^c{}_e, \tag{6}$$

where c is an index in the normal plane $T^\perp\mathcal{S}$, whereas a and b are indices in $T\mathcal{S}$. Given a vector v^a normal to $\mathcal{S}$, we can define the *deformation tensor* $\Theta^{(v)}_{ab}$ as

$$\Theta^{(v)}_{ab} = q^c{}_a q^d{}_b \nabla_c v_d. \tag{7}$$

Then, using expression (4), the *second fundamental tensor* can be expressed as

$$\mathcal{K}^c_{ab} = k^c \Theta^{(\ell)}_{ab} + \ell^c \Theta^{(k)}_{ab} = n^c \Theta^{(n)}_{ab} - s^c \Theta^{(s)}_{ab}. \tag{8}$$

We can express $\Theta^{(v)}_{ab}$ in terms of the variation of the intrinsic metric along v^a. Given a (tensorial) object $A_{a_1 \cdots a_n}{}^{b_1 \cdots b_m}$ tangent to $\mathcal{S}$ we denote by δ_v the operator $(\delta_v A)_{a_1 \cdots a_n}{}^{b_1 \cdots b_m} = q_{a_1}{}^{c_1} \cdots q_{a_n}{}^{c_n} q_{d_1}{}^{b_1} \cdots q_{d_m}{}^{b_m} \mathcal{L}_v A_{c_1 \cdots c_n}{}^{d_1 \cdots d_m}$, where $\mathcal{L}_v$ denote the Lie derivative along (some extension of) v^a. Then, it follows

$$\delta_v q_{ab} = \frac{1}{2}\Theta^{(v)}_{ab}. \tag{9}$$

(a) *Shear and expansion associated with* v^a. Defining the expansion $\theta^{(v)}$ and shear tensor $\sigma^{(v)}_{ab}$ associated with the normal vector v^a as

$$\theta^{(v)} \equiv q^{ab}\nabla_a v_b = \delta_v \ln\sqrt{q}, \quad \sigma^{(v)}_{ab} \equiv \Theta^{(v)}_{ab} - \frac{1}{2}\theta^{(v)} q_{ab}, \tag{10}$$

we express the deformation tensor $\Theta^{(v)}_{ab}$ in terms of his trace and traceless parts

$$\Theta^{(v)}_{ab} = \sigma^{(v)}_{ab} + \frac{1}{2}\theta^{(v)} q_{ab}. \tag{11}$$

(b) *Mean curvature vector* H^a. Taking the trace of $\mathcal{K}^c_{ab}$ on $\mathcal{S}$ we define the *mean curvature vector*[a]

$$H^c \equiv q^{ab}\mathcal{K}^c_{ab} = \theta^{(\ell)} k^c + \theta^{(k)} \ell^c. \tag{12}$$

[a]Note the opposite sign convention with respect to the contribution by J. M. M. Senovilla.[15]

The extrinsic curvature information of $(\mathcal{S}, q_{ab})$ in $(\mathcal{M}, g_{ab})$ is completed by the *normal fundamental forms* associated with normal vectors v^a. In particular[23]

$$\Omega_a^{(n)} = s^c q^d{}_a \nabla_d n_c, \quad \Omega_a^{(s)} = n^c q^d{}_a \nabla_d s_c,$$
$$\Omega_a^{(\ell)} = \frac{1}{k^b \ell_b} k^c q^d{}_a \nabla_d \ell_c, \quad \Omega_a^{(k)} = \frac{1}{k^b \ell_b} \ell^c q^d{}_a \nabla_d k_c. \tag{13}$$

All these normal fundamental forms are related up to a sign and a total derivative on $\mathcal{S}$. Using the normalizations (2) we get[b]: $\Omega_a^{(n)} = -\Omega_a^{(s)}, \Omega_a^{(\ell)} = -\Omega_a^{(k)}, \Omega_a^{(\ell)} = \Omega_a^{(n)} - {}^2D_a \lambda$. We choose to employ the 1-form $\Omega_a^{(\ell)}$ in the following.

2.1.4. *Transformation properties under null normal rescaling*

Under the rescaling (2) $\ell^a \to f\ell^a$, $k^a \to f^{-1}k^a$ the introduced fields transform as

$$\begin{aligned}
&q_{ab} \to q_{ab}, && {}^2D_a \to {}^2D_a, \\
&\mathcal{K}^c_{ab} \to \mathcal{K}^c_{ab}, && H^a \to H^a, \\
&\Theta^{(\ell)}_{ab} \to f\Theta^{(\ell)}_{ab}, && \theta^{(\ell)} \to f\theta^{(\ell)}, \quad \sigma^{(\ell)}_{ab} \to f\sigma^{(\ell)}_{ab}, \\
&\Theta^{(k)}_{ab} \to f^{-1}\Theta^{(k)}_{ab}, && \theta^{(k)} \to f^{-1}\theta^{(k)}, \quad \sigma^{(k)}_{ab} \to f^{-1}\sigma^{(k)}_{ab}. \\
&\Omega^{(\ell)}_a \to \Omega^{(\ell)}_a + {}^2D_a(\ln f),
\end{aligned} \tag{14}$$

Finally, given an axial Killing vector ϕ^a on $\mathcal{S}$, we can write the angular momentum[c]

$$J = \frac{1}{8\pi} \int_{\mathcal{S}} \Omega^{(\ell)}_a \phi^a {}^2\epsilon. \tag{15}$$

The transformation rule of $\Omega_a^{(\ell)}$ in (14) together with the divergence-free property of ϕ^a (following from its Killing character) guarantee that the quantity J does not depend on the choice of null normals ℓ^a, k^a (i.e. J does not change under a boost).

2.2. *Trapping horizons*

2.2.1. *Worldtubes of marginally trapped surfaces*

A *trapping horizon*[14] is (the closure of) a hypersurface $\mathcal{H}$ foliated by closed marginal (outer) trapped surfaces: $\mathcal{H} = \bigcup_{t\in\mathbb{R}} S_t$, with $\theta^{(\ell)}|_{S_t} = 0$. Trapping horizons are classified according to the signs of $\theta^{(k)}$ and $\delta_k\theta^{(\ell)}$. In particular, the sign of $\theta^{(k)}$ controls if the singularity occurs either in the *future* or in the *past* of $\mathcal{S}$, whereas the sign of $\delta_k\theta^{(\ell)}$ controls the (local) *outer-* or *inner-most* character of $\mathcal{H}$. Then, a trapping horizon is said to be: (i) *future* (respectively, *past*) if $\theta^{(k)} < 0$ (respectively, $\theta^{(k)} > 0$), and (ii) *outer* (respectively, *inner*) if there exists[d] ℓ^a and k^a such that $\delta_k\theta^{(\ell)} < 0$ (respectively, $\delta_k\theta^{(\ell)} > 0$).

[b] When using $\ell^a k_a = -e^\sigma$ one gets: $\Omega_a^{(\ell)} = -\Omega_a^{(k)} - {}^2D_a\sigma$. This will be relevant later, in Eq. (39).

[c] The quantity J coincides with the Komar angular momentum in case that ϕ^a can be extended to an axial Killing in the neighborhood of $\mathcal{S}$.

[d] The sign of $\delta_k\theta^{(\ell)}$ is not invariant on the whole $\mathcal{S}$ under a rescaling (2). However, if there exists ℓ^a and k^a such that $\delta_k\theta^{(\ell)} < 0$ on $\mathcal{S}$, then there does not exist any choice of ℓ^a and k^a such that $\delta_k\theta^{(\ell)} > 0$ on $\mathcal{S}$; see Ref. 24 and also the marginally trapped surface stability condition in Ref. 25.

2.2.2. *Future outer trapping horizons*

In a black hole setting the singularity occurs in the future of sections $\mathcal{S}_t$ of $\mathcal{H}$, so that the related trapping horizon is of *future* type, $\theta^{(k)} < 0$. In addition, when considering displacements along k^a (*ingoing* direction) we should move into the trapped region, i.e. $\delta_k \theta^{(\ell)} < 0$, so that the trapping horizon should be *outer*.

The resulting characterization of quasi-local black hole horizons as *Future Outer Trapping Horizons* (FOTHs) is further supported by the following analysis of the area evolution. Hawking's area theorem for event horizons (cf. Sec. 1.2) captures a fundamental feature of classical black holes. It is natural to wonder about a quasi-local version of it. Let us consider an evolution vector h^a along the trapping horizon $\mathcal{H}$, characterized as: (i) h^a is tangent to $\mathcal{H}$ and orthogonal to $\mathcal{S}_t$, and (ii) h^a transports $\mathcal{S}_t$ onto $\mathcal{S}_{t+\delta t}$: $\delta_h t = 1$. We can write h^a and a *dual* vector τ^a orthogonal to $\mathcal{H}$ as

$$h^a = \ell^a - Ck^a, \quad \tau^a = \ell^a + Ck^a. \tag{16}$$

Then $h^a h_a = -\tau^a \tau_a = 2C$, i.e. h^a is spacelike for $C > 0$, null for $C = 0$ and timelike for $C < 0$. The evolution of the area $A = \int_{\mathcal{S}} dA = \int_{\mathcal{S}} {}^2\epsilon$ along h^a is given by

$$\delta_h A = \int_{\mathcal{S}} \theta^{(h)2}\epsilon = \int_{\mathcal{S}} (\theta^{(\ell)} - C\theta^{(k)})^2\epsilon = -\int_{\mathcal{S}} C\theta^{(k)2}\epsilon. \tag{17}$$

Considering for simplicity the spherical symmetric case (C = const; see discussion of Eq. (37) in 3.2.4, for the general case), the *trapping horizon* condition, $\delta_h \theta^{(\ell)} = 0$, writes $\delta_\ell \theta^{(\ell)} - C\delta_k \theta^{(\ell)} = 0$, so that $C = \frac{\delta_\ell \theta^{(\ell)}}{\delta_k \theta^{(\ell)}}$. Applying the *Raychaudhuri* equation for $\delta_\ell \theta^{(\ell)}$ [see later Eq. (21)], together with the $\theta^{(\ell)} = 0$ condition, we find

$$C = -\frac{\sigma^{(\ell)}_{ab}\sigma^{(\ell)ab} + 8\pi T_{ab}\ell^a \ell^b}{\delta_k \theta^{(\ell)}}. \tag{18}$$

Under the null energy and outer horizon conditions, it follows that $C \geq 0$, so that the future condition guarantees the nondecrease of the area in (17). Therefore, FOTHs are *null* or *spacelike* hypersurfaces ($C \geq 0$), satisfying an area law result, and therefore providing appropriate models for quasi-local black hole horizons.

2.3. *Isolated and dynamical horizons*

The distinct geometric structure of null and spatial hypersurfaces suggests different strategies for the study of the stationary and dynamical regimes of quasi-local black holes, modeled as future outer trapping horizons. This has led to the parallel development of the *isolated horizon* and the *dynamical horizon* frameworks.[26–29]

In equilibrium, *Isolated Horizons* (IH) provide a hierarchy of geometric structures constructed on a null hypersurface $\mathcal{H}$ that is foliated by closed (outer) marginally trapped surfaces. They characterize different levels of stationarity for a black hole horizon in an otherwise dynamical environment:

(i) *Non-Expanding Horizons* (NEH). They represent the minimal notion of equilibrium by imposing the stationarity of the intrinsic geometry q_{ab}.

(ii) *Weakly Isolated Horizons* (WIH). They are NEHs endowed with an additional structure needed for a Hamiltonian analysis of the horizon and its related (thermo-)dynamics. They impose no additional constraints on the geometry of the NEH.

(iii) *Isolated Horizons* (IH). These are WIHs whose extrinsic geometry is also invariant along the evolution. They provide the strongest stationarity notion on $\mathcal{H}$.

The nonstationary regime can be characterized by *Dynamical Horizons* (DH), namely spacelike hypersurfaces $\mathcal{H}$ foliated by closed future marginally trapped surfaces, i.e. $\theta^{(\ell)} = 0$ and $\theta^{(k)} < 0$. Introduced in a 3+1 formulation, they provide a complementary perspective to the *dual-null foliation formulation*[14] of trapping horizons, making them naturally adapted for an Initial Value Problem perspective.

2.4. *IHs and DHs as stationary and dynamical sections of FOTHs*

A natural question when considering the transition from equilibrium to the dynamical regime is whether a section $\mathcal{S}_t$ of a FOTH can be partially stationary and partially dynamical. Or, in other words, whether the element of area dA can be non-expanding ($C = 0$) in a part of $\mathcal{S}_t$ whereas it already expands ($C > 0$) in another part. Namely, can h^a be both null and spacelike on a section $\mathcal{S}_t$ of a FOTH?

The answer is in the negative. Transitions between non-expanding and dynamical parts of a FOTH must happen *all at once*. More precisely, assuming the null energy condition, a FOTH can be completely partitioned into non-expanding and dynamical sections. For a section $\mathcal{S}_t$ to be completely dynamical ($C > 0$) it suffices that it has $\delta_\ell \theta^{(\ell)} < 0$ somewhere on it. Otherwise h^a is null ($C = 0$) all over $\mathcal{S}_t$.[30,24]

In more physical terms, it suffices that some *energy* crosses the horizon *somewhere*, and the *whole* horizon instantaneously grows as a whole. This nonlocal behavior is a consequence of the *elliptic* nature of quasi-local horizons. As shown in Sec. 3.2.3, the function C determining the metric type of h^a satisfies an elliptic equation [cf. Eq. (37)]. Under the outer condition $\delta_k \theta^{(\ell)} < 0$ one can apply a *maximum principle* to show that C is non-negative [generalization of Eq. (18)]. Moreover, it suffices that $\delta_\ell \theta^{(\ell)} \neq 0$ somewhere, for having $C > 0$ everywhere.

3. Quasi-Local Horizons: Properties from a 3+1 Perspective

3.1. *Equilibrium regime*

3.1.1. *Null hypersurfaces: Characterization and basic elements*

A hypersurface $\mathcal{H}$ is null if and only if the induced metric is degenerate. Equivalently, if and only if there is a tangent null vector ℓ^a orthogonal to all vectors tangent to $\mathcal{H}$: $\ell^a v_a = 0,\ \forall\, v^a \in T\mathcal{H}$.

Let us introduce some elements on the geometry of $\mathcal{H}$. Choosing a null vector k^a transverse to $\mathcal{H}$, we can write[e] the degenerate metric as $q_{ab} = g_{ab} + k_a \ell_b + \ell_a k_b$.

[e]We abuse notation and employ the same notation employed in sections $\mathcal{S}_t$ of $\mathcal{H}$, cf. Eq. (4).

A projector onto $\mathcal{H}$ can also be constructed as: $\Pi_a{}^b = \delta_a{}^b + \ell_a k^b = q_a{}^b - k_a \ell^b$. As a part of the extrinsic curvature of $\mathcal{H}$, a *rotation* 1-*form* can be introduced[31] on $\mathcal{H}$ as $\omega_a^{(\ell)} = \frac{1}{\ell^b k_b} k^c \nabla_a \ell_c$. This 1-form *lives* on $\mathcal{H}$, i.e. $k^a \omega_a^{(\ell)} = 0$. In particular, we can write $\Pi_a{}^c \nabla_c \ell^b = \omega_a^{(\ell)} \ell^b + \Theta^{(\ell)}{}_a{}^b$, where $\Theta_{ab}^{(\ell)}$ is given by expression (7) [cf. Eq. (5.23) in Ref. 28]. Contracting with ℓ^a we find: $\ell^c \nabla_c \ell^a = \kappa^{(\ell)} \ell^a$, a pre-geodesic equation where the non-affinity coefficient $\kappa^{(\ell)}$ is defined as $\kappa^{(\ell)} = \ell^a \omega_a^{(\ell)}$. If a foliation $\{\mathcal{S}_t\}$ of $\mathcal{H}$ is given, we can write [cf. Eq. (5.35) in Ref. 28]: $\omega_a^{(\ell)} = \Omega_a^{(\ell)} - \kappa^{(\ell)} k_a$.

Vectors ℓ^a and k^a can be completed to a tetrad $\{\ell^a, k^a, (e_1)^a, (e_2)^a\}$, where $(e_i)^a$ are tangent to sections $\mathcal{S}_t$. Normalizations given in (2) are then completed to

$$\ell \cdot (e_i)_a = 0, \quad k^a (e_i)_a = 0, \quad (e_i)^a (e_i)_b = \delta_{ab}. \tag{19}$$

Defining the complex null vector $m^a = \frac{1}{\sqrt{2}}[(e_1)^a + i(e_2)^a]$, the Weyl scalars are defined as the components of the Weyl tensor $C^a{}_{bcd}$ in the null tetrad $\{\ell^a, k^a, m^a, \overline{m}^a\}$

$$\begin{aligned} &\Psi_0 = C^a{}_{bcd}\, \ell_a m^b \ell^c m^d, \quad \Psi_3 = C^a{}_{bcd}\, \ell_a k^b \overline{m}^c k^d, \\ &\Psi_1 = C^a{}_{bcd}\, \ell_a m^b \ell^c k^d, \quad \Psi_4 = C^a{}_{bcd}\, \overline{m}_a k^b \overline{m}^c k^d. \\ &\Psi_2 = C^a{}_{bcd}\, \ell_a m^b \overline{m}^c k^d, \end{aligned} \tag{20}$$

3.1.2. *Null hypersurfaces: Evolution*

It is illustrative to give a 3+1 perspective on $\mathcal{H}$. Given a foliation $\mathcal{H} = \bigcup_{t \in \mathbb{R}} \mathcal{S}_t$ let us evaluate explicitly the evolution along ℓ^a of some quantities defined on sections $\mathcal{S}_t$.

(i) Expansion equation (null Raychaudhuri equation):

$$\delta_\ell \theta^{(\ell)} - \kappa^{(\ell)} \theta^{(\ell)} + \frac{1}{2} \theta^{(\ell)2} + \sigma_{ab}^{(\ell)} \sigma^{(\ell)ab} + 8\pi T_{ab} \ell^a \ell^b = 0. \tag{21}$$

(ii) Tidal equation:

$$\delta_\ell \sigma_{ab}^{(\ell)} = \kappa^{(\ell)} \sigma_{ab}^{(\ell)} + \sigma_{cd}^{(\ell)} \sigma^{(\ell)cd} q_{ab} - q^c{}_a q^d{}_b C_{ecfd} \ell^e \ell^f. \tag{22}$$

(iii) Evolution for Ω_a:

$$\delta_\ell \Omega_c^{(\ell)} + \theta^{(\ell)} \Omega_a^{(\ell)} = 8\pi T_{cd}\, \ell^c q^d{}_a + {}^2D_a \left(\kappa^{(\ell)} + \frac{\theta^{(\ell)}}{2} \right) - {}^2D_c \sigma^{(\ell)c}{}_a. \tag{23}$$

3.1.3. *Non-expanding horizons*

A NEH[32] is a null-hypersurface $\mathcal{H} \approx S^2 \times \mathbb{R}$, on which the expansion associated with ℓ^a vanishes ($\theta^{(\ell)} = 0$), the Einstein equations hold and $-T^a_c \ell^c$ is future directed (*null dominant energy condition*). Note that any foliation $\mathcal{H} = \bigcup_{t \in \mathbb{R}} \mathcal{S}_t$ produces a foliation of $\mathcal{H}$ by MOTS $\mathcal{S}_t$.

(i) *NEH characterization.* Making $\theta^{(\ell)} = 0$ in the Raychaudhuri Eq. (21) we get

$$\sigma_{ab} \sigma^{ab} + 8\pi T_{ab} \ell^a \ell^b = 0. \tag{24}$$

Since the two terms are positive-definite, they vanish independently. This provides an *instantaneous* characterization of a NEH:

$$\theta^{(\ell)} = 0, \quad \sigma^{(\ell)}_{ab} = 0, \quad T_{ab}\ell^a\ell^b = 0. \tag{25}$$

From Eq. (11) with $v^a = \ell^a$, it follows $\Theta^{(\ell)}_{ab} = 0$. The NEH characterization is equivalent, cf. Eq. (9), to the *evolution independence* of the induced metric q_{ab}

$$\delta_\ell q_{ab} = \frac{1}{2}\Theta^{(\ell)}_{ab} = 0. \tag{26}$$

From Eq. (8), we conclude that a NEH fixes half of the degrees of freedom in the second fundamental form $\mathcal{K}^c_{ab}$ of $\mathcal{S}_t$ in $\mathcal{M}$. This will be relevant in Sec. 4.2.1.

(ii) *Connection* $\hat{\nabla}_a$ *on a NEH.* A null hypersurface has no unique (Levi-Civita) connection compatible with the metric. However, on a NEH $\mathcal{H}$ one can introduce a preferred connection as that one induced from the spacetime connection ∇_a: $u^c\hat{\nabla}_c w^a \equiv u^c\nabla_c w^a$, $\forall u^a, w^a \in T\mathcal{H}$. Indeed using NEH characterization (26), $u^c\nabla_c w^a$ is tangent to $\mathcal{H}$: $\ell_d(u^c\nabla_c w^d) = u^c\nabla_c(\ell_d w^d) - u^c w^d \Theta^{(\ell)}_{cd} = 0$.

(iii) *Geometry of a NEH.* We refer (cf. Ref. 33) to the pair $(q_{ab}, \hat{\nabla}_a)$ as the *geometry of a NEH.* Writing the components of the $\hat{\nabla}_a$ connection in terms of quantities on $\mathcal{S}_t$

$$\begin{aligned} q^c{}_a q^b{}_d \hat{\nabla}_c v^d &= {}^2D_a(q^b{}_c v^c), \\ q^c{}_a k_d \hat{\nabla}_c v^d &= {}^2D_a(v^c k_c) - q^c{}_a v^d \Theta^{(k)}_{cd}, \\ \ell^c\hat{\nabla}_c v^a &= \delta_\ell v^a + v^c\omega^{(\ell)}_c \ell^a, \end{aligned} \tag{27}$$

the free data on a NEH are given, from an evolution perspective, by $(q_{ab}|_{\mathcal{S}_t}, \Omega^{(\ell)}_a|_{\mathcal{S}_t}, \kappa^{(\ell)}|_{\mathcal{H}}, \Theta^{(k)}_{ab}|_{\mathcal{S}_t})$, where q_{ab} is time independent.

(iv) *Weyl tensor on a NEH.* Under the rescaling (3), the 1-form $\omega^{(\ell)}_a$ transforms as $\omega^{(\ell)}_a \to \omega^{(\ell)}_a + \hat{\nabla}_a \ln f$. Its exterior derivative $d\omega^{(\ell)}$ provides a gauge invariant object: understanding $\omega^{(\ell)}_a$ as a gauge connection, $d\omega^{(\ell)}$ is its gauge-invariant curvature. Using the NEH condition, $\Theta^{(\ell)}_{ab} = 0$, one can express (cf. Sec. 7.6.2. in Ref. 28)

$$d\omega^{(\ell)} = 2\,\mathrm{Im}\Psi_2{}^2\epsilon. \tag{28}$$

Hence, $\mathrm{Im}\Psi_2$ is gauge invariant on a NEH. Actually the full Ψ_2 is invariant, as it follows from its *boost* transformation rules and the values of Ψ_0 and Ψ_1 on a NEH,[28]

$$\Psi_0|_{\mathcal{H}} = \Psi_1|_{\mathcal{H}} = 0. \tag{29}$$

3.1.4. *Weakly isolated horizons*

A *Weakly Isolated Horizon* (WIH) $(\mathcal{H}, [\ell^a])$ is a NEH together with a class of null normals $[\ell^a]$ such that: $\delta_\ell \omega^{(\ell)}_a = 0$. This condition permits to set a well-posed variational problem for spacetimes containing stationary quasi-local horizons. This enables the development of a Hamiltonian analysis on the horizon $\mathcal{H}$ leading to the construction of conserved quantities under WIH-symmetries.[31] In particular, the expression for the angular momentum in Eq. (15) is recovered

$$J_{\mathcal{H}} = \frac{1}{8\pi}\int_{\mathcal{S}_t} \omega_c^{(\ell)}\phi^c\, {}^2\epsilon = \frac{1}{8\pi}\int_{\mathcal{S}_t} \Omega_c^{(\ell)}\phi^c\, {}^2\epsilon = -\frac{1}{4\pi}\int_{\mathcal{S}_t} f\mathrm{Im}\Psi_2\, {}^2\epsilon, \tag{30}$$

with $\phi^a = {}^2D_c f\, {}^2\epsilon^{ac}$ (ϕ^a is an axial Killing vector, in particular divergence-free).

The WIH structure is relevant for the discussion of IH thermodynamics (cf. A. Nielsen's contribution[22]). We do not address this issue here and just comment on the equivalence of the WIH condition with a thermodynamical *zeroth law*. Reminding $\omega_a^{(\ell)} = \Omega_a^{(\ell)} - \kappa^{(\ell)}k_a$, the (vacuum) evolution equation (23) for $\Omega_a^{(\ell)}$ leads to $\mathcal{L}_\ell\Omega_a^{(\ell)} = {}^2D_a\kappa^{(\ell)}$. More generally, $\delta_\ell\omega_a^{(\ell)} = \hat{\nabla}\kappa^{(\ell)}$ (cf. for example Eq. (8.5) in Ref. 28). That is, on WIHs the non-affinity coefficient (*surface gravity*) is constant: $\kappa^{(\ell)} = \kappa_o$.

WIHs and NEH geometry. WIHs do not constrain the underlying NEH geometry. In other words, every NEH admits a WIH structure. In fact, given $\kappa^{(\ell)} \neq$ const., the rescaling $\ell' = \alpha\ell$, with $\kappa_o =$ const. $= \nabla_\ell\alpha + \alpha\kappa^{(\ell)}$, leads to a constant $\kappa^{(\ell')} = \kappa_o$. Finally, free data for a WIH are again $(q_{ab}|_{\mathcal{S}_t}, \Omega_a^{(\ell)}|_{\mathcal{S}_t}, \kappa^{(\ell)}|_{\mathcal{H}}, \Theta_{ab}^{(k)}|_{\mathcal{S}_t})$, but now $q_{ab}|_{\mathcal{S}_t}$, $\Omega_a^{(\ell)}|_{\mathcal{S}_t}$ and $\kappa^{(\ell)}|_{\mathcal{H}} = \kappa_o$ are time-independent.

3.1.5. *(Strongly) isolated horizons*

An isolated horizon (IH) is a WIH on which the whole extrinsic geometry is time-invariant: $[\delta_\ell, \hat{\nabla}_a] = 0$. This condition can be characterized[33,28] as $\delta_\ell\Theta^{(k)} = 0$, that leads to the geometric constraint

$$\kappa^{(\ell)}\Theta_{ab}^{(k)} = \frac{1}{2}({}^2D_a\Omega_b^{(\ell)} + {}^2D_b\Omega_a^{(\ell)}) + \Omega_a^{(\ell)}\Omega_b^{(\ell)} - \frac{1}{2}{}^2R_{ab} + 4\pi\left(q^c{}_a q^d{}_b T_{cd} - \frac{T}{2}q_{ab}\right). \tag{31}$$

With Eq. (26), this fixes completely the second fundamental form $\mathcal{K}^c_{ab}$. Free data of an IH, $(q_{ab}|_{\mathcal{S}_t}, \Omega_a^{(\ell)}|_{\mathcal{S}_t}, \kappa^{(\ell)}|_{\mathcal{H}} = \kappa_o)$, are time independent. Their geometric (gauge-invariant) content can be encoded in the pair[f]: $({}^2R, \mathrm{Im}\Psi_2)$. On the one hand, ${}^2R$ accounts for the gauge-invariant part of q_{ab}. Regarding $\Omega_a^{(\ell)}$, from $d\omega^{(\ell)} = 2\mathrm{Im}\Psi_2\,{}^2\epsilon$ and $\kappa^{(\ell)} =$ const., it follows that $d\Omega^{(\ell)} = 2\mathrm{Im}\Psi_2\,{}^2\epsilon$. On a sphere $\mathcal{S}_t$ we can write $\Omega_a^{(\ell)} = \Omega_a^{\mathrm{div-free}} + \Omega_a^{\mathrm{exact}}$, so that $\Omega_a^{\mathrm{exact}} = {}^2D_a g$ is gauge-dependent [cf. (14)]. From $d\Omega_a^{\mathrm{div-free}} = 2\mathrm{Im}\Psi_2$, the gauge-invariant part of $\Omega_a^{(\ell)}$ is encoded in $\mathrm{Im}\Psi_2$.

IH multipoles of axially symmetric horizons. On an axially symmetric IH, the gauge-invariant part of the geometry, $({}^2R, \mathrm{Im}\Psi_2)$, can be decomposed onto spherical harmonics. On an axially symmetric section $\mathcal{S}_t$ of $\mathcal{H}$, a coordinate system can be *canonically* constructed,[36,37] such that [with $A_{\mathcal{H}} = 4\pi(R_{\mathcal{H}})^2$]

$$q_{ab}dx^a \otimes dx^b = (R_{\mathcal{H}})^2(F^{-1}\sin^2\theta d\theta \otimes d\theta + Fd\phi \otimes d\phi). \tag{32}$$

[f]Note the relation with the complex scalar $\mathcal{K}$ in Refs. 34 and 35.

In particular, $dA = (R_{\mathcal{H}})^2 \sin\theta d\theta d\phi$ (*round sphere* area element). We can then use standard spherical harmonics $Y_{\ell m}(\theta)$, with $m = 0$ in this axisymmetric case

$$\int_{\mathcal{S}_t} Y_{\ell 0}(\theta) Y_{\ell' 0}(\theta) d^2A = (R_{\mathcal{H}})^2 \delta_{\ell\ell'}, \tag{33}$$

to define the *IH geometric multipoles*[36] I_n and L_n

$$I_n = \frac{1}{4} \int_{\mathcal{S}_t} {}^2\!R\, Y_{n0}(\theta)\, d^2A,$$

$$L_n = - \int_{\mathcal{S}_t} \mathrm{Im}\Psi_2\, Y_{n0}(\theta)\, d^2A. \tag{34}$$

Then, *mass* M_n and *angular momentum* J_n multipoles are defined[36–39] by adequate dimensional rescalings of I_n and L_n.

3.1.6. *Gauge freedom on a NEH: Non-uniqueness of the foliation*

Before proceeding to the dynamical case, we underline the existence of a fundamental gauge freedom in the equilibrium (null) case: *any foliation* $\{\mathcal{S}_t\}$ *of a NEH* $\mathcal{H}$ *provides a foliation of* $\mathcal{H}$ *by marginally trapped surfaces.* This is equivalent to the rescaling freedom of the null normal $\ell^a \to f\ell^a$. Therefore, the amount of gauge freedom in the equilibrium case is encoded in one arbitrary function f on $\mathcal{S}_t$.

Note that in this equilibrium horizon context, the relevant spacetime geometric object (the hypersurface $\mathcal{H}$) is *unique*, whereas the gauge-freedom enters in its evolution description due to the *non-uniqueness* of its possible foliation by MOTS.

3.2. *Dynamical case*

3.2.1. *Existence and foliation uniqueness results*

Let us introduce two fundamental results following from the application of geometric analysis techniques to the study of dynamical trapping horizons.

Property 1 (Dynamical Horizon foliation uniqueness).[40] *Given a dynamical FOTH* $\mathcal{H}$, *the foliation by marginally trapped surfaces is unique.*

This first result identifies an important *rigidity property* of DHs: the uniqueness of its evolution description. This is in contrast with the equilibrium null case, with its freedom in the choice of the foliation. In particular, on a dynamical FOTH the evolution vector is completely determined: h^a is unique up to *time* reparametrization.

Property 2 (Existence of DHs).[30,41] *Given a marginally trapped surface* $\mathcal{S}_0$ *satisfying an appropriate stability condition on a Cauchy hypersurface* Σ, *to each* 3+1 *spacetime foliation* $(\Sigma_t)_{t\in\mathbb{R}}$ *there corresponds a unique dynamical FOTH* $\mathcal{H}$ *containing* $\mathcal{S}_0$ *and sliced by marginally trapped surfaces* $\{\mathcal{S}_t\}$ *such that* $\mathcal{S}_t \subset \Sigma_t$.

This second result addresses the Initial Value Problem of DHs, in particular the existence of an evolution for a given MOTS into a dynamical FOTH. The result requires a stability condition (namely, $\mathcal{S}_0$ is required to be *stably outermost*[30,41,25,42]), so that the sign of the variation of $\theta^{(\ell)}$ in the inward (outward) direction is under control. This is essentially the *outer* condition[14] in the FOTH characterization.

3.2.2. *"Gauge" freedom: Non-uniqueness of dynamical horizons*

The evolution of an AH into a DH is non-unique, as a consequence of combining Properties 1 and 2 above. Let us consider an initial AH $\mathcal{S}_0 \subset \Sigma_0$ and two different 3+1 slicings $\{\Sigma_{t_1}\}$ and $\{\Sigma_{t_2}\}$, compatible with Σ_0. From Property 2 there exist DHs $\mathcal{H}_1 = \bigcup_{t_1} \mathcal{S}_{t_1}$ and $\mathcal{H}_2 = \bigcup_{t_2} \mathcal{S}_{t_2}$, with $\mathcal{S}_{t_1} = \mathcal{H}_1 \cap \Sigma_{t_1}$ and $\mathcal{S}_{t_2} = \mathcal{H}_2 \cap \Sigma_{t_2}$ marginally trapped surfaces. Let us consider now the sections of $\mathcal{H}_1$ by $\{\Sigma_{t_2}\}$, i.e. $\mathcal{S}'_{t_2} = \mathcal{H}_1 \cap \Sigma_{t_2}$, so that $\mathcal{H}_1 = \bigcup_{t_2} \mathcal{S}'_{t_2}$. In the generic case, slicings $\{\mathcal{S}'_{t_2}\}$ and $\{\mathcal{S}_{t_1}\}$ of $\mathcal{H}_1$ are different (one can consider a deformation of the slicing $\{\Sigma_{t_2}\}$, if needed). Therefore, from the foliation uniqueness of Property 1, sections $\mathcal{S}'_{t_2}$ cannot be marginally trapped surfaces. It follows then that $\mathcal{H}_1$ and $\mathcal{H}_2$ are different as hypersurfaces in $\mathcal{M}$: if $\mathcal{H}_1 = \mathcal{H}_2$, sections $\mathcal{S}_{t_2}$ (MOTSs) and $\mathcal{S}'_{t_2}$ (non-MOTSs) would coincide by construction, leading to a contradiction. In addition to this non-uniqueness, DHs *interweave* in spacetime due to the existence of causal constraints[40]: a DH $\mathcal{H}_1$ cannot lie completely in the causal past of another DH $\mathcal{H}_2$ (cf. Fig. 1).

Comparing with the discussion in Sec. 3.1.6 on the uniqueness and gauge-freedom issues in the equilibrium case, we conclude from the previous geometric considerations that the dynamical and equilibrium cases contain the same amount of gauge freedom, namely a function on $\mathcal{S}$, although *dressed* in a different form. More specifically, whereas in the NEH case there is a fixed horizon, with a rescaling

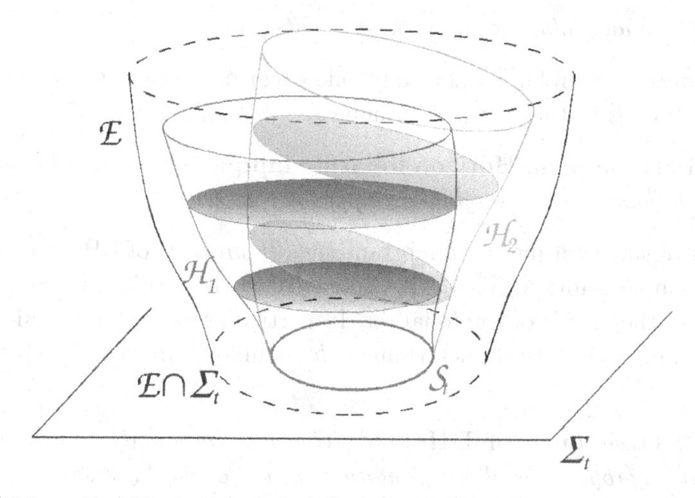

Figure 1. (Color online) Illustration of the DH non-uniqueness. Dynamical horizons $\mathcal{H}_1$ and $\mathcal{H}_2$ represent evolutions from a given initial MOTS corresponding to different spacetime 3+1 slicings.

freedom ($\ell^a \to f\ell^a$, f function on $\mathcal{S}_t$), in the DH case the foliation is fixed, but a (gauge) freedom appears in the choice of the evolving horizon (lapse function N on $\mathcal{S}_t$). In other words, in the dynamical case the choice is among *distinct* spacetime geometric objects, $\mathcal{H}_1$ and $\mathcal{H}_2$, whereas in the equilibrium case the choice concerns the *description* (foliation) of a single spacetime geometric object $\mathcal{H}$.

3.2.3. *FOTH characterization*

As discussed in Subsec. 2.2, a FOTH with evolution vector $h^a = \ell^a - Ck^a$ is characterized by: (i) a *trapping horizon condition*: $\theta^{(\ell)} = 0, \delta_h\theta^{(\ell)} = 0$, (ii) a future condition $\theta^{(k)} < 0$, and (iii) an outer condition: $\delta_k\theta^{(\ell)} < 0$. These conditions can be made more explicit in terms of the variations[24,43]

$$\begin{aligned}\delta_{\alpha\ell}\theta^{(\ell)} &= -\alpha(\sigma^{(\ell)}_{ab}\sigma^{(\ell)ab} + 8\pi T_{ab}\ell^a\ell^b),\\ \delta_{\beta k}\theta^{(\ell)} &= \beta\left[-{}^2D^c\Omega^{(\ell)}_c + \Omega^{(\ell)}_c\Omega^{(\ell)c} - \frac{1}{2}{}^2R + 8\pi T_{ab}k^a\ell^b\right] + {}^2\Delta\beta - 2\Omega^{(\ell)}_c{}^2D^c\beta,\end{aligned} \tag{35}$$

with α and β functions on $\mathcal{S}_t$. Making $\beta = 1$, the outer condition writes

$$\delta_k\theta^{(\ell)} = -{}^2D^c\Omega^{(\ell)}_c + \Omega^{(\ell)}_c\Omega^{(\ell)c} - \frac{1}{2}{}^2R + 8\pi T_{ab}k^a\ell^b < 0, \tag{36}$$

for some ℓ^a and k^a, whereas the trapping horizon condition (with $\alpha = 1, \beta = C$) is

$$\delta_h\theta^{(\ell)} = \delta_\ell\theta^{(\ell)} - \delta_{Ck}\theta^{(\ell)} = \delta_\ell\theta^{(\ell)} - C\delta_k\theta^{(\ell)} - {}^2\Delta C + 2\Omega^{(\ell)}_c{}^2D^cC = 0, \tag{37}$$

that is

$$-{}^2\Delta C + 2\Omega^{(\ell)}_c{}^2D^cC - C\left[-{}^2D^c\Omega^{(\ell)}_c + \Omega^{(\ell)}_c\Omega^{(\ell)c} - \frac{1}{2}{}^2R\right] = \sigma^{(\ell)}_{ab}\sigma^{(\ell)ab} + 8\pi T_{ab}\tau^a\ell^b. \tag{38}$$

This elliptic condition on C, in particular through the application of a maximum principle relying on the outer condition $\delta_k\theta^{(\ell)} < 0$, is at the heart of the nonlocal behavior of the worldtube $\bigcup_{t\in\mathbb{R}}\mathcal{S}_t$ discussed in Sec. 2.4.

Remark on the variation/deformation/stability operator $\delta_v\theta^{(\ell)}$. Before proceeding further, Eq. (35) requires some explanation. In Sec. 2.1.3, we have introduced δ_v in terms of the Lie derivative on a tensorial object. However, the expansion $\theta^{(\ell)}$ is not a scalar quantity in the sense of a point-like (tensorial) field defined on the manifold $\mathcal{M}$. The expansion is a quasi-local object whose very definition at a point $p \in \mathcal{M}$ requires the choice of a (portion of a) surface $\mathcal{S}$ passing through p. In this sense, $\delta_{\gamma v}$ (with γ a function on $\mathcal{S}$) cannot be in general evaluated as a Lie derivative. Consider a displacement of the surface $\mathcal{S}_t$ by a vector γv^a. The surface $\mathcal{S}_{t+\delta t}$ and therefore $\theta^{(\ell)}|_{t+\delta t}$ depend on the angular dependence of γ, so that $\delta_{\gamma v}\theta^{(\ell)} \neq \gamma\delta_v\theta^{(\ell)}$. The operator δ_v still satisfies a linear property for constant linear combinations, $\delta_{av+bw}\theta^{(\ell)} = a\delta_v\theta^{(\ell)} + b\delta_w\theta^{(\ell)}$ $(a, b \in \mathbb{R})$, and the Leibnitz rule, $\delta_v(\gamma\theta^{(\ell)}) = (\delta_v\gamma)\theta^{(\ell)} + \gamma\delta_v\theta^{(\ell)}$. Details about this operator can be found in Refs. 30, 24, 43.[g] Here we rather exploit a *practical trick* for the evaluation of $\delta_{\gamma v}\theta^{(\ell)}$, based on the remark that given the vector v^a normal to $\mathcal{S}$, and *not* multiplied by a function on $\mathcal{S}$, it still holds formally $\delta_v\theta^{(\ell)} = \mathcal{L}_v\theta^{(\ell)}$. Then, we can evaluate $\delta_{\gamma v}\theta^{(\ell)}$ as $\delta_{\gamma v}\theta^{(\ell)} = \delta_{\tilde{v}}\theta^{(\ell)} = \mathcal{L}_{\tilde{v}}\theta^{(\ell)}$, with $\tilde{v}^a = \gamma v^a$. In particular, the application of this strategy to the second line of (35) goes as follows. We write $\tilde{k}^a = \beta k^a$ and calculate $\delta_{\tilde{k}}\theta^{(\ell)}$ through a Lie derivative evaluation. This results in

$$\delta_{\tilde{k}}\theta^{(\ell)} = (-\tilde{k}^c\ell_c)\left[{}^2D^c\Omega^{(\tilde{k})}_c + \Omega^{(\tilde{k})}_c\Omega^{(\tilde{k})^c} - \frac{1}{2}{}^2R\right] + 8\pi T_{ab}\tilde{k}^a\ell^b. \tag{39}$$

Using $(-\tilde{k}^c\ell_c) = \beta$, $\Omega^{(\tilde{k})}_a = \Omega^{(k)}_a + {}^2D_a\ln\beta$ and $\Omega^{(k)}_a = -\Omega^{(\ell)}_a$ the expression for $\delta_{\tilde{k}}\theta^{(\ell)}$ in (35) follows (cf. footnote b).

3.2.4. *Generic properties of dynamical FOTHs*

We review some generic properties of dynamical trapping horizons.[14, 44, 26, 32, 24]

(i) *Topology law*: under the dominant energy condition, sections $\mathcal{S}_t$ are topological spheres. This can be shown by integrating $\delta_k\theta^{(\ell)} < 0$ on $\mathcal{S}_t$. Under the assumed energy condition, the Euler characteristic χ

$$\chi = \frac{1}{4\pi}\int_{\mathcal{S}} {}^2R\, {}^2\epsilon = \frac{1}{2\pi}\int_{\mathcal{S}} (-\delta_k\theta^{(\ell)} + \Omega^{(\ell)}_c\Omega^{(\ell)^c} + 8\pi T_{ab}k^a\ell^b)\, {}^2\epsilon,$$

is positive and, being $\mathcal{S}_t$ a closed 2-surface, its spherical topology follows.

(ii) *Signature law*: under the null energy condition, $\mathcal{H}$ is completely partitioned into null worldtube sections (where $\delta_\ell\theta^{(\ell)} = 0$) and space-like worldtube sections (where $\delta_\ell\theta^{(\ell)} \neq 0$ at least on a point). Applying a maximum principle to the trapping horizon constraint condition, Eq. (37), it follows that either $C =$ const ≥ 0, or C is a function $C > 0$ everywhere on $\mathcal{S}$ (cf. discussion in Sec. 2.4).

(iii) *Area law*: under the null energy condition, if $\delta_\ell\theta^{(\ell)} \neq 0$ somewhere on $\mathcal{S}_t$, the area grows locally everywhere on $\mathcal{S}_t$. Otherwise the area in constant along the evolution. This follows from applying the future condition, $\theta^{(k)} < 0$, and the signature law to $\delta_h{}^2\sqrt{q} = -C\theta^{(k)}\sqrt{q}$ [cf. Eq. (17)].

(iv) *Preferred choice of null tetrad on a DH*. According to the *foliation uniqueness* and *existence* results discussed in Sec. 3.2.1, there is a unique evolution vector

[g] See also the treatment in terms of Lie derivatives in the *double null foliations* treatment in Refs. 14 and 23.

h^a tangent to $\mathcal{H}$ and orthogonal to $\mathcal{S}_t$, such that h^a transports $\mathcal{S}_t \in \Sigma_t$ onto $\mathcal{S}_{t+\delta t} \in \Sigma_{t+\delta t}$: that is, $\delta_h t = 1$, for a given function t defining a 3+1 spacetime foliation $\{\Sigma_t\}$. Denoting the unit timelike normal to Σ_t by n^a, the lapse function by N, i.e. $n_a = -N\nabla_a t$, and the normal to $\mathcal{S}_t$ tangent to Σ_t by s^a, we can write on the horizon $\mathcal{H}_N$

$$h^a = Nn^a + bs^a, \tag{40}$$

for some b fixed from N and C in (16), as $2C = (b+N)(b-N)$. The expression of the evolution vector as $h^a = \ell^a - Ck^a$ [cf. Eq. (16)] links the scaling of ℓ^a and k^a to that of h^a. In particular, ℓ^a is singled out as the only null normal to $\mathcal{S}_t$ such that $h^a \to \ell^a$ as the trapping horizon is driven to stationarity ($C \to 0 \Leftrightarrow \delta_\ell \theta^{(\ell)} \to 0$). Writing generically the null normals at $\mathcal{H}_N$ as $\ell^a = f \cdot (n^a + s^a)$ and $k^a = (n^a - s^a)/(2f)$, Eqs. (40) and (16) lead to a preferred scaling of null normals on the DH $\mathcal{H}_N$

$$\ell^a_N = \frac{N+b}{2}(n^a + s^a), \quad k^a_N = \frac{1}{N+b}(n^a - s^a). \tag{41}$$

3.2.5. *Geometric balance equations*

One of the main motivations for the development of quasi-local horizon formalisms is the extension of the laws of black hole thermodynamics to dynamical regimes. This involves in particular finding balance equations to control the rate of change of *physical quantities* on the horizon, in terms of appropriate fluxes through the hypersurface. This is an extensive subject whose review is beyond our scope. In the spirit of the present discussion, we restrain ourselves to comment on the balance equations for two geometric quantities on $\mathcal{S}_t$: the area $A = \int_{\mathcal{S}} dA = \int_{\mathcal{S}} {}^2\epsilon$ and the angular momentum $J[\phi]$ in Eq. (15), for an axial Killing (or, more generally, divergence-free) vector ϕ^a. That is, we aim at writing

$$\frac{dA}{dt} = \int_{\mathcal{S}_t} F^A \, dA, \quad \frac{dJ[\phi]}{dt} = \int_{\mathcal{S}_t} F^J \, dA, \tag{42}$$

for appropriate area F^A and angular momentum F^J fluxes, with d/dt associated to the foliation Lie-transported by h^a. Eventually, one would aim at writing a *first law of thermodynamics* by appropriately combining the previous balance equations

$$\kappa_t \frac{dA}{dt} + \Omega_t \frac{dJ[\phi]}{dt} = \int_{\mathcal{S}_t} F^E \, dA, \tag{43}$$

for some functions κ_t and Ω_t on $\mathcal{S}_t$, so that F^E is interpreted as an energy flux.[44, 26, 45–47, 24, 48–50] As a first step towards (42) we write evolution equations for the expansion $\theta^{(h)}$ and the 1-form $\Omega^{(\ell)}_a$ along the evolution vector h^a. These equations are given by the projection of some of the components of the Einstein equations onto $\mathcal{H}$. Introducing a 4-*momentum current density* $p_a = -T_{ab}\tau^b$, with τ^a

the vector orthogonal to $\mathcal{H}$ defined in (16), such equations provide three of the components of p_a. The fourth is given by the trapping horizon condition (38). In brief:

(i) Evolution element of area[51,52] ($p_a h^a = -T_{ab}\tau^b h^a$):

$$(\delta_h + \theta^{(h)})\theta^{(h)} = -\kappa^{(h)}\theta^{(h)} + \sigma^{(h)}_{ab}\sigma^{(\tau)ab} + \frac{(\theta^{(h)})^2}{2} - {}^2D^a Q_a + 8\pi T_{ab}\tau^a h^b - \frac{\theta^{(k)}}{8\pi}\delta_h C, \tag{44}$$

with $Q_a = \frac{1}{4\pi}[C\Omega^{(\ell)}_a - 1/2\,{}^2D_a C]$ and $\kappa^{(h)} = -h^b k^c \nabla_b \ell_c$.

(ii) Evolution normal (rotation) form[23,52] $\Omega^{(\ell)}_a$ ($p_b q^b{}_a = -T_{bc}\tau^c q^b{}_a$):

$$(\delta_h + \theta^{(h)})\Omega^{(\ell)}_a = {}^2D_a\kappa^{(h)} - {}^2D^c\sigma^{(\tau)}_{ac} - {}^2D_a\theta^{(h)} + 8\pi q^b{}_a T_{bc}\tau^c - \theta^{(k)}\,{}^2D_a C. \tag{45}$$

(iii) Normal component ($p_a\tau^a = -T_{ab}\tau^b\tau^a$): linear combination, using $\tau^a = 2\ell^a - h^a$, of $T_{ab}\tau^a h^b$ (area element evolution) and $T_{ab}\tau^a\ell^b$ [trapping horizon constraint (38)].

In order to derive the evolution equation for A, we write $A = \int_{\mathcal{S}} dA = \int_{\mathcal{S}} {}^2\epsilon$ so that, using the transport of $\mathcal{S}_t$ into $\mathcal{S}_{t+\delta t}$ by h^a, we have $\frac{dA}{dt} = \int_{\mathcal{S}} \delta_h(dA) = \int_{\mathcal{S}} \theta^{(h)} dA$ and $\frac{d^2A}{dt^2} = \int_{\mathcal{S}} (\delta_h\theta^{(h)} + (\theta^{(h)})^2) dA$. From Eq. (44) it then follows

$$\frac{d^2A}{dt^2} + \bar{\kappa}'\frac{dA}{dt} = \int_{\mathcal{S}_t}\left[8\pi T_{ab}\tau^a h^b + \sigma^{(h)}_{ab}\sigma^{(\tau)ab} + \frac{(\theta^{(h)})^2}{2} + (\bar{\kappa}' - \kappa')\theta^{(h)}\right]{}^2\epsilon, \tag{46}$$

where $\kappa' \equiv \kappa - \delta_h \ln C$ and $\bar{\kappa}' = \bar{\kappa}(t) \equiv A^{-1}\int_{\mathcal{S}_t} \kappa'^2\epsilon$. Note that this is a second-order equation for the area.[51] Near equilibrium, the second time derivative as well as higher-order terms can be neglected leading to the Hawking and Hartle expression[53]

$$\bar{\kappa}'\frac{dA}{dt} = \int_{\mathcal{S}_t} [8\pi T_{ab}\ell^a\ell^b + \sigma^{(\ell)}_{ab}\sigma^{(\ell)ab}]dA.$$

Regarding the evolution equation for $J[\phi]$, we make use of Eq. (45) together with a divergence-free condition on ϕ^a (that relaxes the Killing condition) and the condition that ϕ^a is Lie-dragged by the evolution vector h^a. Then[23,54,55]

$$\frac{d}{dt}J(\phi) = -\int_{\mathcal{S}_t} T_{ab}\tau^a\phi^b\,{}^2\epsilon - \frac{1}{16\pi}\int_{\mathcal{S}_t}\sigma^{(\tau)}_{ab}\delta_\phi q^{ab}\,{}^2\epsilon, \tag{47}$$

with the second term on the right-hand side accounting for a non-Killing ϕ^a. Interestingly in dynamical (spacelike) horizons $\mathcal{H}$, the conditions ${}^2D_a\phi^a = 0$ and $\delta_h\phi^a$ completely fix[54] the form of the vector ϕ^a: $\phi^a = {}^2\epsilon^{ab}\,{}^2D_b\theta^{(h)}$.

3.2.6. *Open geometric issues and physical remarks*

To close this generic section on geometric aspects of dynamical horizons, we list some relevant open geometric problems:

(i) *Canonical choice of dynamical trapping horizon.* DHs are highly non-unique in a given black hole spacetime. A natural question concerns the possibility of making a canonical choice. There has been some attempts in this direction based on *entropic* arguments.[51,56–58] A very interesting avenue lies in the recently introduced notion of the *core of the trapped region*[21] (see also J. M. M. Senovilla's contribution[15]).
(ii) *Asymptotics of dynamical horizons to the event horizon.* One would expect DHs to asymptote generically to the event horizon at late times. This is indeed a topic of active research.[59–61,16]
(iii) *Black hole singularity* covering *by dynamical horizons.* In addition to the asymptotics of DHs to the event horizon, it is also of interest to assess their behaviour at the *birth* of the black hole singularity, in particular their capability to separate (*dress*) singularities from the rest of the spacetime (see Sec. 5.4.4).

DHs as physical surfaces. Dynamical horizons are objects with very interesting geometric properties for the study of black hole spacetimes. In addition, from a physical perspective it is remarkable that they admit a nontrivial thermodynamical description (cf. A. Nielsen's contribution[22]). However, it is also important to underline that, if thought of as boundaries of compact physical objects (in the sense we think, say, of the surface of a neutron star), then they have nonstandard physical properties:

(a) They are *non-unique.* From an Initial Value Problem perspective, the question about the evolution of a given AH is not well-posed, since it depends on the 3+1 slicing choice (such non-uniqueness in evolution is typical in gauge dynamics).
(b) Dynamical trapping horizons are *superluminal*, something difficult to reconciliate with the physical surface of an object.
(c) DHs show a *nonlocal behavior.* For instance, they grow globally (reacting *as a whole*) when *energy* crosses them at a given local region (even a point). This is a consequence of their intrinsic *elliptic*, rather than *hyperbolic*, behavior.

4. Black Hole Spacetimes in an Initial-Boundary Value Problem Approach

In the context of an Initial-Boundary Value Problem approach to the construction of spacetimes, dynamical trapping horizons play a role at two levels: (i) first, as an *a priori* ingredient to be incorporated into a given PDE formulation of Einstein equations, and (ii) as an *a posteriori* tool to extract information of the constructed spacetimes. In this section we address their application as an *a priori* ingredient.

4.1. *The initial value problem in general relativity: 3+1 formalism*

Our general basic problem is the control[62] of the qualitative and quantitative aspects of *generic* solutions to Einstein equations in dynamical scenarios involving a black hole spacetime. The Initial-Boundary Value Problem approach provides a powerful avenue to it. Such a strategy is well suited, on the one hand, to the use of global analysis and Partial Differential Equations (PDE) tools for controlling the qualitative aspects of the problem and, on the other hand, to the employment of numerical techniques to assess the quantitative ones. In particular, we focus here on the Cauchy (and hyperboloidal) Initial Value Problem.

4.1.1. *Einstein equations: Constraint and evolution system*

General Relativity is a geometric theory in which not all the fields constitute physical degrees of freedom (gauge theory), so that constraints among the fields are present. In the passage from the geometric formulation of the theory to an analytic problem in the form of a specific PDE system, several PDE subsystems enter into scene.[63] First, the *constraint system* is determined by the (Gauss–Codazzi) conditions that data on a three-dimensional Riemannian manifold must satisfy to be considered as initial data on a spacetime slice. The Hamiltonian and momentum constraints are determined by the $G_{ab}n^b$ components of the Einstein equation, where n^a is a unit timelike vector normal to the initial slice. Second, the *evolution system* is built from the rest of Einstein equation, including possible auxiliary fields. The *gauge system* determines the dynamical choice of coordinates in the spacetime. Finally, a *subsidiary system* controls the internal consistency of the previous systems.

4.1.2. *3+1 formalism*

We introduce some notation regarding the 3+1 formalism.[64] As in Sec. 3.2.4, given a 3+1 slicing of spacetime by spacelike hypersurfaces $\{\Sigma_t\}$, the unit timelike normal to Σ_t is denoted by n^a and the lapse function as N, $n_a = -N\nabla_a t$, with t the scalar function defining the 3+1 slicing. The 3+1 evolution vector is denoted by $t^a = Nn^a + \beta^a$, where β^a is the shift vector. The induced metric on Σ_t is denoted by γ_{ab}, i.e. $\gamma_{ab} = g_{ab} + n_a n_b$. We choose the following sign convention for the extrinsic curvature of Σ_t in $\mathcal{M}$: $K_{ab} = -\gamma^c{}_a\nabla_c n_b = -\frac{1}{2}\mathcal{L}_n\gamma_{ab}$. In particular, we can write $K_{ij} = \frac{1}{2N}(\gamma_{ik}D_j\beta^k + \gamma_{jk}D_i\beta^k - \dot{\gamma}_{ij})$, where the dot denotes the derivative $\mathcal{L}_t$. Indices $i, j, k, \ldots$ are used for objects leaving on Σ_t. For concreteness, we focus on a particular 3+1 decomposition of Einstein equations, namely involving the following conformal decomposition (*conformal Ansatz*[65]) for data (γ_{ij}, K^{ij}) on Σ_t:

$$\gamma_{ij} = \Psi^4\tilde{\gamma}_{ij}, \quad K_{ij} = \Psi^\zeta\tilde{A}_{ij} + \frac{1}{3}K\gamma_{ij}, \tag{48}$$

for several ζ choices. Denoting by $\tilde{D}_i$ the Levi-Civita connection associated with $\tilde{\gamma}_{ij}$ and inserting (48) into Einstein equations leads to a coupled elliptic–hyperbolic PDE system on the variables Ψ, β^i, N and $\tilde{\gamma}_{ab}$. The elliptic part has the form

$$\begin{aligned}
\tilde{D}_k\tilde{D}^k\Psi - \frac{{}^3\tilde{R}}{8}\Psi &= S_\Psi[\Psi, N, \beta^i, K, \tilde{\gamma}, \ldots], \\
\tilde{D}_k\tilde{D}^k\beta^i + \frac{1}{3}\tilde{D}^i\tilde{D}_k\beta^k + {}^3\tilde{R}^i_k\beta^k &= S_\beta[\Psi, N, \beta^i, K, \tilde{\gamma}, \ldots], \\
\tilde{D}_k\tilde{D}^kN + 2\tilde{D}_k\ln\Psi\tilde{D}^kN &= S_N[N, \Psi, \beta^i, K, \tilde{\gamma}, \dot{K}, \ldots],
\end{aligned} \tag{49}$$

where the equation on Ψ follows from the Hamiltonian constraint, the equation on β^i follows from the momentum constraint and the third equation on N follows from a (gauge) condition imposed on $\dot{K}$. If only solved on an initial slice with $\tilde{\gamma}_{ij}$, $\dot{\tilde{\gamma}}^{ij}$, K and $\dot{K}$ as free data, this system constitutes the *Extended Conformal Thin Sandwich* approach to initial data.[66,67] If we solve it during the whole evolution, together with

$$\frac{\partial^2\tilde{\gamma}^{ij}}{\partial t^2} - \frac{N^2}{\Psi^4}\Delta\tilde{\gamma}^{ij} - 2\mathcal{L}_\beta\frac{\tilde{\gamma}^{ij}}{\partial t} + \mathcal{L}_\beta\mathcal{L}_\beta\tilde{\gamma}^{ij} = S^{ij}_{\tilde{\gamma}}[N, \Psi, \beta^i, K, \tilde{\gamma}, \ldots], \tag{50}$$

for $\tilde{\gamma}_{ij}$, it defines a particular constrained evolution formalism.[68–70]

4.2. *Initial data: Isolated horizon inner boundary conditions*

There are two standard approaches to ensure that initial data on a slice Σ_0 correspond to a black hole spacetime. The *punctures* approach exploits the nontrivial topology[71,72] of Σ_0, whereas the *excision* approach removes a sphere from the initial slice and enforces it to be inside the black hole region. In a sense, they both reflect the *global* versus *quasi-local* discussion in Sec. 1. Here we discuss the use of inner boundary conditions derived from the IH formalism, when constructing initial data of black holes *instantaneously in equilibrium* in an *excision approach*.

4.2.1. *Non-expanding horizon conditions*

The NEH condition $\Theta^{(\ell)}_{ab} = 0$ in Eq. (26) [or (25)] provides three inner boundary conditions for the elliptic system (49). In particular, they enforce the excised surface $\mathcal{S}_0$ to be a section of a quasi-local horizon instantaneously in equilibrium.

For a given choice of free initial data in system (49), the geometric NEH inner boundary conditions, $\Theta^{(\ell)}_{ab} = 0$, must be complemented with two additional inner boundary (gauge) conditions. Denoting by s^i the normal vector to $\mathcal{S}_t$ tangent to Σ_t, we write $\beta^i = \beta^\perp s^i + \beta^i_\parallel$, with $\beta^\perp = \beta^i s_i$ and $\beta^i_\parallel s_i = 0$. Adapting the coordinate system to the horizon (i.e. $t^a = \ell^a + \beta^a_\parallel \Leftrightarrow \beta^\perp = N$) supplies a fourth gauge condition that, together with the $\theta^{(\ell)} = 0$ and $\sigma^{(\ell)}_{ab} = 0$ NEH conditions, reads[73–75,28]

$$\tilde{s}^i \tilde{D}_i \Psi + \tilde{D}_i \tilde{s}^i \Psi + \Psi^{-1} K_{ij} \tilde{s}^i \tilde{s}^j - \Psi^3 K = 0,$$
$$^2\tilde{D}_a \tilde{\beta}^{\|}_b + {}^2\tilde{D}_b \tilde{\beta}^{\|}_a - ({}^2\tilde{D}_c \beta^c_{\|}) \tilde{q}_{ab} = 0, \quad \beta^{\perp} = N, \tag{51}$$

where $\tilde{q}_{ab} = \Psi^4 q_{ab}$ and $\tilde{\beta}^{\|}_a = \tilde{q}_{ab} {\beta^{\|}}^b$. A fifth boundary condition, namely for N, can be obtained by choosing a slicing inner boundary condition. The (gauge) *weakly isolated horizon* structure can be used in this sense.[76,28]

4.2.2. *(Full) isolated horizon conditions*

The next geometric quasi-equilibrium horizon structure is a (full) IH (cf. Secs. 3.1.4 and 3.1.5). This involves three additional conditions that cannot be accommodated in system (49) for fixed free initial data. However, we can revert the argument and employ IH conditions to determine improved quasi-equilibrium free initial data $\tilde{\gamma}_{ab}$ and $\dot{\tilde{\gamma}}_{ab}$ by solving the full set of Einstein equations (49) and (50) under a *quasi-equilibrium Ansatz.* Namely, we can set $\partial_t \tilde{\gamma}^{ab}$ and $\frac{\partial^2 \tilde{\gamma}^{ab}}{\partial t^2}$ in (50) to prescribed functions f^{ab}_1 and f^{ab}_2 and consider the elliptic system formed by (49) together with

$$-\frac{N^2}{\Psi^4} \tilde{\Delta} \tilde{\gamma}^{ab} + \mathcal{L}_\beta \mathcal{L}_\beta \tilde{\gamma}^{ab} = S^{ab}_{\tilde{\gamma}} - f^{ab}_2 + 2\mathcal{L}_\beta f^{ab}_1. \tag{52}$$

This extended elliptic system is solved for ten fields: (Ψ, β^a, N) and the five $\tilde{\gamma}^{ab}$. Geometrically, we need to impose four gauge inner conditions, leaving exactly six inner conditions to be fixed. Remarkably, this fits exactly the six IH conditions[77]

$$\Theta^{(\ell)}_{ab} = 0, \quad \Theta^{(k)}_{ab} = \Theta^{(k)}_{ab}(\kappa_o, \tilde{q}_{ab}, \Omega^{(\ell)}_a) \Leftrightarrow F^{\Theta^{(k)}}_{ab}(\kappa_o, \Psi, \beta^a, N, \tilde{\gamma}_{ab}) = 0, \tag{53}$$

where $F^{\Theta^{(k)}}_{ab}$ is determined by the expression for $\Theta^{(k)}_{ab}$ in Eq. (31), fixed up to the value of the constant κ_o. It is interesting to remark that this IH prescription[77] completely fixes (up to a κ_o one-parameter family) the extrinsic curvature tensor $\mathcal{K}^c_{ab} = k^c \Theta^{(\ell)}_{ab} + \ell^c \Theta^{(k)}_{ab}$ [cf. Eq. (8)] of $\mathcal{S}_0$ as embedded in the spacetime $\mathcal{M}$.

4.3. *Constrained evolutions: Trapping horizon inner boundary conditions*

The elliptic–hyperbolic system (49)–(50) provides a constrained evolution scheme for the dynamical construction of the spacetime. Adopting an excision approach to black holes, we need five inner boundary conditions for the elliptic part of the system. In principle, dynamical trapping horizon conditions on the inner boundary worldtube $\mathcal{H} = \cup_t \mathcal{S}_t$ provide a geometric prescription guaranteeing that $\mathcal{H}$ remains in the black hole region. However, imposing FOTH conditions on $\mathcal{H}$ can be *too stringent* in generic evolutions. The reason is that the constructed worldtube of MOTS $\mathcal{H}$, regarded as a hypersurface in spacetime, can change signature.

This is in conflict with the outer condition in Sec. 2.2 (something related to *jumps* occurring generically[78–80] in AH evolutions; see Sec. 5.1.1) so that the resulting PDE system can become ill-posed. In this context, trapping horizon conditions, together with the requirement of recovering NEH inner conditions at the equilibrium limit, provide an appropriate relaxed set of inner boundary conditions.[81] More specifically, trapping horizon conditions provide two geometric conditions $\theta^{(\ell)} = 0$ and $\delta_h\theta^{(\ell)} = 0$, whereas three additional gauge conditions guarantee the recovery of NEH at equilibrium.

As a first step, as in Sec. 4.2.1, we choose a coordinate system adapted to the horizon. This means that spacetime evolution t^a is tangent to $\mathcal{H}$. Decomposing the shift as $\beta^a = \beta^\perp s^a + \beta^a_\parallel$, then t^a is written as $t^a = Nn^a + \beta^a = (Nn^a + bs^a) + \beta^a_\parallel + (\beta^\perp - b)s^a = h^a + \beta^a_\parallel + (\beta^\perp - b)s^a$. Therefore t^a is tangent to $\mathcal{H}$ if and only if $\beta^\perp = b$.

(i) *Geometric trapping horizon conditions.* Condition $\theta^{(\ell)} = 0$ leads, in terms of the 3+1 quantities in Sec. 4.1.2, to the expression in the first line of Eq. (51). Condition $\delta_h\theta^{(\ell)} = 0$ in Eq. (38), using the adapted coordinate system $\beta^\perp = b$, leads to

$$[-{}^2D_a{}^2D^a - 2L^a{}^2D_a + A](\beta^\perp - N) = B(\beta^\perp + N), \tag{54}$$

where $L_a = K_{ij}s^i q^j{}_a$, $A = \frac{1}{2}{}^2R - {}^2D_aL^a - L_aL^a - 4\pi T_{ab}(n^a + s^a)(n^b - s^b)$, and $B = \frac{1}{2}\sigma^{(\hat{\ell})}_{ab}\sigma^{(\hat{\ell})ab} + 4\pi T_{ab}(n^a + s^b)(n^b + s^b)$, with $\hat{\ell}^a = n^a + s^a$.

(ii) *Gauge boundary conditions* I. Aiming at recovering NEH boundary conditions for $\beta^a_\parallel$, we first express $\delta_h q_{ab} = \theta^{(h)}q_{ab} + 2\sigma^{(h)}_{ab}$ in adapted coordinates ($h^a = t^a - \beta^a_\parallel$)

$$2\sigma^{(h)}_{ab} = \left(\frac{\partial q_{ab}}{\partial t} - \frac{\partial}{\partial t}\ln\sqrt{q}\, q_{ab}\right) - ({}^2D_a\beta^\parallel_b + {}^2D_b\beta^\parallel_a - {}^2D_c\beta^c_\parallel q_{ab}). \tag{55}$$

Then, the coordinate choice $\partial_t q_{ab} - \partial_t \ln\sqrt{q}\; q_{ab} = 0$ leads to the condition on $\beta^\parallel_a$

$${}^2D_a\beta^\parallel_b + {}^2D_b\beta^\parallel_a - {}^2D_c\beta^c_\parallel q_{ab} = -2\sigma^{(h)}_{ab}, \tag{56}$$

that is completed by using the evolution equation for $\sigma^{(h)}_{ab}$ on $\mathcal{H}$

$$\begin{aligned}\delta_h\sigma^{(h)}_{ab} &= -q^d{}_a q^f{}_b C^c{}_{def}\ell_c\ell^e - C^2 q^d{}_a q^f{}_b C^c{}_{def}k_c k^e \\ &\quad - 8\pi C\left[q^c{}_a q^d{}_b T_{cd} - \frac{1}{2}(q^{cd}T_{cd})q_{ab}\right] + \cdots. \end{aligned}\tag{57}$$

(iii) *Gauge boundary conditions* II. The slicing condition for N is essentially free. However, from Properties 1 and 2 in Sec. 3.2.1, such a choice is equivalent to choosing particular a dynamical horizon $\mathcal{H}$. Since each $\mathcal{H}$ is a genuine geometric object, this suggests the possibility of recasting into *geometric* terms the gauge choice of inner boundary condition for N, by selecting a trapping horizon $\mathcal{H}$ satisfying some specific *geometric criterion* for $\mathcal{H}$. As an example of this, maximizing the area growth rate $\dot{A}$ of $\mathcal{H}$ leads[51,81] to the condition $\beta^{\perp} - N = -\text{const.} \cdot \theta^{(\hat{k})}$, with $\hat{k}^a = n^a - s^a$.

5. *A posteriori* Analysis of Black Hole Spacetimes

We address here the application of dynamical trapping horizons to the *a posteriori* analysis of spacetimes, their main application in the Initial Value Problem approach.

5.1. *"Tracking" the* black hole region*: AH finders*

As discussed in Sec. 1.2, event horizons cannot be located during the spacetime evolution. However, in applications such as numerical relativity, assessing if a region of spacetime lies inside the black hole region can be crucial during the evolution. Under the assumption of cosmic censorship, the location of AHs in spatial sections Σ_t and the worldtubes constructed by piling them up (see Sec. 1.3.1) are extremely useful to determine the evolutive properties of the black hole. In this sense, *AH finders* prove to be extraordinary practical tools. These are algorithms for searching surfaces $\mathcal{S}_t \subset \Sigma_t$ that satisfy the MOTS condition $\theta^{(\ell)} = 0$. There are many approaches to this problem,[82] but all of them aim at solving the condition $D_i s^i - K + K_{ij} s^i s^j = 0$. For instance, assuming spherical topology, we can characterize the surface in an adapted (spherical) coordinate system as $F(r, \theta, \varphi) = r - h(\theta, \varphi)$ with $F = \text{const}$, so that the normal vector to $\mathcal{S}_t$ is given by $s_i = \frac{1}{\sqrt{D^i F \cdot D_i F}} D_i F$ with $D_i F = (1, -\partial_\theta h, -\partial_\varphi h)$ in the spherical coordinate system. The MOTS condition becomes then a nonlinear elliptic equation on h that can be solved very efficiently.

5.1.1. *Understanding AH jumps*

Noncontinuous *jumps* of AHs occur generically in 3+1 black hole evolutions. The dynamical trapping horizon framework sheds light[79,80,83] on these *AH jumps*, suggesting a spacetime picture where the jumps are understood as multiple spatial cuts of a single underlying spacetime MOTS worldtube. Jumps are associated with the change of metric type of the horizon hypersurface (see Fig. 2). This is particularly dramatic in binary black hole simulations, where at a given time t the two individual nonconnected horizons *jump* to a common one. A specific prediction of the dynamical horizon picture is that new (common) horizons form in *pairs*[37,83]: the outermost (apparent) horizon growing in area and a *dual* inner one whose area decreases in the time t. Apart from providing a better understanding of the underlying geometry of the trapped region, this spacetime picture can be of use in the study of flows interpolating between a given MOTS and the eventual event horizon, something of potential interest for studies of the Penrose inequality (see Sec. 5.2.2).

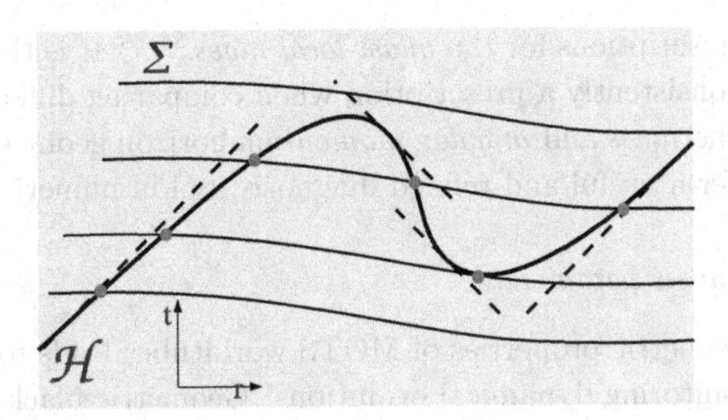

Figure 2. (Color online) Illustration of AH jumps as multiple cuts of a single spacetime MOTS-worltube $\mathcal{H}$. In particular, timelike sections of $\mathcal{H}$ produce jumps (null hypersurfaces are represented with 45°).

5.2. *Horizon analysis parameters*

Assigning parameters to (individual) black holes can offer crucial insight into the dynamical evolution. These can be physical parameters like the mass or the angular momentum, or diagnosis parameters informing of relevant dynamical properties. Given the generic absence of background rigid structures, first-principles parameters are often out of reach and one must follow nonrigorous or *pragmatic* approaches.

5.2.1. *Mass and angular momentum. IH and DH multipoles*

In our discussion we have avoided entering into first-principles physical issues, stressing rather the geometric properties of dynamical trapping horizons and their applications. However, mass and angular momentum estimates for individual black holes, either fundamental or effective, are extremely important in the modeling of astrophysical systems involving matter or binary systems. The problem has two aspects. First, one must identify a surface to be associated with the black hole boundary. Discussion in Sec. 1 shows that this is a delicate question. In any case, AHs provide surfaces $\mathcal{S}_t \in \Sigma_t$ tracking the black hole region, that can be employed as preferred choices for pragmatic estimations. The second problem refers to the ambiguities in the quasi-local characterization of the gravitational field mass and angular momentum in General Relativity.[84,85] Regarding the angular momentum, the Komar expression (15) characterizes appropriately the axisymmetric case. Effective prescriptions[86–88] exist for generic horizons. Regarding the mass, the irreducible mass M_{irred} $A \equiv 16\pi M_{\mathrm{irred}}^2$ provides a purely geometric estimation in terms of the area. Its physical interpretation as the portion of the black hole mass that cannot be extracted by a Penrose process, together with its equivalence with the Hawking energy, $M_{\mathrm{Hawking}} = \sqrt{A/(16\pi)}(1+1/(8\pi)\oint \theta^{(\ell)}\theta^{(k)}dA)$ for MOTSs, makes it useful in numerical applications and in thermodynamical treatments.[45,46] Given A and J one can also consider[32] the Christodoulou expression for the Kerr mass

$$M_{\mathrm{Chris}} = \left(\frac{A}{16\pi} + \frac{4\pi J^2}{A}\right)^{\frac{1}{2}}. \tag{58}$$

There are many prescriptions for the *quasi-local mass*.[84,85] It is therefore crucial to choose and keep consistently a prescription when comparing different solutions. In this latter sense, the *mass* and *angular momentum* horizon geometric multipoles I_n and L_n in (34) offer a useful and refined diagnosis tool in numerical studies.[37,89]

5.2.2. *Useful diagnosis parameters*

Insight into the geometric properties of MOTS worldtubes leads to useful diagnosis parameters for monitoring dynamical evolutions. Geometric black hole inequalities provide a particular avenue. In particular, the conjectured Penrose's inequality $A \leq 16\pi M^2_{\rm ADM}$ for asymptotically flat spacetimes provides a bound to the AH area (strictly speaking, the bound is on the area of a minimal surface enclosing the AH). A violation of $\epsilon_{\rm Penrose} \equiv A/(16\pi M^2_{\rm ADM}) \leq 1$ *indicates* a more exterior MOTS. In the axially symmetric case this can be refined in terms of a so-called[90,91] *Dain number*

$$\epsilon_{\rm Dain} \equiv \frac{A}{8\pi(M^2_{\rm ADM} + \sqrt{M^4_{\rm ADM} - J^2})} \leq 1. \tag{59}$$

Moreover, the rigidity part of the conjecture provides an extremely simple characterization of Kerr as satisfying $\epsilon_{\rm Dain} = 1$. In the same spirit, the geometric inequality[92] $J \leq M^2_{\rm ADM}$ provides a characterization of (sub)extremality of black holes. However, these inequalities involve total quantities such as the ADM mass. It is remarkable that the dynamical horizon structure (actually the *outer* trapping horizon condition) provides exactly the needed conditions to prove the quasi-local inequality[93,94,42]

$$A \geq 8\pi|J|, \tag{60}$$

in generic spacetimes with matter satisfying the dominant energy condition. The validity of the area-angular momentum inequality (60) is equivalent to the non-negativity of the surface gravity κ of isolated and dynamical horizons,[32] supporting the internal consistency of the first law of black hole thermodynamics. Inequality (60) provides a quasi-local characterization of black hole (sub)extremality, that is directly related to changes in the horizon metric type[80] and jumps discussed in Sec. 5.1.1. This is also the context of the *Booth & Fairhurst extremality parameter*[80,95]

$$e \equiv 1 + \frac{1}{4\pi}\int_{\mathcal{S}} dA\, \delta_k \theta^{(\ell)} \leq 1. \tag{61}$$

5.3. *Heuristic and effective approaches in a posteriori spacetime analysis*

Hitherto we have discussed analysis tools to be applied in numerically constructed spacetimes, but related to sound geometric structures. However, when developing a qualitative understanding of the underlying dynamics, involving, e.g. a comparison with Newtonian or Special Relativity scenarios, the available geometric notions are often not enough. This is manifest in astrophysical contexts requiring estimations for linear, orbital angular momentum or binding energies. In some cases, a choice must be done between saying nothing at all or rather adopting a *heuristic* approach.

An example of the latter is the following *heuristic* proposal[96] for a quasi-local black hole linear momentum. Given a vector ξ^a transverse to a MOTS $\mathcal{S}$, applying on $\mathcal{S}$ the linear momentum ADM prescription at spatial infinity leads to

$$P(\xi) = \frac{1}{8\pi}\int_{\mathcal{S}_t}(K_{ab} - K\gamma_{ab})\xi^a s^b \, {}^2\epsilon. \qquad (62)$$

In spite of its *ad hoc* nature, this quantity has been successfully applied in the analysis[96] of linear and orbital angular momentum in binary black hole orbits and in the recoil dynamics of the black hole resulting from asymmetric binary mergers.

5.4. *An effective correlation approach to the analysis of spacetime dynamics*

The qualitative and quantitative understanding of strong-field spacetime dynamics represents a challenge in gravitational physics both at a fundamental level and in applications. In astrophysical settings a natural strategy consists in extending to general relativistic scenarios the Newtonian *celestial mechanics* approach. This has indeed led to fundamental achievements in the understanding of the physics of compact objects. However, the focus on the properties of individual objects, in particular in multi-component systems, also meets fundamental obstacles in a gravitational theory (i) without *a priori* rigid structures providing canonical structures, and (ii) with global aspects playing a crucial role. The latter encompasses global causal issues and also the in-built elliptic character of certain objects, both aspects relevant in the characterization of black holes. In this context, an approach to spacetime analysis that explicitly emphasizes the global/quasi-local properties of the relevant fields, at the price of renouncing a detailed tracking of the geometry and *trajectories* of small compact regions, can offer complementary insights to the *celestial mechanics* approach. Such a *coarse-grained effective* description is much in the spirit of the *correlation* approach in the analysis of complex condensed-matter systems or in quantum/statistical-field theory, where the functional structure of the (local) dynamical fields is encoded in the associated *n-point correlation functionals.*[h] Such an approach underlines the *relational aspects* of the theory, as a complementary methodology to the isolation of the dynamical properties of compact parts of the system. In sum, we can paraphrase the strategy as aiming at *a functional and coarse-grained description of the spacetime geometry, by importing functional tools for the analysis of condensed matter and quantum/statistical field theory systems.*

[h] *N*-point correlation functions encode the functional structure of the local fields. A coarse-grained description appear as a truncation to a finite number of *n*-point functions.

5.4.1. *Cross-correlations of geometric quantities at test screens*

The strategy outlined above is admittedly vague. We sketch now a particular implementation[97] of some of its aspects in a *cross-correlation* approach to the analysis of spacetime dynamics. Aiming at studying the gravitational dynamics in a given spacetime region $\mathcal{R}$, we consider an *outer* $\mathcal{B}_o$ and an *inner* $\mathcal{B}_i$ hypersurfaces lying in the causal future of $\mathcal{R}$. These hypersurfaces are taken as outer and inner boundaries of the bulk spacetime region of interest. The geometry of $\mathcal{B}_o$ and $\mathcal{B}_i$ is causally affected by the dynamics in $\mathcal{R}$, so that $\mathcal{B}_o$ and $\mathcal{B}_i$ can be understood as *balloon probes* into the spacetime geometry. In other words, $\mathcal{B}_o$ and $\mathcal{B}_i$ provide *test screens* (they do not *back-react* on the bulk dynamics) on which we can construct *geometric quantities* h_o and h_i to be cross-correlated. Choosing causally disconnected screens $\mathcal{B}_o$ and $\mathcal{B}_i$, a nontrivial correlation between h_o and h_i encodes geometric information about the common past region $\mathcal{R}$. We can think of this as the reconstruction of the interaction region from the debris in a *scattering* experiment (*inverse scattering* picture). Let us now restrict ourselves to the study of near-horizon spacetime dynamics.[97] In an (asymptotically flat) black hole spacetime setting, null infinity $\mathscr{I}^+$ and the (event) black hole horizon $\mathscr{E}$ provide *canonical* choices for $\mathcal{B}_o$ and $\mathcal{B}_i$, respectively (cf. Fig. 3). Retarded and advanced null coordinates u and v provide good parameters for quantities h_o and h_i calculated as integrals on sections $\mathcal{S}_u \subset \mathscr{I}^+$ and $\mathcal{S}_v \subset \mathcal{E}$. A meaningful notion for the cross-correlation between $h_o(u)$ and $h_i(v)$, considered as time series, requires the introduction of a (gauge-dependent) mapping between u and v at $\mathscr{I}^+$ and $\mathcal{E}$. We refer to this point as the *time-stretching issue.*

5.4.2. *Cross-correlations in an Initial Value Problem approach: Dynamical horizons as canonical inner probe screens*

The adopted Initial Value Problem approach has a direct impact in the *cross-correlation* picture above. In particular, the event horizon is not available during the evolution.[i] Instead, the (outermost) DH $\mathcal{H}$ fixed by the chosen 3+1 foliation stands as a natural spacetime inner boundary $\mathcal{B}_i$. Although *any* hypersurface *covering* the black hole singularity could be envisaged for the present cross-correlation purposes, the DH $\mathcal{H}$ provides a natural geometric prescription. Regarding the *time-stretching issue*, the time function t defining the 3+1 spacetime slicing automatically implements a (gauge) mapping between *retarded* and *advanced* times u and v. Cross-correlations between geometric quantities at $\mathcal{H}$ and $\mathscr{I}^+$ can then be calculated as standard time-series $h_i(t)$ and $h_o(t)$ (cf. Fig. 4). Due to the gauge nature of t, the geometric information in quantities $h_i(t)$ and $h_2(t)$ is not encoded in their *local* (arbitrary) time dependence, but rather in the *global* structure of successive maxima and minima. The calculation of cross-correlations must take this into account.[97] This means, in particular, that quantities to be correlated must be scalars.

---

[i]Regarding $\mathscr{I}^+$, a pragmatic choice in a Cauchy approach consists in substituting it by a timelike worldtube of large radii spheres. However, $\mathscr{I}^+$ can be kept if using a hyperboloidal foliation.

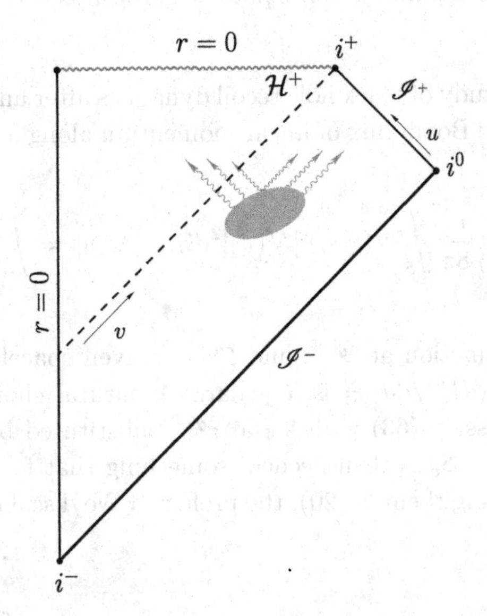

Figure 3. (Color online) Carter–Penrose diagram representing a generic (spherically symmetric) collapse and illustrating the *cross-correlation* approach to near-horizon gravitational dynamics.

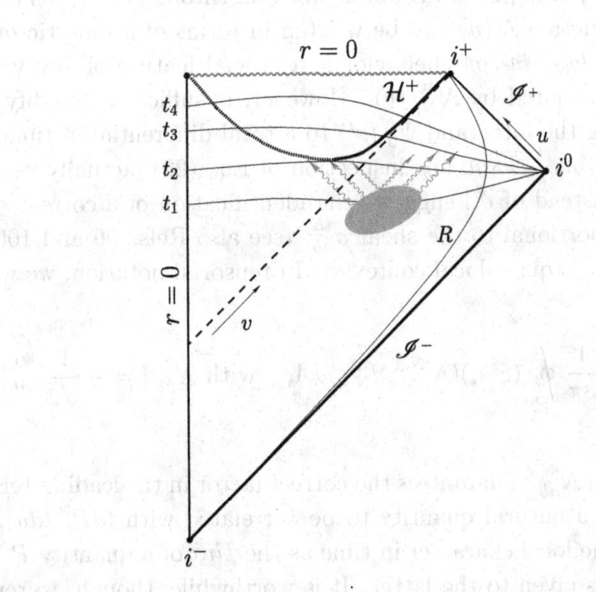

Figure 4. (Color online) Carter–Penrose diagram for the *cross-correlation* picture in a Cauchy IVP approach.

5.4.3. *Application to black hole recoil dynamics: Towards DH news functions*

In the context of the study of black hole recoil dynamics after an asymmetric merger, let us take $h_o(u)$ as the Bondi flux of linear momentum along a (preferred) direction

$$\frac{dP^{\mathrm{B}}[\xi]}{du}(u) = \lim_{(u,r\to\infty)} \frac{r^2}{8\pi} \oint_{\mathcal{S}_{u,r}} (\xi^i s_i)\, |\mathcal{N}(u)|^2 d\Omega, \quad \mathcal{N}(u) = \int_{-\infty}^{u} \Psi_4(u')du'. \tag{63}$$

Here $\mathcal{N}$ is the news function at $\mathscr{I}^+$, and ξ^a is a given spacelike transverse direction to $\mathcal{S}_{u,r}$, so that $(dP^{\mathrm{B}}/du)[\xi]$ is a scalar. A natural choice[j] for $h_i(v)$ would be given by the expression (63) with Ψ_4 at $\mathscr{I}^+$ substituted by some Ψ_0 at $\mathcal{H}$. A preferred null tetrad on $\mathcal{S}_v$ is then needed, something that for DHs is provided by ℓ^a_N and k^a_N in (41). Using them in (20), the preferred Weyl scalar Ψ_0^N is employed to construct

$$\tilde{K}^N[\xi](v) = -\frac{1}{8\pi} \oint_{\mathcal{S}_v} (\xi^i s_i)|\tilde{\mathcal{N}}_N^{(0)}(v)|^2 dA, \quad \text{with } \tilde{\mathcal{N}}_N^{(0)}(v) = \int_{v_0}^{v} \Psi_0^N(v')dv'. \tag{64}$$

In spite of the formal similarity between (63) and (64) there is a fundamental difference: whereas $(dP^{\mathrm{B}}/du)[\xi]$ is an instantaneous flux through $\mathscr{I}^+$, this is not true for $\tilde{K}^N[\xi](v)$. The function $\mathcal{N}(u)$ can be written in terms of geometric quantities on sections $\mathcal{S}_u$. This *local-in-time* behavior is a crucial feature of any valid *news function* and it is not shared by $\tilde{\mathcal{N}}_N^{(0)}(v)$. However, it suffices to modify $\tilde{\mathcal{N}}_N^{(0)}(v)$ with terms completing the integrand $\Psi_0^N(v')$ to a total differential in time. Noting $q^c_a q^d_b C_{lcfd}\ell^l \ell^f = \Psi_0 \overline{m}_a \overline{m}_b + \overline{\Psi}_0 m_a m_b$, inspection of Eq. (22) [actually its dynamical version with h^a instead of ℓ^a] suggests the identification of a correct *news-like* function at $\mathcal{H}$ as proportional to the shear $\sigma^{(h)}_{ab}$ (see also Refs. 99 and 100 for the discussion of the news in quasi-local contexts). In tensorial notation, we write

$$\frac{dP^N}{dv}[\xi](v) = -\frac{1}{8\pi} \oint_{\mathcal{S}_v} (\xi^i s_i)(\mathcal{N}^{\mathrm{N,g}}_{ab} \mathcal{N}^{ab}_{\mathrm{N,g}})dA, \quad \text{with } \mathcal{N}^{\mathrm{N,g}}_{ab} = -\frac{1}{\sqrt{2}}\sigma^{(h)}_{ab}, \tag{65}$$

where the coefficient in $\mathcal{N}^{\mathrm{N,g}}_{ab}$ guarantees the correct factor in the leading-term. This $(dP^N/dv)[\xi]$ provides a natural quantity to be correlated with $(dP^{\mathrm{B}}/du)[\xi]$. The notation underlines the local character in time as the *flux* of a quantity $P^N[\xi]$, but no physical meaning is given to the latter. It is worthwhile, though, to remark on

[j] We also mention an effective curvature vector[97,98] constructed from the Ricci scalar ${}^2R$ on sections $\mathcal{S}_v$ of $\mathcal{H}$, that provides an intrinsic prescription for $h_i(v)$ leading to nontrivial[97] cross-correlations with $(dP^{\mathrm{B}}/du)[\xi]$.

the formal similarity of the monopolar part of the square of the news $\mathcal{N}_{ab}^{\mathrm{N,g}}$, i.e.

$$\frac{dE^N}{dv}(v) = \frac{1}{16\pi}\oint_{\mathcal{S}_v} \sigma^{(h)}_{ab}\sigma^{ab}_{(h)} dA$$
$$= \frac{1}{16\pi}\oint_{\mathcal{S}} [\sigma^{(\ell)}_{ab}\sigma^{(\ell)ab} - 2C\sigma^{(\ell)}_{ab}\sigma^{(k)ab} + C^2\sigma^{(k)}_{ab}\sigma^{(k)ab}] dA \tag{66}$$

with the expression of the flux of gravitational energy[44,26] through a DH, in particular with its *transverse* part.[45,46] The identification of $\sigma^{(h)}_{ab}$ as a news-like function suggests a further step, by introducing a heuristic notion of *Bondi-like* 4-momentum flux through $\mathcal{H}$. Considering the unit normal $\hat{\tau}^a$ to $\mathcal{H}$ ($\hat{\tau}^a = \tau^a/\sqrt{|\tau^b\tau_b|} = (\ell^a + Ck^a)/\sqrt{2C} = (bn^a + Ns^a)/\sqrt{2C}$), and for a generic spacetime vector η^a

$$\frac{dP^N_\tau}{dv}[\eta] = -\frac{1}{16\pi}\oint_{\mathcal{S}_v} (\eta^a\hat{\tau}_a)\sigma^{(h)}_{ab}\sigma^{ab}_{(h)} dA, \tag{67}$$

has formally the expression of a *Bondi-like* 4-momentum.[k] The flux of energy associated with an Eulerian observer n^a would be

$$\frac{dE^N_\tau}{dv}(v) \equiv \frac{dP^N_\tau}{dv}[n^a] = \frac{1}{16\pi}\oint_{\mathcal{S}} \frac{b}{\sqrt{2C}}(\sigma^{(h)}_{ab}\sigma^{(h)ab}) dA, \tag{68}$$

where $\frac{b}{\sqrt{2C}} = \sqrt{1 + N^2/2C}$. The flux of linear momentum for $\xi^a \in T\Sigma_t$ would be

$$\frac{dP^N_\tau}{dv}[\xi] = -\frac{1}{16\pi}\oint_{\mathcal{S}_v} \frac{N}{\sqrt{2C}}(\xi^a s_a)(\sigma^{(h)}_{ab}\sigma^{(h)ab}) dA. \tag{69}$$

Near equilibrium ($C \to 0$), we have $\sigma^{(h)}_{ab}\sigma^{ab}_{(h)} \sim C$ on DHs [cf. Eq. (17)] so that expressions (68) and (69) are regular ($O(\sqrt{C})$). Integrating (69) in time would lead to a Bondi-like counterpart[l] of the heuristic *ADM-like* linear momentum in (62).

Before finishing this section, let us mention that the present discussion on horizon news-like functions can be related[97] to a *viscous fluid analogy* for quasi-local horizons.[23,51] In particular, geometric *decay* and *oscillation timescales* (respectively, τ and T) can be constructed on the horizon[97] from the expansion $\theta^{(h)}$ and shear $\sigma^{(h)}_{ab}$, respectively related to bulk and shear viscosity terms. In the context of black hole recoil dynamics, this provides an instantaneous geometric prescription for a *slowness parameter*[101] $P = T/\tau$ controlling the qualitative aspects of the dynamics.

---

[k] An alternative expression would follow by using in (67), instead of $\sigma^{(h)}_{ab}\sigma^{(h)ab}$, the integrand in the DH energy flux,[44,26,45,46] that would also include the *longitudinal* part $\Omega^{(\ell)}_a\Omega^{(\ell)a}$.

[l] A related prescription for a DH linear momentum flux would be given by angular integration of the appropriate components in the *effective gravitational-radiation energy-tensor* of Ref. 46.

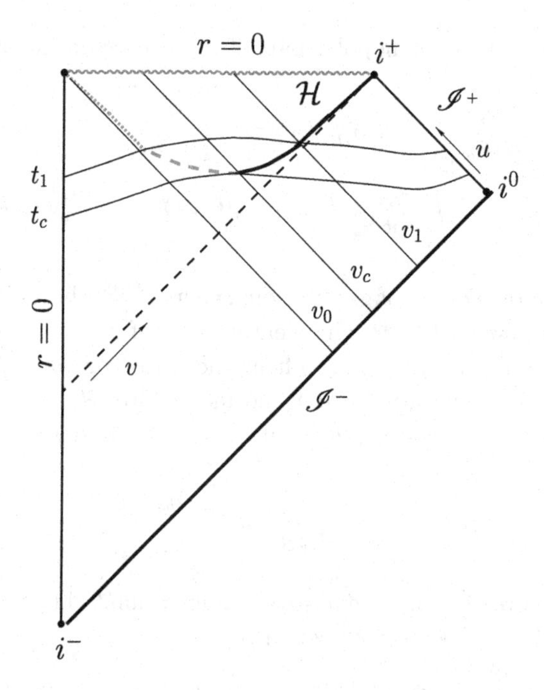

Figure 5. (Color online) Illustration of the splitting of a DH into *internal* and *external* sections by a 3+1 slicing.

5.4.4. *The role of the inner horizon in the integration of fluxes along* $\mathcal{H}$

Flux integrations along $\mathcal{H}$ require appropriate parametrizations of $\mathcal{H}$, such as an *advanced time* v. Then, given the flux $F_Q(v)$ of a quantity $Q(v)$, we can write[m]

$$Q(v) = Q(v_0) + \mathrm{sign}(C) \int_{v_0}^{v} F_Q(v')dv', \tag{70}$$

this requiring an initial value $Q(v_0)$. However, such coordinate v is not natural in an Initial Value Problem approach. As discussed in Sec. 5.1.1, the 3+1 slicing $\{\Sigma_t\}$ induces a splitting of the DH into *internal* and *external* sections. The integration in (70) can then be split into *external* and *internal* horizon parts (cf. Fig. 5)

$$Q(t) = Q(v_0) + \mathrm{sign}(C) \int_{t_c}^{t} (F_Q)^{\mathrm{int}}(t')dt' + \mathrm{sign}(C) \int_{t_c}^{t} (F_Q)^{\mathrm{ext}}(t')dt' + \mathrm{Res}(t), \tag{71}$$

where the error $\mathrm{Res}(t)$ is $\mathrm{Res}(t) = \mathrm{sign}(C) \int_t^{\infty} (F_Q)^{\mathrm{int}}(t')dt'$.

If the growth of Q is understood as ultimately associated with some flow into the black hole singularity, the actual essential role of the horizon $\mathcal{H}$ would be that of capturing the associated fluxes. This assumes that the worldtube $\mathcal{H}$ *begins* at the *formation* of the singularity. More complex singularity structures (as those coming from a binary merger) would require a more detailed analysis of this point. From

[m]The coefficient $\mathrm{sign}(C)$, $+1$ for spacelike $\mathcal{H}$ and -1 for timelike $\mathcal{H}$, takes into account the possible integration of fluxes happening when timelike sections of $\mathcal{H}$ occur; cf. Fig. 2.

this perspective, there is nothing intrinsically special about dynamical horizons: *any* hypersurface separating the black hole singularity from past null infinity $\mathscr{I}^-$ (e.g. the event horizon) would be appropriate for fluxes evaluation. However, from a quasi-local perspective, if DHs are shown to *cover* systematically the black hole singularity (or, more generally, the inner Cauchy horizon), then they actually provide excellent geometric prescriptions for such test screens (this is the motivation for the point (iii) in Sec. 3.2.6).

5.4.5. *Auxiliary test-field evolutions in curved backgrounds*

In Sec. 5.4.3 we have considered cross-correlations between different contractions of the Weyl tensor at distinct hypersurfaces. It is legitimate to question if such cross-correlations are meaningful at all, given the *a priori* different geometric content of the involved functions. Let us consider the following approach to this issue: evolve, together with the gravitational degrees of freedom in Einstein equations, an auxiliary (set of) scalar field(s) Φ_i without back-reaction on the geometry (i.e. *test fields*) and whose evolution on the dynamically evolving background spacetime closely tracks[n] its relevant geometric features. Then, the correlation approach outlined in Subsec. 5.4 for a (coarse-grained) extraction of geometric content, can be applied directly on Φ_i. We can paraphrase this approach as *pouring sand on a transparent surface.* On the one hand, this removes the ambiguity in the choice of quantities h_i and h_o at inner and outer hypersurfaces. On the other hand, and more importantly, it also permits to extend to the bulk spacetime the (cross-)correlation strategy between spacetime boundaries.

6. General Perspective

We have presented an introduction to some aspects of quasi-local black holes in an Initial Value Problem approach to the spacetime construction. From a fundamental perspective, quasi-local black hole horizons provide crucial insights into the geometry of the black hole and trapped regions and a sound avenue to black hole physics in generic scenarios. However, quasi-local black holes also meet challenges when considered as physical surfaces of a compact object. We have adopted a *pragmatic* or effective approach in which quasi-local black hole horizons are understood as hypersurfaces with remarkable geometric properties that provide worldtubes of canonical surfaces in a given 3+1 slicing of the spacetime. We have shown how they can be used as an *a priori* ingredient in evolution schemes to Einstein equations, where they provide inner boundary conditions for black hole spacetimes. Then we have illustrated their use as *a posteriori* analysis tools tracking and characterizing quasi-locally the black hole properties and providing, through their *rigidity*

[n]See Ref. 102 for a discussion of a similar approach in a binary black hole context, and Ref. 35 for a methodology sharing part of the spirit but directly tracking spacetime curvature quantities.

properties, excellent *test-screen* probes into the near-horizon black hole spacetime geometry.

Acknowledgments

I thank the organizers of the 2011 Shanghai Asia-Pacific School, especially C.-M. Cheng, S. A. Hayward and J. Nester, for their kind invitation and hospitality. I thank E. Gourgoulhon, B. Krishnan, R. P. Macedo, P. Mösta, A. Nielsen and L. Rezzolla for the interactions during the writing of these notes. I acknowledge the support of the Alexander von Humboldt Foundation, the Spanish MICINN (FIS2008-06078-C03-01) and the Junta de Andalucía (FQM2288/219).

References

1. R. Penrose, *Annals N. Y. Acad. Sci.* **224** (1973) 125.
2. R. Penrose, *Phys. Rev. Lett.* **14** (1965) 57.
3. S. Hawking, *Proc. R. Soc. London, Ser. A* **300** (1967) 182.
4. S. Hawking and R. Penrose, *Proc. R. Soc. London, Ser. A* **314** (1970) 529.
5. S. W. Hawking and G. F. R. Ellis, *The Large Scale Structure of Space-Time* (Cambridge University Press, 1973).
6. R. Penrose, *Riv. Nuovo Cim.* **1** (1969) 252.
7. M. Heusler, *Liv. Rev. Relat.* **1** (1998) 6.
8. S. W. Hawking, *Phys. Rev. Lett.* **26** (1971) 1344.
9. S. W. Hawking, *Commun. Math. Phys.* **25** (1972) 152.
10. J. M. Bardeen, B. Carter and S. W. Hawking, *Commun. Math. Phys.* **31** (1973) 161.
11. A. B. Nielsen and V. Faraoni, *Class. Quant. Grav.* **28** (2011) 175008.
12. M. Kriele and S. A. Hayward, *J. Math. Phys.* **38** (1997) 1593.
13. P. T. Chruściel, in *The Conformal Structure of Spacetime: Geometry, Analysis, Numerics*, J. Frauendiener and H. Friedrich (eds.), Lecture Notes in Physics (Springer, 2002), p. 61.
14. S. Hayward, *Phys. Rev. D* **49** (1994) 6467.
15. J. M. M. Senovilla, *Int. J. Mod. Phys. D* **22** (2011) 2139.
16. A. B. Nielsen, *Class. Quant. Grav.* **27** (2010) 245016.
17. D. M. Eardley, *Phys. Rev. D* **57** (1998) 2299.
18. E. Schnetter and B. Krishnan, *Phys. Rev.* **D73** (2006) 021502.
19. I. Ben-Dov, *Phys. Rev. D* **75** (2007) 064007.
20. J. E. Aman, I. Bengtsson and J. M. M. Senovilla, *J. Phys. Conf. Ser.* **229** (2010) 012004.
21. I. Bengtsson and J. M. M. Senovilla, (2010) arXiv:1009.0225.
22. A. B. Nielsen, *Int. J. Mod. Phys. D* **22** (2011) 2205.
23. E. Gourgoulhon, *Phys. Rev. D* **72** (2005) 104007.
24. I. Booth and S. Fairhurst, *Phys. Rev. D* **75** (2007) 084019.
25. I. Racz, *Class. Quant. Grav.* **25** (2008) 162001.
26. A. Ashtekar and B. Krishnan, *Phys. Rev. D* **68** (2003) 104030.
27. I. Booth, *Can. J. Phys.* **83** (2005) 1073.
28. E. Gourgoulhon and J. L. Jaramillo, *Phys. Rep.* **423** (2006) 159.
29. B. Krishnan, *Class. Quant. Grav.* **25** (2008) 114005.
30. L. Andersson, M. Mars and W. Simon, *Phys. Rev. Lett.* **95** (2005) 111102.

31. A. Ashtekar, C. Beetle and J. Lewandowski, *Phys. Rev. D* **64** (2001) 044016.
32. A. Ashtekar and B. Krishnan, *Liv. Rev. Relat.* **7** (2004) 10.
33. A. Ashtekar, C. Beetle and J. Lewandowski, *Class. Quant. Grav.* **19** (2002) 1195.
34. R. Penrose and W. Rindler, *Spinors and Space-Time*, Vol. 1: *Two-Spinor Calculus and Relativistic Fields* (Cambridge University Press, 1984).
35. R. Owen, J. Brink, Y. Chen, J. D. Kaplan, G. Lovelace, K. D. Matthews, D. A. Nichols, M. A. Scheel, F. Zhang, A. Zimmerman and K. S. Thorne, *Phys. Rev. Lett.* **106** (2011) 151101.
36. A. Ashtekar, J. Engle, T. Pawlowski and C. Van Den Broeck, *Class. Quant. Grav.* **21** (2004) 2549.
37. E. Schnetter, B. Krishnan and F. Beyer, *Phys. Rev. D* **74** (2006) 024028.
38. M. Jasiulek, *Class. Quant. Grav.* **26** (2009) 245008.
39. R. Owen, *Phys. Rev. D* **80** (2009) 084012.
40. A. Ashtekar and G. J. Galloway, *Adv. Theor. Math. Phys.* **9** (2005) 1.
41. L. Andersson, M. Mars and W. Simon, *Adv. Theor. Math. Phys.* **12** (2008) 853.
42. J. L. Jaramillo, M. Reiris and S. Dain, *Phys. Rev. D* **84** (2011) 121503.
43. L. M. Cao, *JHEP* **03** (2011) 112.
44. A. Ashtekar and B. Krishnan, *Phys. Rev. Lett.* **89** (2002) 261101.
45. S. Hayward, *Phys. Rev. Lett.* **93** (2004) 251101.
46. S. A. Hayward, *Phys. Rev. D* **70** (2004) 104027.
47. I. Booth and S. Fairhurst, *Phys. Rev. Lett.* **92** (2004) 011102.
48. S. A. Hayward, arXiv:0810.0923.
49. Y. H. Wu and C. H. Wang, *Phys. Rev. D* **80** (2009) 063002.
50. Y. H. Wu and C. H. Wang, *Phys. Rev. D* **83** (2011) 084044.
51. E. Gourgoulhon and J. L. Jaramillo, *Phys. Rev. D* **74** (2006) 087502.
52. E. Gourgoulhon and J. L. Jaramillo, *New Astron. Rev.* **51** (2008) 791.
53. S. W. Hawking and J. B. Hartle, *Commun. Math. Phys.* **27** (1972) 283.
54. S. A. Hayward, *Phys. Rev. D* **74** (2006) 104013.
55. S. A. Hayward, arXiv:gr-qc/0607081.
56. I. Booth, M. P. Heller and M. Spalinski, *Phys. Rev. D* **83** (2011) 061901.
57. A. B. Nielsen, M. Jasiulek, B. Krishnan and E. Schnetter, *Phys. Rev. D* **83** (2011) 124022.
58. I. Booth, M. P. Heller, G. Plewa and M. Spalinski, *Phys. Rev. D* **83** (2011) 106005.
59. C. Williams, *Annales Henri Poincare* **9** (2008) 1029.
60. C. Williams, *Commun. Math. Phys.* **293** (2010) 589.
61. C. Williams, arXiv:1005.5401.
62. H. Friedrich, *Ann. Phys. (Berlin)* **15** (2005) 84.
63. J. L. Jaramillo, J. A. Valiente Kroon and E. Gourgoulhon, *Class. Quant. Grav.* **25** (2008) 093001.
64. E. Gourgoulhon, Lectures delivered at Institut Henri Poincaré in 2006, arXiv:gr-qc/0703035.
65. A. Lichnerowicz, *J. Math. Pures Appl.* **23** (1944) 37. Reprinted in A. Lichnerowicz, *Choix d'œuvres mathématiques* (Hermann, Paris, 1982).
66. J. W. York Jr., *Phys. Rev. Lett.* **82** (1999) 1350.
67. H. P. Pfeiffer and J. W. York Jr., *Phys. Rev. D* **67** (2003) 044022.
68. S. Bonazzola, E. Gourgoulhon, P. Grandclément and J. Novak, *Phys. Rev. D* **70** (2004) 104007.
69. I. Cordero-Carrion, J. M. Ibanez, E. Gourgoulhon, J. L. Jaramillo and J. Novak, *Phys. Rev. D* **77** (2008) 084007.
70. I. Cordero-Carrion *et al.*, *Phys. Rev. D* **79** (2009) 024017.

71. D. Gannon, *J. Math. Phys.* **16** (1975) 2364.
72. D. Gannon, *Gen. Rel. Grav.* **7** (1976) 219.
73. J. L. Jaramillo and E. Gourgoulhon and G. A. Mena Marugán, *Phys. Rev. D* **70** (2004) 124036.
74. G. B. Cook and H. P. Pfeiffer, *Phys. Rev. D* **70** (2004) 104016.
75. S. Dain, J. L. Jaramillo and B. Krishnan, *Phys. Rev. D* **71** (2005) 064003.
76. J. L. Jaramillo, M. Ansorg and F. Limousin, *Phys. Rev. D* **75** (2007) 024019.
77. J. L. Jaramillo, *Phys. Rev. D* **79** (2009) 087506.
78. A. B. Nielsen and M. Visser, *Class. Quant. Grav.* **23** (2006) 4637.
79. I. Booth, L. Brits, J. A. Gonzalez and C. Van Den Broeck, *Class. Quant. Grav.* **23** (2006) 413.
80. I. Booth and S. Fairhurst, *Phys. Rev.* **D77** (2008) 084005.
81. J. L. Jaramillo, E. Gourgoulhon, I. Cordero-Carrion and J. M. Ibanez, *Phys. Rev. D* **77** (2008) 047501.
82. J. Thornburg, *Liv. Rev. Relat.* **10** (2007) 3.
83. J. L. Jaramillo, M. Ansorg and N. Vasset, *AIP Conf. Proc.* **1122** (2009) 308.
84. L. B. Szabados, *Liv. Rev. Relat.* **12** (2009) 4.
85. J. L. Jaramillo and E. Gourgoulhon, *Mass and motion in general relativity*, Springer Series on Fundamental Theory of Physics Vol. 162 (2011), L. Blanchet, A. Spallicci, B. Whiting (Eds.).
86. O. Dreyer, B. Krishnan, D. Shoemaker and E. Schnetter, *Phys. Rev. D* **67** (2003) 024018.
87. G. B. Cook and B. F. Whiting, *Phys. Rev. D* **76** (2007) 041501.
88. M. Korzynski, *Class. Quant. Grav.* **24** (5944) 5935.
89. N. Vasset, J. Novak and J. L. Jaramillo, *Phys. Rev. D* **79** (2009) 124010.
90. S. Dain, C. O. Lousto and R. Takahashi, *Phys. Rev. D* **65** (2002) 104038.
91. J. L. Jaramillo, N. Vasset and M. Ansorg, *EAS Publications Series* **30** (2008) 257.
92. S. Dain, *Phys. Rev. Lett.* **96** (2006) 101101.
93. J. Hennig, C. Cederbaum and M. Ansorg, *Commun. Math. Phys.* **293** (2010) 449.
94. S. Dain and M. Reiris, *Phys. Rev. Lett.* **107** (2011) 051101.
95. I. Booth, *Can. J. Phys.* **86**, (2008) 669–673.
96. B. Krishnan, C. O. Lousto and Y. Zlochower, *Phys. Rev. D* **76** (2007) 081501.
97. J. L. Jaramillo, R. P. Macedo, P. Moesta and L. Rezzolla, *Phys. Rev. D* **85** (2012) 084030; *Phys. Rev. D* **85** (2012) 084031.
98. L. Rezzolla, R. P. Macedo and J. L. Jaramillo, *Phys. Rev. Lett.* **104** (2010) 221101.
99. S. A. Hayward, *Class. Quantum Grav.* **11** (1994) 3037.
100. S. A. Hayward, *Phys. Rev. D* **68** (2003) 104015.
101. R. H. Price, G. Khanna and S. A. Hughes, *Phys. Rev. D* **83** (2011) 124002.
102. E. Bentivegna, D. M. Shoemaker, I. Hinder and F. Herrmann, *Phys. Rev. D* **77** (2008) 124016.

Chapter 2

Physical Aspects of Quasi-Local Black Hole Horizons*

Alex B. Nielsen
Max-Planck-Institut für Gravitationsphysik,
Am Mühlenberg 1, D-14476 Golm, Germany
alex.nielsen@aei.mpg.de

We discuss some of the physical aspects expected to be associated with black holes. These include Hawking radiation, horizon entropy and cosmic censorship. In particular we focus on whether these properties are more naturally associated to causally defined horizons or quasi-local horizons.

1. Introduction

There are a number of physical effects expected to be associated with black holes. Black holes are now a standard paradigm for explaining a large number of astrophysical phenomena such as quasars, gamma-ray bursts and X-ray binaries.[1] Such astrophysical black holes are expected to be very simple, described to a good approximation by the Kerr stationary vacuum solution. Much smaller black holes may also be produced in high energy particle collisions, in cosmic rays[2] or in particle colliders.[3]

From a theoretical viewpoint black holes have provided the background for a great deal of speculation about extensions to the known laws of physics. First, they are expected to contain singularities, or at least regions where the usual description of gravity in terms of general relativity breaks down. Second, the similarities between the classical laws of thermodynamics and the laws of black hole mechanics have led to the conjecture that black holes carry a type of entropy related to their surface area. This idea has motivated the search for the microscopic gravitational degrees of freedom that can explain this entropy through a statistical mechanical state counting argument. Third, black holes are expected to be unstable due to the emission of Hawking radiation. This Hawking radiation should transport energy away from the black hole and lead to the mass of the black hole diminishing. This effect may even lead to the black hole disappearing entirely and this poses a challenge to the accepted description of the universe in terms of unitary quantum evolution.

*This chapter was previously published in International Journal of Modern Physics D, Vol. 20, No. 11 (2011) 2205–2221

It is the hope of many researchers to use the properties of black holes to provide clues about the features a quantum theory of gravity would exhibit. The properties of black holes mentioned above have already led to a number of novel physics concepts that are at least in part inspired by their study. We will examine here a number of ideas that are particularly relevant to the discussion of black hole horizons. Several of the ideas are closely related and connected themes run through them. Several of these ideas have been extended to other types of horizons such as cosmological horizons or the Rindler horizons of accelerated observers, with varying degrees of success, but here we will focus only on those associated with gravitational collapse and the formation of black holes.

A key question in all of this and a question that is also a central theme in the contributions of José Luis Jaramillo and José Senovilla, is the question of which type of horizon, if any, should these properties be applied to. There are a great many different types of horizons discussed in the literature. These can be classified into two main approaches to defining the horizon; the global causal event horizon and the quasi-local geometrical horizon. Precise definitions of these concepts can be found in the contributions of José Luis Jaramillo and José Senovilla.

Another issue relevant in the context of gravity is to what extent the various concepts rely on the validity of the Einstein equations. The Einstein equations may receive corrections from a number of sources including compactifications of higher-dimensions, supersymmetry or quantum loop corrections. In Einstein gravity there is a direct equivalence between the term $R_{ab}l^a l^b$ appearing in the Raychaudhuri equation and the null energy term $T_{ab}l^a l^b$. In Einstein gravity in addition, the horizon-entropy is given by $S = A/4$ but this is not so in more general theories. Most quasi-local horizon definitions are based on the change of area in null directions. We will see that this may have to be changed to accommodate more general gravitational theories.

Perhaps one of the first to suggest black hole thermodynamics be applied to quasi-local horizons such as apparent horizons rather than event horizons was Hájíček[4] who conjectured that Hawking radiation originates from the region close to the apparent horizon independently of whether an event horizon exists or not. This idea was further examined by Hiscock[5] who proposed identifying the entropy with one quarter the area of the apparent horizon and Collins[6] who obtained a $T\mathrm{d}S = \mathrm{d}Q$ like relation for apparent horizons. An important contribution was made by Hayward[7] who defined an outer condition for apparent horizons and was able to show that the area of a future outer trapping horizon is non-decreasing if the null energy condition is satisfied locally on the horizon. These ideas received further attention when it was shown that the microstates of black hole entropy can be counted in loop quantum gravity for a constant area isolated horizon[8] and in analogue models where a true event horizon is not necessary for the production of Hawking radiation.[9]

Part of black hole physics is founded on rigorously proved mathematical theorems, such as the Penrose singularity theorem or the Hawking area theorem. Part of the work is founded on less rigorously formulated conjectures such as the generalized second law or holography and much is founded on analogy, such as black hole thermodynamics. Perhaps the central "miracle" of black hole thermodynamics, the result that gives most credence to the thermodynamics analogy, is the fact that the temperature arising in the laws of black hole mechanics turns out to be exactly the same temperature as the thermal spectrum computed for Hawking radiation. Without this agreement much of the foundation of further results and ideas would be weakened if not entirely removed. A number of standard textbook results are derived for Killing horizons in asymptotically flat spacetimes and part of the original impetus for studying quasi-local horizons was to extend some of these results to more general situations, without the need to assume a global Killing vector field or a certain structure at infinity. Thus the isolated horizons of Ashtekar et al.[10] need not be Killing horizons, even though they are constant area null horizons and thus still causal horizons in the narrower sense.

But the problem goes deeper than this. Many of the conjectured extensions to the laws of physics are fundamentally rooted in the question of what it means to be a horizon and are intricately tied up with the fundamental features of these horizons. Do the properties refer to the causal structure of the spacetime or to the geometry? Are the properties local and measurable or non-local and idealized?

One of the reasons given originally for preferring event horizons over apparent horizons and the related trapping horizons is the foliation dependence of trapping horizons. Examples of collapse spacetimes are known for which an infinite number of intersecting trapping horizons are found, each at different locations,[11] whereas the location of the event horizon is known to be foliation independent and unique. This problem forms part of the contributions by José Senovilla. If physical properties are to be associated with quasi-local geometrical horizons, to what extent do the physical properties of black holes require a resolution to this uniqueness problem?

Another reason for favoring the event horizon over apparent horizons is that for asymptotically flat spacetimes that are asymptotically predictable and satisfy the null energy condition it is known that the apparent horizon should lie behind the event horizon. Therefore whatever physical processes are associated with the apparent horizon would be forever concealed from asymptotic observers. However, there does not appear to be any physical reason to expect the effective null energy condition to hold forever in the future of a black hole. In this case the apparent horizon is expected to appear outside the event horizon.[12] Nevertheless, there have been statements in the literature that future outer trapping horizons must be spacelike or null.[13] A counterexample involving quantum fields would not be terribly surprising, although an explicitly worked example would be very interesting.

2. The Motivating Physical Properties

The use of horizons and related concepts span a great number of fields in theoretical physics from mathematical and geometrical relativity to quantum gravity through numerical relativity and analogue gravity models. We will look at a number of these below, by no means an exhaustive list. Many of the concepts presented below are related and inspired by one another.

Two celebrated results form the basis for most of this work, the Penrose singularity theorem and the Hawking area theorem.

2.1. *Penrose singularity theorem*

For spacetimes satisfying the weak energy condition and admitting a noncompact Cauchy surface, if the spacetime contains a trapped surface then there will exist inextendible (ending at a singularity) null geodesics and the spacetime will be null incomplete.[14] This theorem effectively states that Einstein gravity predicts its own downfall. If a trapped surface exists somewhere a singularity will form and general relativity will break down. Most researchers expect that general relativity will break down before the formation of a singularity due to quantum gravity effects but there is still no completely understood theory of how this will happen. The trapped surfaces that are used in the proof of the theorem are quasi-local. It is only through the unproved cosmic censorship conjecture that singularities are related to event horizons.

2.2. *Hawking area theorem*

The area of the event horizon is non-decreasing in spacetimes that satisfy the weak energy condition and for which no singularities are visible from infinity.[15] This growth of the event horizon includes the merger of multiple black holes. The result, together with speculation about the verifiability of the second law of thermodynamics, led Bekenstein to postulate that the area of a black hole is a measure of its entropy.[16] This in turn led to the formulation of the laws of black hole mechanics in an analogous form to the classical laws of thermodynamics.[17] We turn now to these laws, paying particular attention to their formulation for quasi-local horizons. Further details are available elsewhere.[18,10,19,20]

2.3. *Zeroth law*

The zeroth law states that the surface gravity is constant over a horizon in equilibrium. At its most basic the law just requires a function that is constant over a stationary horizon. The zeroth law can either define this function, or define what is meant by equilibrium. Any two functions that are constant over the horizon and thus satisfy a zeroth law can be combined together to give a third function that satisfies the zeroth law.

It is relatively easy to prove a zeroth law of black hole mechanics once one has a definition of surface gravity. But there are different definitions of surface gravity.[21,22] For a static, spherically symmetric metric of the form

$$ds^2 = -e^{-2\phi(r)}\Delta r dt^2 + \frac{dr^2}{\Delta(r)} + r^2 d\Omega^2 , \tag{1}$$

the surface gravity given by the usual Killing vector method is

$$\kappa_{Kil} = e^{-\phi}\frac{\Delta'}{2} , \tag{2}$$

where all functions are evaluated on the horizon. The surface gravity defined in terms of the Kodama vector[18] is

$$\kappa_{Kod} = \frac{\Delta'}{2} , \tag{3}$$

again with all functions evaluated on the horizon. If the value of the metric function ϕ at the horizon is not equal to one these two definitions will not be equal, although both will satisfy a zeroth law. There are many static black hole solutions for which ϕ cannot be set to one at the horizon.[23] For the definitions given for isolated horizons[10] there are in fact an infinite number of different surface gravities all of which satisfy the zeroth law.

2.4. *First law*

There are various interpretations of the first law. In ordinary thermodynamics the first law relates the change in the internal energy to the heat flow and work done. If energy is to appear in an equivalent black hole version of the first law then we must face the problem of defining energy in general relativity, a notoriously difficult problem in itself. In the textbook definitions this problem is solved either by using the asymptotic spacetime mass, the ADM mass,[17] or using an intergated horizon flux and defining something resembling an integrated Bondi flux at the horizon.[24]

For the static spherically symmetric Reissner–Nordström solution we have that the areal radius of both the event horizon and trapping horizon is given by $r_H = M + \sqrt{M^2 - Q^2}$ where M is the ADM mass and Q the asymptotically measured charge. Thus the area difference between two Reissner–Nordström black holes whose ADM masses differ by ΔM and whose charges differ by ΔQ will be, to leading order in ΔM and ΔQ,

$$\Delta M \approx \frac{1}{2\pi}\frac{r_H - M}{r_H^2}\Delta\left(\frac{A}{4}\right) + \frac{Q}{r_H}\Delta Q . \tag{4}$$

The term $(r_H - M)/r_H^2$ is the surface gravity of the static Reissner–Nordström black hole. This result has been extended to included rotation and stationary matter shells.[17] One remarkable thing about these results is that they relate quantities measured at infinity and relating to the whole spacetime, such as the ADM mass, to properties solely related to the black hole and computed on its horizon, such as the area. A physical process version for this result can be given in terms of small perturbations of a background stationary spacetime.[24]

In the fully dynamical but still spherically symmetric case, a very simple relation is obtainable relating the change of area of the trapping horizon and the change of the Misner–Sharp quasi-local mass. This just uses the relation satisfied at the trapping horizon, $r = 2m$, and the areal radius definition $A = 4\pi r^2$.

$$\delta_h m = \frac{1}{2\pi}\frac{1}{2r}\delta_h\left(\frac{A}{4}\right). \tag{5}$$

Here and in what follows the derivatives δ_h refer to derivatives along the horizon. This is a special case of a more general result[10] and is exact in spherical symmetry without needing to assume a stationary background or slow evolution. The formula gives a simple relation between the change in mass and the change in one quarter of the area, both quasi-local properties, without appeal to any asymptotically flat region. However, the term relating these two, $1/4\pi r$, does not vanish for finite area, extremal solutions, such as occur in Reissner–Nordström spacetimes. A modification due to Hayward[18] gives

$$\delta_h m = \frac{1}{2\pi}\kappa\delta_h\left(\frac{A}{4}\right) + w\delta_h V\,, \tag{6}$$

again involving the Misner–Sharp mass in spherical symmetry, where the second term on the right hand side can be interpreted as a work term, with $w = T_{ab}(l^a n^b + n^a l^b)/2$ and V is the volume on a spatially flat hypersurface, $V = 4\pi r^3/3$. The relation between the change in mass and change in area, denoted κ, does now vanish in extremal situations and is given by the expression (3). The price to pay for this is that the heat supply and work terms cannot be varied independently unlike the situation in ordinary thermodynamics. In ordinary thermodynamics the temperature is an intensive parameter and independent of the extensive volume, but for black holes the temperature and entropy are both related to this volume.

By restricting to slowly evolving horizons, Booth and Fairhurst[25] obtained a first law of the form

$$\delta_h m \approx \frac{1}{2\pi}\kappa_o\delta_h\left(\frac{A}{4}\right) \approx \frac{1}{8\pi}\int_S \left[\sigma_l^2 + G_{cd}l^c l^d\right]\varepsilon_{ab}\,. \tag{7}$$

In this case κ_o denotes a constant surface gravity which can be chosen to reduce to the isolated Killing horizon value in the static limit. The first approximation is obtained only for situations near spherical symmetry allowing the Misner–Sharp definition of mass to be used, while the second approximation holds in the general case. The right hand side can be interpreted as a flux of matter via the Einstein equations from $T_{ab}l^a l^b$ and the σ_l^2 term can be attributed to gravitational waves. As the authors point out,[19] Eq. 7 can be interpreted in general cases as a first law without any definition of horizon energy at all.

2.5. *Second law*

The second law is perhaps the most important of the thermodynamic/mechanics relations. For event horizons it is simply the area theorem of Hawking, supplemented in the case of non-Einstein gravity by a number of similar arguments.[26,27] An energy condition, via the equations of motion, guarantees the increase of area and this energy condition must hold also in the future. If the energy condition is violated somewhere in the future, as it is expected to be in Hawking radiation, then the event horizon entropy can be decreasing even if the energy condition is satisfied locally.[12] For trapping horizons there is a simple quasi-local relation relating the area increase and the energy condition. For a congruence of curves we have

$$\delta_h \varepsilon_{ab} = \theta_h \varepsilon_{ab} \,, \tag{8}$$

where δ_h denotes the variation along the congruence and ε_{ab} is the area two form of the horizon normal space (see the notes of José Luis Jaramillo for further details.) This equation implies that the cross sectional area will be increasing if the expansion, θ_h, is positive. When h represents the generators of the event horizon a proof by contradiction establishes that θ_h cannot be negative, and hence the area is non-decreasing. Taking the derivative of the area two form along a null direction and using the Raychaudhuri equation we have[28,29]

$$\delta_l \delta_l \varepsilon_{ab} = \left(\kappa_l \theta_l + \frac{\theta_l^2}{2} - \sigma_l^2 - G_{cd} l^c l^d \right) \varepsilon_{ab} \,. \tag{9}$$

In the perturbative limit where $\theta_l^2 \approx 0$ and the transition is between two approximately non-expanding states, this equation can be integrated to obtain the Hartle–Hawking flux law for a black hole. An event horizon cannot make a true transition between one non-expanding state and another non-expanding state because by the Raychaudhuri equation and the energy condition, θ_l along a null causal horizon is always decreasing.

In the case where the horizon is not necessarily null, such as the trapping horizon case, the θ_h can be split into a linear combination of the two normal directions θ_l and θ_n. For a future trapping horizon $\theta_l = 0$ by definition and θ_n is required to be negative. Since the trapping horizon is required to satisfy $\delta_h \theta_l = 0$ by definition, we have

$$\delta_h \varepsilon_{ab} = \frac{\theta_n}{\delta_n \theta_l} \left(\sigma_l^2 + G_{cd} l^c l^d \right) \varepsilon_{ab} \,. \tag{10}$$

Again this condition will hold regardless of the Einstein equations. To convert the $G_{ab} l^a l^b$ into something involving the stress–energy tensor and the energy conditions one can make use of the Einstein equations $G_{ab} = 8\pi T_{ab}$. However, in general modified theories the Einstein equations do not hold. In addition, in a general diffeomorphism invariant theory the horizon-entropy of a stationary horizon is not necessarily equal to one quarter of the area. The horizon-entropy can be computed via integration of a surface entropy density s_{ab} given by the formula[30]

$$s_{ab} = -2\pi \frac{\partial \mathscr{L}}{\partial R_{cdef}} \hat{\varepsilon}_{cd}\hat{\varepsilon}_{ef}\varepsilon_{ab}\,, \tag{11}$$

where $\mathscr{L}$ denotes the Lagrangian density, R_{cdef} is the Riemann tensor, and $\hat{\varepsilon}_{cd}$ is the antisymmetric binormal form. In the case of ordinary Einstein gravity this just gives one quarter of the area two-form, but in scalar-tensor theories and $f(R)$ gravity this gives the area two-form scaled by a non-trivial positive function, $s_{ab} = s\varepsilon_{ab}$. In a dynamical setting the change of this horizon-entropy is given by

$$\delta_h s_{ab} = (\delta_h s + s\theta_h)\,\varepsilon_{ab}\,. \tag{12}$$

The term $(\delta_h s + s\theta_h)$ needs to be positive in order for the generalized entropy to be increasing. In explicit examples, such as when s represents the Brans–Dicke scalar field,[31] the value of $s = \phi$ both increases and decreases along the horizon, although always remaining positive. For a causal horizon in the Brans–Dicke case we find

$$\delta_l\left(\theta_l + \frac{l^a\nabla_a\phi}{\phi}\right) = \kappa_l\left(\theta_l + \frac{l^a\nabla_a\phi}{\phi}\right) - \frac{\theta_l^2}{2} - \sigma_l^2 - \frac{(\omega+1)}{\phi^2}\left(l^a\nabla_a\phi\right)^2 - \frac{8\pi}{\phi}T_{ab}l^al^b. \tag{13}$$

If the causal horizon settles down at late times to a static horizon with $(\delta_l\phi + \phi\theta_l) = 0$ then at previous times its value cannot ever be negative because to get from a negative vale to zero its derivative must be somewhere positive, which is forbidden by (13), provided the null energy condition holds $T_{ab}l^al^b \geq 0$ and l is scaled such that $\kappa_l = 0$.

A similar result can be obtained for quasi-local horizons, but to do this we must introduce the following requirements:[12,32] let H be a three-dimensional hypersurface foliated by closed spacelike two-surfaces whose null normals l and n satisfy

$$\begin{aligned} \varepsilon^{ab}\delta_l s_{ab} &= 0\,, \\ \varepsilon^{ab}\delta_n s_{ab} &< 0\,, \\ \delta_n\left(\varepsilon^{ab}\delta_l s_{ab}\right) &< 0\,. \end{aligned} \tag{14}$$

These conditions are similar to those for a future outer trapping horizon and indeed reduce to them in the case of Einstein gravity where $s_{ab} = \varepsilon_{ab}/4$. But in general these conditions are different and do not describe a surface foliated by marginally trapped surfaces. For the case of Brans–Dicke theory the first condition corresponds to $\theta_l\phi + \delta_l\phi = 0$. Where $\delta_l\phi \neq 0$ this is not the apparent horizon. The change of the entropy is now

$$\delta_h s_{ab} = \frac{\varepsilon^{cd}\delta_n s_{cd}}{\delta_n(\varepsilon^{ef}\delta_l s_{ef})}\left(\frac{\theta_l^2}{2} + \sigma_l^2 + \frac{\omega+1}{\phi^2}\left(\delta_l\phi\right)^2 + \frac{8\pi}{\phi}T_{gh}l^gl^h\right)\phi\varepsilon_{ab}\,. \tag{15}$$

From this we see that the horizon-entropy is increasing for a horizon satisfying (14) and matter obeying the null energy condition. A horizon-entropy second law holds for these quasi-local horizons in modified gravity theories, even though they are not foliated by marginally trapped surfaces.

From the very beginning[16] the association of entropy to black holes was meant as a measure of ignorance about the internal state of the black hole. As such, this interpretation of the entropy is most naturally associated with causal horizons as they are true boundaries to causal information flow. Timelike quasi-local horizons are certainly no boundary to two-way information flow. If instead entropy is interpreted geometrically as in[30] then an association with the conditions (14) seems natural as an entropic trapping horizon.

3. The Conjectured Physical Properties

The third celebrated result in black hole physics is the Hawking radiation.[33] Although the Hawking result is backed by a wide range of different calculations, the result is still the subject of some controversy and alternative interpretations. Current technologies seem a long way from observing this effect for astrophysical black holes.

The prediction of Hawking radiation led to the possibility that black holes could lose mass and eventually disappear. This in turn led to the black hole information paradox[34] which in turn spawned a wide number of possible resolutions from modifying quantum mechanics to allowing information to fall into baby universes.

3.1. *Hawking radiation*

The original derivation of Hawking radiation relied on the use of null rays that approach the event horizon but still escape to infinity.[33] A novel way to view the Hawking process was through the tunneling argument.[35] This tunneling argument lends itself rather naturally to application on trapping horizons[36,37] and recent work has focused on the use of Kodama vector fields to understand spherically symmetric situations.[38] But there appear to be problems with this approach. As a very simple example, consider the following metric

$$ds^2 = -dt^2 + \frac{dr^2}{\Delta^2} + \frac{r^2 d\Omega^2}{\Delta}, \tag{16}$$

with $\Delta = 1 - 2M/r$. This metric is just a conformal transformation of the familiar Schwarzschild metric. It has exactly the same local null cone structure and future directed light rays propagate from near $r = 2M$ to infinity. However, the metric also has a $\theta_l = 0$ trapping horizon at $r = 3M$ even though light rays are able to cross this surface from inside to outside without impedance. The metric is static with a timelike Killing vector field and since it is also spherically symmetric it admits a Kodama vector field that is parallel to the Killing vector field. Although the norm of the Kodama vector is tending to zero as one approaches $r = 3M$, the Kodama vector field itself vanishes at $r = 3M$ which strongly suggests that using the Kodama vector field to probe the structure of this metric is likely to run into problems.

In the context of "veiled" general relativity[39] this metric can even be viewed as physically equivalent to the Schwarzschild metric, with the same properties of perihelion precession, Shapiro time delay and gravitational redshift. It is therefore somewhat disconcerting that the trapping horizon and Kodama vector techniques are unable to recover the expected behavior at $r = 2M$.

It has recently been argued that no horizon at all is necessary for Hawking radiation as long as one has the necessary structure at infinity[40,41] and that "it 'looks like' a horizon might form in the not too distant future". This occurs because of a key adiabatic approximation used in deriving the Hawking effect that the spacetime should be changing slowly with respect to the peak frequency of the radiation and allows the possibility that a spacetime may get very close to forming a horizon without one ever actually forming. It is also argued that Hawking-like radiation will occur in fuzzball geometries[42] where no horizon is formed at all due to long-range quantum gravity effects.

An analogue of Hawking radiation may have been observed in the laboratory as a classical stimulated process involving analogue white hole horizons in water surface waves.[43] This is a classical stimulated process rather than a quantum spontaneous process with long surface waves being converted into short deep water waves. In this case the location of the horizon is frequency dependent and is thus smeared out for a range of frequencies. The water flow that is creating the analogue geometry can be modified or switched off externally and hence there is no true event horizon. Nonetheless, a good fit to a Boltzmann factor distribution for the frequencies of the emitted waves is observed. This opens up the possibility of testing experimentally key assumptions that go into the Hawking result.

3.2. *Cosmic censorship*

The cosmic censorship conjecture[44,45] is the idea that singularities of the gravitational field should not be visible to observers. The weak version forbids the existence of inextendible past directed null curves from future null infinity. In other words if a singularity forms it is covered by an event horizon and the spacetime is strongly asymptotically predictable. This builds on the Penrose singularity theorem which gives sufficient grounds for the formation of a singularity in the presence of trapped surfaces. In this way there is a link between trapped surfaces and the formation of event horizons. The cosmic censorship conjecture is very handy in classical gravity, especially without knowledge of how quantum gravity should behave, in that it allows one to define Cauchy surfaces and initial data surfaces for controlled evolution. The conjecture has been refined over the years and while there is much evidence supporting it in various guises,[46] there are also known exceptions to other formulations. For example, weak cosmic censorship is violated for slow Vaidya collapse,[47] although such solutions are not asymptotically flat. Weak cosmic censorship is also violated for asymptotically-flat finely-tuned self-similar collapse.[48]

3.3. *Penrose conjecture*

The Penrose conjecture[49] states that the areal-mass of black holes cannot be greater than the total mass of a given spacetime.

$$M \geq \sqrt{\frac{A}{16\pi}}\,. \tag{17}$$

This is related to and inspired by cosmic censorship. A counter-example to the Penrose conjecture is likely to involve a violation of cosmic censorship. In order to be logically independent of cosmic censorship and independent of the future evolution of the spacetime, the conjecture is usually formulated in terms of marginally trapped surfaces or at least the infimum of surfaces enclosing a marginally trapped surface. Assuming the energy conditions and Einstein equations, the minimum area needed to enclose the apparent horizon is always less than the area of the event horizon. However, it was noticed[50] that the area of the trapping horizon can be greater than the area of the event horizon even though the event horizon fully encloses the trapping horizon. The Penrose inequality does not hold for the outermost apparent horizon.[13]

The Penrose conjecture is also actually a form of positive mass relation although it can be expressed in purely geometric terms. A related condition has been proved for time-symmetric zero extrinsic curvature initial data hypersurfaces,[51] based on a idea by Geroch[52] where a suitable surface is expanded outward at a rate inversely proportional to the surface's mean curvature, so-called inverse mean curvature flow. But it remains an open question as to whether the conjecture holds in general.

Related inequalities can be derived for spinning black holes in terms of their angular momentum.[53,54] In particular it was proved[54] that the relation

$$A \geq 8\pi|J|\,, \tag{18}$$

holds for axially symmetric apparent horizons in dynamical spacetimes satisfying the dominant energy condition.

3.4. *Generalized second law*

The Generalized Second Law states loosely that the intrinsic entropy of horizons plus the entropy of matter outside the horizons is a non-decreasing function towards the future. The idea builds off of the area increase law for event horizons and its extensions. As we saw in the section on the second law, horizon-entropy increase results typically require an energy condition assumption. This energy condition is known to be violated in the case of Hawking radiation.

When emitting Hawking radiation the area of the black hole can actually decrease and correspondingly the horizon-entropy of the black hole decreases too. However the Hawking process produces large amounts of entropy outside of the black hole in the form of radiation and so the Generalized Second Law is expected to hold even in these cases. The Generalized Second Law can be given either as a total change

$$\Delta S_{horizon} + \Delta S_{outside} \geq 0\,, \tag{19}$$

in which case the change can be between a present configuration and a later equilibrium configuration, or it can be given as a more immediate differential change

$$\frac{\mathrm{d}}{\mathrm{d}t}\left(S_{horizon} + S_{outside}\right) \geq 0\,, \tag{20}$$

for a suitably defined time parameter t defining evolutions. The first version allows one to sidestep the issue of defining entropy in non-equilibrium situations. However, some form of non-equilibrium entropy should be defined in order to retain the Markovian property of Cauchy evolution.[55] Knowledge of a possibly equilibrium past is not needed to predict the future from a present non-equilibrium state. Entropy counters such as the black hole area are certainly definable in dynamical situations. The relation between energy flux and horizon growth is more immediate for quasi-local horizons than it is for causal horizons since causal horizons can be growing, even in entirely flat space regions with no matter.

A number of proofs of the Generalized Second Law have been proposed over the years.[56] Some of them can be applied equally well to causal horizons and quasi-local horizons as they depend on small perturbative deformations of isolated horizons or *gedanken* experiments involving transitions from one stationary state to another.

Recently it has been argued that the Generalized Second Law will only hold for null causal horizons such as the event horizon.[57] In fact a proof is given there that it cannot hold in general for trapping horizons or any other kind of horizon that is not a causal horizon. The result relies on a $1+1$ dimensional conformal transformation of the Hartle–Hawking vacuum under the linear approximation $\theta_l^2 = 0$.

3.5. *Entropy bounds*

Entropy bounds are the idea that regions of spacetime can only contain a maximum amount of entropy corresponding to the equivalent entropy held in a black hole.[58]

$$S \leq 2\pi E R\,, \tag{21}$$

with E the energy of the system and R the areal radius, or in the so-called covariant formulation of Bousso[59]

$$S \leq \frac{A}{4}\,. \tag{22}$$

The validity of this bound of course depends critically on what is meant by S, E and R.[60] Although this idea is not directly related to issue of horizon definitions, similar definitional problems occur in interpreting the terms in thermodynamic relations for black holes. The bound is set by a Schwarzschild black hole for which both event horizon and apparent horizon coincide. However, it is explicitly motivated by the assumption that black holes have entropy and this entropy is related to their area. Therefore it can be seen as a proposed law of nature deriving from the association of entropy to black hole horizons. Similar, although weaker bounds, exist for uncollapsed objects.[61]

3.6. *Holographic principle*

The holographic principle is the conjecture that physical properties of regions of spacetime can be entirely read off from just their boundary properties. The holographic principle is in part motivated by considerations that the microstates of a black hole are encoded in its horizon. One can even argue that the origin of black hole entropy *must* be sought in the holographic principle.[62] The most completely understood example of the holographic principle and somewhat independent of the black hole motivation is the AdS-CFT duality in its particular manifestation of type IIB string theory on a ten-dimensional background that is asymptotically $AdS_5 \times S_5$ and an $N = 4$ supersymmetric Yang–Mills conformal field theory that lives on the four-dimensional boundary of the AdS_5.

In a more general setting, the gauge–gravity duality relates the plasma phase of strongly interacting non-Abelian gauge theories, similar to the quark–gluon phase seen at RHIC and LHC, to gravitational systems. For example, transport properties of the plasma can be studied by looking at quasi-normal models of the corresponding black holes. A related project, fluid-gravity duality,[63] relates non-linear hydrodynamics and solutions of the Navier–Stokes equation to long wavelength perturbations of higher dimensional black holes or black branes.

The fluid–gravity correspondence has been used recently[62] to relate information about the behavior of the black hole to entropy currents defined for the fluid. In this case the acausal nature of the event horizon is a definite problem. In static situations it has been argued that the association of thermodynamic properties of the dual quantum field theories with the apparent horizon is more robust than with the event horizon.[64]

Another aspect of the fluid–gravity correspondence is the possibility of associating fluid properties to the black hole horizon itself. This is epitomized by the membrane paradigm[65] in which the horizon of the black hole is replaced by a time-like stretched horizon just outside the black hole horizon which can be thought of as a surface containing all the relevant physical properties to compute electromagnetic and quantum effects in the exterior region. In early applications[65] the stretched horizon was located just outside the absolute event horizon. Using the Einstein equations one can derive evolution equations for the event horizon as a viscous fluid by a two-dimensional Navier–Stokes-like equation. More recent work[66,28,29] has extended this to quasi-local horizons too. One interesting result of this extension is that the bulk viscosity becomes positive for quasi-local horizons, rather than negative as with event horizons, allowing a fully "physical" evolution of the fluid equations.

3.7. *Black hole complementarity*

Black hole complementarity[67] is the idea that states falling into black holes are duplicated (somehow) on the horizon. This means that a full copy of whatever falls into the black hole is preserved outside, while allowing for "nothing special" to be noticed by observers falling into the hole. In this way the model resolves the black hole information paradox by keeping an entire account of all states outside of the horizon.

As the focus here is on the loss of information and the irretrievable loss of this information, this argument is most naturally associated to the global event horizon (this is stated explicitly in[67]), or rather the stretched horizon, which is somewhat loosely defined as a timelike surface a small distance outside the global event horizon, such that its area is a small value greater than the area of the global event horizon when compared along a past directed null rays (in four dimensions there are many such past null rays normal to the event horizon but because the authors are working in two dimensions which past directed null rays is not specificed).

A key point of the black hole complementarity proposal is that the duplication of information at the horizon cannot be observed by any single observer. The event horizon plays a key role in the argument because this is where the duplication is meant to take place. The teleological nature of the event horizon means this duplication process must also be teleological. A number of studies have attempted to show that there are black hole configurations for which the duplication can be observed by a single observer.[68,69] For black hole complementarity to be effective these configurations cannot occur in nature. On the other hand black hole complementarity may just be wishful thinking, an attempt to solve a problem that nature solves in a completely different way, if at all.

4. Summary

To what extent are these properties physical, in the sense that they can be observed and tested by physicists with finite measuring apparatus? Since Hawking radiation can be detected by local observers carrying particle detectors it is perhaps the most physically measureable indication of a black hole. Hawking radiation may have a quasi-local origin, as suggested by Hájíček[4] and by the local nature of quantum field theory. As the example of veiled Schwarzschild shows, the existence of a trapping horizon may not even be sufficient for the production of Hawking radiation. The fact that null expansions are conformal frame dependent is one fact mitigating against trapping horizons having a physical role. If physical properties are to be conformal frame independent[39] then equations like $\theta_l = 0$ cannot qualify by themselves. One possible way to maintain the quasi-local nature but dispense with null expansions is given by the conditions (14).

In principle it should be possible to measure the temperature of the Hawking radiation of a black hole, either from a true gravitational black hole or an analogue model. This means the disagreement between the Killing-defined temperature (1) and the Kodama-defined temperature (3) is in principle decidable by experiment. Whether the non-uniqueness of the trapping horizon has any role to play in the calculation of Hawking radiation remains an open question. The relationship between the energy flux and horizon area change might also be measurable with sufficient experimental control or possibly on a purely formal level using the dual fluid-gravity correspondence. This also opens the door to the possibility that the non-uniqueness of the trapping horizon may have a counterpart in the dual fluid sector.[62]

The event horizon has a definite role in black hole complementarity. An event horizon in an otherwise locally flat region of spacetime must be able to duplicate infalling information, keeping some of it on the outside of the black hole. That this may involve highly non-local physics makes it hard to evaluate with current quantum field theoretic tools. The complementarity conjecture is by design impossible to verify by a single observer, unless certain regular black holes are constructable.[68] The causal event horizon also has some mathematical role to play in the Generalized Second Law as its teleological nature may be essential to satisfying this rule when applied over a sufficient length of time. The very direct relationship between the behavior of a quasi-local horizon and matter fields means that the Generalized Second Law applied to trapping horizons may fail for quantum fields.

There remains much work to be done in understanding these issues. The literature currently contains a number of contradictory statements. The use of the AdS-CFT correspondence led Booth et al.[62] to state that

> "[The result[64]] *strongly suggests that the causal boundary of a black hole is not the relevant entropy carrier...*"

and consideration of the $1 + 1$ dimensional conformally coupled matter sector led Wall[57] to conclude

> *"...the causal horizon is the only sort of horizon for which the semiclassical* [Generalized Second Law] *can hold."*

Somewhat less spectacularly Barcelo et al.[40] demonstrate that

> *"...any collapsing compact object (regardless of whether or not any type of horizon ever forms) will, provided the exponential approximation and adiabatic condition hold, emit a slowly evolving Planckian flux of quanta."*

whereas Hayward et al.[38] state that

> "[the] *method* [Hamilton–Jacobi tunneling for deriving Hawking radiation] *works precisely for future outer trapping horizons."*

To make many of the ideas we have seen above compatible may require a weakening of our current interpretations of black hole physics. The first law surface gravity relating the change in energy to the change in entropy may not be the same as the temperature of Hawking radiation. The entropy of ignorance may be carried by causal horizons while a geometric or holographic entropy is carried by quasi-local horizons. The necessary conditions for a Hawking flux of energy may not refer to any of these horizons. Further work is needed to settle which of these statements is correct, if any. Full calculations in curved space quantum field theory or quantum gravity would of course be definitive. But these theories are still incomplete and untested. The hope that we can use the physical properties of black holes to motivate ideas in quantum gravity seems problematic when there is still so much uncertainty about what it means to be a black hole.

5. Acknowledgments

The author is very grateful for support from the Alexander von Humboldt Foundation and hospitality at the Max Planck Institute for Gravitational Physics in Potsdam-Golm. The author would also like to thank the organizers of the 2011 APCTP-NCTS school on gravitation.

References

1. M. Begelman and R. Rees *"Black Holes in the Universe"* Cambridge University Press (2010).
2. J. L. Feng and A. D. Shapere, *Phys. Rev. Lett.* **88** (2002) 021303. [hep-ph/0109106].
3. S. B. Giddings and S. D. Thomas, *Phys. Rev.* **D65** (2002) 056010. [hep-ph/0106219].
4. P. Hajicek, *Phys. Rev.* **D36** (1987) 1065.
5. W. A. Hiscock, *Phys. Rev.* **D40** (1989) 1336.
6. W.Collins, *Phys. Rev.* **D45** (1992) 495.
7. S. A. Hayward, *Phys. Rev.* **D49** (1994) 6467.
8. A. Ashtekar, J. Baez, A. Corichi and K. Krasnov, *Phys. Rev. Lett.* **80** (1998) 904.
9. M. Visser, *Int. J. Mod. Phys.* **D12** (2003) 649
10. A. Ashtekar and B. Krishnan, Living Rev. Rel. **7** (2004) 10.
11. A. B. Nielsen, M. Jasiulek, B. Krishnan and E. Schnetter, *Phys. Rev.* **D83** (2011) 124022.
12. A. B. Nielsen, *Class. Quant. Grav.* **27** (2010) 245016.
13. I. Ben-Dov, *Phys. Rev.* **D70**, 124031 (2004).
14. R. Penrose, *Phys. Rev. Lett.* **14** (1965) 57–59.
15. S. W. Hawking, *Phys. Rev. Lett.* **26** (1971) 1344
16. J. D. Bekenstein, *Phys. Rev.* **D7** (1973) 2333–2346.
17. J. M. Bardeen, B. Carter and S. W. Hawking, *Commun. Math. Phys.* **31** (1973) 161–170.
18. S. A. Hayward, *Class. Quant. Grav.* **15** (1998) 3147–3162.
19. I. Booth, S. Fairhurst, *Phys. Rev.* **D75** (2007) 084019.
20. A. B. Nielsen, *Gen. Rel. Grav.* **41** (2009) 1539.
21. A. B. Nielsen and J. H. Yoon, *Class. Quant. Grav.* **25** (2008) 085010.
22. M. Pielahn, G. Kunstatter and A. B. Nielsen, arXiv:1103.0750 [gr-qc].
23. A. B. Nielsen, *Phys. Rev.* **D74** (2006) 044038. [gr-qc/0603127].
24. R. M. Wald, *Quantum Field Theory in Curved Spacetimes and Black Hole Thermodynamics*, University of Chicago Press, Chicago (1994).
25. I. Booth, S. Fairhurst, *Phys. Rev. Lett.* **92** (2004) 011102.
26. Jacobson T, Kang G and Myers R C 1995 *Phys. Rev. D* **52** 3518.

27. G. Kang, *Phys. Rev.* **D54** (1996) 7483–7489.
28. E. Gourgoulhon and J. L. Jaramillo, *Phys. Rev.* **D74** (2006) 087502.
29. E. Gourgoulhon and J. L. Jaramillo, *New Astron. Rev.* **51** (2008) 791–798.
30. R. M. Wald, *Phys. Rev.* **D48** (1993) 3427.
31. M. A. Scheel, S. L. Shapiro and S. A. Teukolsky, *Phys. Rev.* **D51** (1995) 4236.
32. V. Faraoni and A. B. Nielsen, arXiv:1103.2089 [gr-qc].
33. S. W. Hawking, *Commun. Math. Phys.* **43** (1975) 199–220.
34. J. Preskill, [hep-th/9209058].
35. M. K. Parikh and F. Wilczek, *Phys. Rev. Lett.* **85** (2000) 5042–5045.
36. R. Di Criscienzo, M. Nadalini, L. Vanzo, S. Zerbini and G. Zoccatelli, *Phys. Lett.* **B657** (2007) 107–111. [arXiv:0707.4425 [hep-th]].
37. A. B. Nielsen and D. h. Yeom, *Int. J. Mod. Phys. A* **24** (2009) 5261.
38. S. A. Hayward, R. Di Criscienzo, L. Vanzo, M. Nadalini and S. Zerbini, *Class. Quant. Grav.* **26** (2009) 062001.
39. Deruelle N and Sasaki M, arXiv:1007.3563 [gr-qc].
40. C. Barcelo, S. Liberati, S. Sonego and M. Visser, *Phys. Rev.* **D83** (2011) 041501.
41. C. Barcelo, S. Liberati, S. Sonego and M. Visser, *JHEP* **1102** (2011) 003.
42. B. D. Chowdhury and S. D. Mathur, *Class. Quant. Grav.* **26** (2009) 035006.
43. S. Weinfurtner et al., *Phys. Rev. Lett.* **106** (2011) 021302.
44. R. M. Wald, In Iyer, B.R. (ed.) et al.: *Black holes, gravitational radiation and the universe* 69-85. [gr-qc/9710068].
45. H. Ringstrom, *Living Rev. Rel.* **13** (2010) 2.
46. S. Hod, *Phys. Rev. Lett.* **100** (2008) 121101.
47. W. A. Hiscock, L. G. Williams and D. M. Eardley, *Phys. Rev.* **D26** (1982) 751–760.
48. M. W. Choptuik, *Phys. Rev. Lett.* **70** (1993) 9–12.
49. M. Mars, *Class. Quant. Grav.* **26** (2009) 193001.
50. P. S. Jang and R. M. Wald, *J. Math. Phys.*, **18** (1977) 41.
51. G. Huisken and T. Ilmanen, *J. Diff. Geom.* **59** (2001) 353.
52. R. Geroch, *Annals N.Y.Acad. Sci.* ,**224** (1973) 108.
53. I. Booth and S. Fairhurst, *Phys. Rev.* **D77** (2008) 084005.
54. J. L. Jaramillo, M. Reiris and S. Dain, [arXiv:1106.3743 [gr-qc]].
55. A. Corichi and D. Sudarsky, *Mod. Phys. Lett.* **A17** (2002) 1431–1444.
56. A. C. Wall, *JHEP* **0906** (2009) 021.
57. A. C. Wall, [arXiv:1105.3520 [gr-qc]].
58. J. D. Bekenstein, *Phys. Rev.* **D23** (1981) 287.
59. R. Bousso, *JHEP* **9907** (1999) 004.
60. D. N. Page, *JHEP* **0810** (2008) 007.
61. G. Abreu, M. Visser, *JHEP* **1103** (2011) 056.
62. I. Booth, M. P. Heller, G. Plewa and M. Spalinski, *Phys. Rev.* **D83** (2011) 106005.
63. S. Bhattacharyya, V. EHubeny, S. Minwalla and M. Rangamani, *JHEP* **0802** (2008) 045.
64. P. Figueras, V. E. Hubeny, M. Rangamani and S. F. Ross, *JHEP* **0904** (2009) 137.
65. R. H. Price and K. S. Thorne, *Phys. Rev.* **D33** (1986) 915–941.
66. E. Gourgoulhon, *Phys. Rev.* **D72** (2005) 104007.
67. L. Susskind, L. Thorlacius and J. Uglum, *Phys. Rev.* **D48** (1993) 3743–3761.
68. D. -h. Yeom and H. Zoe, *Phys. Rev.* **D78** (2008) 104008.
69. S. E. Hong, D. -i. Hwang, D. -h. Yeom and H. Zoe, *JHEP* **0812** (2008) 080.

Chapter 3

On Uniqueness Results for Static, Asymptotically Flat Initial Data Containing MOTS

Alberto Carrasco[1] and Marc Mars[2]

Dept. Física Fundamental and Instituto de Física Fundamental y Matemáticas (IUFFyM)
Universidad de Salamanca,
Plaza de la Merced s/n, 37008 Salamanca, Spain
[1]*acf@usal.es,* [2]*marc@usal.es*

Marginally outer trapped surfaces are widely believed to be useful quasi-local, dynamical replacements for black holes. It is also expected that in a stationary situation both concepts should essentially concide. In this contribution we explore this expectation by studying whether the uniqueness theorems for static black holes extend to static spacetimes contaning a spacelike hypersurface, possibly with boundary, which includes an asymptotically flat end and a marginally outer trapped surface (MOTS) which is bounding with respect to that end. We conclude that, under suitable conditions which involve no global-in-time assumptions, the answer to this question is affirmative. The matter model is arbitrary except for the condition that it satisfies the null energy condition and it admits a static black hole uniqueness proof using the doubling method of Bunting and Masood-ul-Alam.

1. Introduction

Trapped surfaces are defined in terms of local properties of light pulses emitted orthogonally from them and, hence, are quasi-local in nature. Black holes, on the other hand, are defined as spacetimes containing regions from which no light ray can escape. The concept of "non-escape" is necessarily a subtle one because one needs to have an *a priori* notion of which region the light ray should not be able to reach. Also, "non-escape" means that the light ray should not enter such regions no matter how far/long it travels in spacetime. Thus, black holes are global objects both in space and in time. It is clear that trapped surfaces and black holes are in principle very different in nature. However, both definitions are devised to indicate the presence of a strong gravitational field and both involve bending of light in one way or another. Thus, it is conceivable that these two objects are related to each other. In fact, it is widely believed that this is the case and that hypersurfaces defined by the evolution of (suitable types of) trapped surfaces can serve as quasi-local replacements for the concept of the black hole event horizon. This belief is supported by several facts, some rigorous and some conjectural. From the rigorous side, and restricting the discussion to the usual definition of black holes for asymptotically flat spacetimes, it is known that trapped surfaces are necessarily contained within the black hole region (i.e. the region not visible from infinity) in any black hole spacetime satisfying the null energy condition (see e.g. Chapter 9.2 of[34] and Chapter 12.2 of[61]). Thus, the global notion of strong gravitational field captures the quasi-local notion in a satisfactory way. The converse, namely whether spacetimes containing trapped surfaces are necessarily black hole spacetimes, is much more subtle and at present largely unknown. However, it would follow as a simple corollary of the singularity theorems of Hawking, Penrose and others (see[34,57]) provided the Penrose weak cosmic censorship conjecture[51] holds true. This conjecture is supported by physical arguments of predictability, but as a rigorous mathematical problem is still wide open[62,1] except for very special cases with high degree of symmetry. The present situation is, therefore, that certain classes of trapped surfaces (or rather their evolutions) are widely expected to be useful as suitable dynamical and quasi-local replacements of black holes, but no rigorous proof of this fact is available. Thus, in addition to exploring the physical properties of the horizons generated by the evolution of trapped surfaces, it also becomes necessary to explore situations where the relationship between such horizons and event horizons can be made more precise.

From a physical perspective, one of the simplest situations where this relationship should be possible to analyze involves stationary spacetimes. The reason is obviously that in a non-evolving spacetime, the dynamical, quasi-local concept of black holes should be simply the same as the global notion. This question is, however, much more subtle than one may think *a priori*. To start with, there is the issue of which type of trapped surfaces should be used when exploring this equivalence. It is clear that surfaces which show a borderline behaviour among trapped surfaces should be used. The two most natural candidates are marginally trapped surfaces, defined by the condition that one of their future null expansions vanishes identically while the other is non-positive, and marginally outer trapped surfaces (MOTS for short), where only the vanishing of the outer null expansion is required (of course this requires a proper definition of "outer", which may not be a simple issue). These two types of surfaces are the borderline case, respectively, of weakly trapped surfaces and weakly outer trapped surfaces where the first class is defined by the condition that the mean curvature vector of the surface is future causal, while the other is defined by the condition that the outer future null expansion is non-positive. A conjecture by Eardley[32] states that the boundary of the spacetime region containing weakly outer trapped surfaces in a black hole spacetime coincides with the event horizon. This is known to be true in all stationary black holes for which the event horizon is a Killing horizon[a]. In dynamical situations, this conjecture has been proved by Ben-Dov[12] in the particular case of the Vaidya spacetime. On the other hand, Bengtsson and Senovilla[13] have proved that that the region containing weakly trapped surfaces in the Vaidya spacetime does not extend to the event horizon. These results suggest that the concept of weakly outer trapped surface does capture the essence of a black hole better than that of weakly trapped surfaces, and hence that MOTS are the most suitable quasi-local replacements for black holes.

From the discussion above it follows that a fundamental problem regarding MOTS is how they relate to event horizons in stationary spacetimes. The approach one would like to take is to assume as few global conditions on spacetime as possible (specially global-in-time). This is because, if we know *a priori* that the spacetime is a black hole, then many properties of MOTS follow at once (e.g. that they must be enclosed by the event horizon) and it becomes impossible to check in which sense are MOTS *quasi-local* replacements of black holes. On the other hand, if one only assumes the existence of Killing vectors which is timelike at spatial infinity and makes no further hypothesis on the spacetime, then the study of MOTS becomes a non-trivial problem, and this equivalence issue can be explored. It is natural to start with the simplest possible situation, namely static spacetimes. A natural approach to analyze the relationship between MOTS and event horizons in this situation is to study whether the uniqueness theorems for static black holes can be extended to static spacetimes containing MOTS. Obviously, an extension of the classic static black uniqueness theorems to static spacetimes with MOTS would imply, *a posteriori*, the equivalence between both concepts. This has been the aim of a series of papers[18–20] and has been the main subject of one of the authors' Ph.D. thesis.[17] In this article we will summarise these results and we will discuss the main ideas behind their proofs.

Our main result (Theorem 11) shows that, at least as far as uniqueness of static black holes is concerned, event horizons and MOTS do coincide, provided the spacetime satisfies suitable conditions. In accordance with the discussion above, we do not wish to make any *a priori* global-in-time assumptions in the spacetime. In fact we have tried to work as much as possible directly at the initial data level (i.e. assuming only initial data but not a spacetime containing it) although in some cases we also need to assume the existence of a small slab of spacetime around the initial data set. Notice that the latter case is more restrictive than the former one because we do not make any specific assumptions on the matter model and we only impose energy inequalities in our analysis. Consequently, we cannot directly invoke the gravitational field equations in order to generate a spacetime from the initial data. Working at the initial data level (or on small slabs of spacetime containing these data) not only complies with the previous requirement of studying the relationship

[a]This is because sections of Killing horizons have vanishing null expansion along the Killing generator of the horizon.

between MOTS and black holes without assuming *a priori* that a black hole exists, but it is also in good agreement with the general tendency of studying stationary and static spacetimes with as few global assumptions in time as possible, and try to understand these global properties as a consequence of the "evolution" (i.e. to understand the global properties of the maximal Cauchy development of, say, a vacuum initial data set containing a static Killing vector assuming only information on the initial data set).

Our uniqueness result for static spacetimes containing MOTS is also interesting independently of its relationship with black holes. It proves that static configurations are indeed very rigid. This type of result has several implications. For instance, in any evolution of a collapsing system, it is expected that an equilibrium configuration is eventually reached. The uniqueness theorems of black holes are usually invoked to conclude that the spacetime is one of the stationary black holes compatible with the uniqueness theorem. However, this argument assumes implicitly that one has sufficient information on the spacetime to be able to apply the uniqueness theorems, which is far from obvious since the spacetime is being constructed during the evolution. In our setting, as long as the evolution has a MOTS on each time slice, if the spacetime reaches a static configuration, then it is unique.

1.1. *Uniqueness theorems for black holes*

In the late 60s and early 70s the properties of equilibrium states of black holes were extensively studied. The first uniqueness theorem for black holes was found by W. Israel in 1967,[38] who found the surprising result that a static, topologically spherical vacuum black hole is described by the Schwarzschild solution. In the following years, several works[49, 54, 16] established that the Schwarzschild solution indeed exhausts the class of static vacuum black holes with non-degenerate horizons. The method of proof in[38, 49, 54] consisted in constructing two integral identities which were used to investigate the geometric properties of the level surfaces of the norm of the static Killing. This method proved uniqueness under the assumption of connectedness and non-degeneracy of the event horizon. The hypothesis on the connectedness of the horizon was dropped by Bunting and Masood-ul-Alam[16] who devised a new method, known as the *doubling method*, based on finding a suitable conformal rescaling which allowed using the rigidity part of the positive mass theorem[56] to conclude uniqueness. The result of Bunting and Masood-ul-Alam states the following.

Theorem 1 (Bunting, Masood-ul-Alam, 1987[16]). *Let $(M, g^{(4)})$ be a vacuum spacetime with a static Killing vector $\vec{\xi}$. Assume that $(M, g^{(4)})$ contains a connected, asymptotically flat spacelike hypersurface (Σ, g, K) which is time-symmetric (i.e. with vanishing second fundamental form and orthogonal to the Killing vector $\vec{\xi}$), has non-empty compact boundary $\partial\Sigma$ and is such that the static Killing vector $\vec{\xi}$ is causal on Σ and null only on $\partial\Sigma$.*

Then (Σ, g) *is isometric to* $\left(\mathbb{R}^3 \setminus B_{M_{Kr}/2}(0), (g_{Kr})_{ij} = \left(1 + \frac{M_{Kr}}{2|x|}\right)^4 \delta_{ij}\right)$ *for some* $M_{Kr} > 0$, *i.e. the* $\{t = 0\}$ *slice of the Kruskal spacetime with mass* M_{Kr} *outside and including the horizon. Moreover, there exists a neighbourhood of* Σ *in* M *which is isometrically diffeomorphic to the closure of the domain of outer communications of the Kruskal spacetime.*

The hypothesis on the non-degeneracy of the event horizon was dropped by Chruściel[24] in 1999 who applied the doubling method across the non-degenerate components and applied the positive mass theorem for complete manifolds with one asymptotically flat end[9] to conclude uniqueness (the Bunting and Masood-ul-Alam conformal rescaling transforms the degenerate components into cylindrical ends). The developments in the uniqueness of static electro-vacuum black holes went in parallel to the developments in the vacuum case. Some remarkable works which played an important role in the general proof of the uniqueness of static electro-vacuum black holes are,[39, 50, 58, 55, 59, 45, 25, 30] where again the doubling method played a fundamental role. Uniqueness of static black holes using the doubling method has also been proved for other matter models, as for instance the Einstein–Maxwell-dilaton model.[46, 44] The Bunting and Masood-ul-Alam method is in principle capable of dealing with a large variety of matter models and at present it is the most powerful method to prove uniqueness of black holes in the static case.

During the late 60s, uniqueness of *stationary* black holes also started to take shape. The works of Israel, Hawking, Carter and Robinson, between 1967 and 1975, gave an almost complete proof that the Kerr black hole was the only possible stationary vacuum black hole. The first step was given by Hawking (see[34]) who proved that the intersection of the event horizon with a Cauchy hypersurface has $\mathbb{S}^2$-topology. The next step, also due to Hawking,[34] was the demonstration of the so-called Hawking Rigidity Theorem, which states that a stationary black hole must be static or axisymmetric. Finally, the work of Carter[22] and Robinson[53] succeeded in proving that the Kerr solutions are the only possible stationary axisymmetric black holes. Nevertheless, due to the fact that the Hawking Rigidity Theorem requires analyticity of all objects involved, uniqueness was proved only for analytic spacetimes. The recent work by Chruściel and Lopes Costa[29] has contributed substantially to reduce the hypotheses and to fill several gaps present in the previous arguments. Similarly, uniqueness of stationary electro-vacuum black holes has also been proved. Some remarkable works for the stationary electro-vacuum case are,[23, 47] and, more recently[40] where weaker hypotheses are assumed for the proof. Uniqueness of stationary and axisymmetric black holes has also been proved for non-linear σ-models in.[15] It is also worth remarking that, in the case of matter models modeled with Yang–Mills fields, uniqueness of stationary black holes is not true in general and counterexamples exist.[10]

The first answer to the question of whether the uniqueness theorems for static black holes extend to static spacetimes containing MOTS was given by Miao in

2005,[48] who proved uniqueness for the particular case of time-symmetric, asymptotically flat and vacuum spacelike hypersurfaces possessing a minimal compact boundary (note that, in a time-symmetric slice, compact minimal surfaces are MOTS and vice versa). This result generalizes the uniqueness result of Bunting and Masood-ul-Alam for vacuum static black holes discussed above and reads as follows.

Theorem 2 (Miao, 2005[48]). *Let $(M, g^{(4)})$ be a vacuum spacetime with a static Killing vector $\vec{\xi}$. Assume that $(M, g^{(4)})$ possesses a connected, asymptotically flat spacelike hypersurface (Σ, g, K) which is time-symmetric and such that $\partial\Sigma$ is a (non-empty) compact minimal surface.*
Then (Σ, g) is isometric to $\left(\mathbb{R}^3 \setminus B_{M_{Kr}/2}(0), (g_{Kr})_{ij} = \left(1 + \frac{M_{Kr}}{2|x|}\right)^4 \delta_{ij}\right)$ for some $M_{Kr} > 0$, i.e. the $\{t = 0\}$ slice of the Kruskal spacetime with mass M_{Kr} outside and including the horizon. Moreover, there exists a neighbourhood of Σ in M which is isometrically diffeomorphic to the closure of the domain of outer communications of the Kruskal spacetime.

A key ingredient in Miao's proof was to show that the existence of a closed minimal surface implies the existence of an asymptotically flat end Σ^∞ with smooth topological boundary[b] $\partial^{top}\Sigma^\infty$ such that $\vec{\xi}$ is timelike on Σ^∞ and vanishes on $\partial^{top}\Sigma^\infty$. Miao then proved that $\partial^{top}\Sigma^\infty$ coincides in fact with the minimal boundary $\partial\Sigma$ of the original manifold. Hence, the strategy of Miao was to reduce Theorem 2 to Theorem 1.

As a consequence of the static vacuum field equations, the set of points where the Killing vector vanishes in a time-symmetric slice is known to be a totally geodesic surface. Totally geodesic surfaces are of course minimal and in this sense Theorem 2 is a generalization of Theorem 1. In fact, Theorem 1 allows us to rephrase Miao's theorem as follows:

No minimal surface can penetrate in the exterior region where the Killing vector is timelike in any time-symmetric and asymptotically flat slice of a static vacuum spacetime.

In this sense, Miao's result can be regarded as a confinement result for MOTS in time-symmetric slices of static vacuum spacetimes. As mentioned before, a general confinement result of this type was known for black hole spacetimes (i.e. under suitable global-in-time hypotheses). Consequently, Theorem 2 can also be viewed as an extension of this result to the initial data setting for the particular case of time-symmetric, static vacuum slices.

It is clear that this confinement result can be thought to be extended at least in three different directions. Firstly, the matter model may be more general than

[b]Throughout this work the sign ∂^{top} will denote the topological boundary of a set and ∂ the boundary of a manifold with boundary.

vacuum. In fact, for most part of this contribution, the matter model will be general except for the requirement that it satisfies the null energy condition. Secondly, the slices need not be taken as time-symmetric. Recall that we intend to work at the initial data level or, at most, on small slabs of spacetime around them, so we cannot simply select a time-symmetric hypersurface on our spacetime (such a hypersurface would exist around each spacetime point but no *a priori* information of its global-in-space properties would be available). And finally, the condition of asymptotic flatness may be relaxed to just assuming the presence of an outer untrapped surface. The proof given by Miao relies strongly on the vacuum field equations, so these generalizations require different methods. A fundamental step is a proper understanding in static spacetimes of MOTS and of the topological boundary of the exterior region where the Killing field is timelike.

Once these non-penetration properties of MOTS are properly understood, a convenient strategy to prove uniqueness of static spacetimes with MOTS is to try to reduce the problem to the classic uniqueness theorem. This requires finding reasonable conditions on initial data sets which allow us to recover the basic ingredients of the doubling method of Bunting and Masood-ul-Alam.

We finish the Introduction with a brief summary of this contribution. In Section 2 we will introduce the definitions and summarize useful previous results. In Section 3 we will describe the results involving properties of Killing vector fields on stable, strictly stable and locally outermost MOTS. These results, together with a number of properties of the topological boundary of the set where a static Killing vector field is timelike on an initial data set, will allow us to conclude that, under suitable conditions, no bounding MOTS can penetrate in the exterior region where the Killing vector is timelike. This result, which extends Miao's theorem as a confinement statement, will be discussed in Section 4. In Section 5 we will present and discuss our main result which, under reasonable assumptions, establishes the equivalence of MOTS and (spacelike sections of) event horizons for static spacetimes. Finally, in Section 6 we will present a list of possible open problems.

2. Preliminaries

2.1. *Trapped surfaces*

We will consider 4-dimensional spacetimes, namely smooth manifolds[c] M without boundary endowed with a Lorentzian metric $g^{(4)}$ of signature $(-,+,+,+)$. We will also assume that $(M, g^{(4)})$ is time-oriented. The Levi–Civita covariant derivative of $g^{(4)}$ is denoted by $\nabla^{(4)}$ and the corresponding Ricci and scalar curvature tensors are $R^{(4)}_{\mu\nu}$ and $R^{(4)}$, respectively (where $\mu, \nu = 0, 1, 2, 3$). We follow the sign conventions of.[61]

Throughout this work, we will often impose the *null energy condition (NEC)*.

[c]All manifolds in this contribution are connected unless otherwise stated.

Definition 1. *A spacetime* $(M, g^{(4)})$ *satisfies the* **null energy condition (NEC)** *if the Einstein tensor* $G^{(4)}_{\mu\nu} \stackrel{\text{def}}{=} R^{(4)}_{\mu\nu} - \frac{1}{2} R^{(4)} g^{(4)}_{\mu\nu}$ *satisfies* $G^{(4)}_{\mu\nu} k^\mu k^\nu|_{\mathfrak{p}} \geq 0$ *for any null vector* $\vec{k} \in T_{\mathfrak{p}} M$ *and all* $\mathfrak{p} \in M$.

Our conventions for submanifolds are as follows. Let Σ be a smooth manifold, possibly with boundary, of dimension $n < 4$ and $\Phi : \Sigma \rightarrow M$ be an immersion, then we call $\Phi(\Sigma)$ an immersed submanifold. We use the term **submanifold** whenever Φ is, in addition, injective. In this case we often use the same symbol to denote the abstract manifold Σ and its image in M. The same convention is used for contravariant tensors. Note, however, that $\Phi(\Sigma)$ admits two topologies which are in general different: the induced topology as a subset of M and the manifold topology defined by Φ from Σ. When referring to topological concepts in injectively immersed submanifolds we will always use the subset topology unless otherwise stated. If a immersion Φ is such that these two topologies coincide then Φ is an embedding and $\Phi(\Sigma)$ is an **embedded submanifold**.

From now on, Σ will denote a spacelike hypersurface, namely a codimension one embedded submanifold, possibly with boundary, with positive definite first fundamental form (denoted by g). We always assume Σ to be orientable. We will use K to denote the second fundamental form[d] of Σ with respect to the future directed unit normal. Indices on Σ are Latin i, j, etc. and run from 1 to 3.

Definition 2. *A* **surface** S *is a smooth, orientable, codimension one, embedded, not necessarily connected submanifold of* (Σ, g).

The induced metric of S is denoted by γ and we use capital Latin indices A, B, etc. for tensors defined in S. As a submanifold of Σ, S will have a second fundamental form vector $\vec{\kappa}$ and a mean curvature vector $\vec{p} = \text{tr}_S \vec{\kappa}$, where tr_S denotes the trace on S. The second fundamental form vector of S as a submanifold of M will be denoted by $\vec{\Pi}$ and we use $\vec{H}$ for the corresponding mean curvature vector $\vec{H} = \text{tr}_S \vec{\Pi}$.

Let $\vec{n}$ be the unit, future-directed, normal vector to Σ and assume that S admits a notion of *exterior* in Σ (in the next subsection we describe a specific situation where this notion is well defined) so that one of its two unit normals in Σ, denoted by $\vec{m}$, points in this exterior direction. Then, we can define the *outer* and *inner* null vectors, denoted by $\vec{l}_+$ and $\vec{l}_-$, as

$$\vec{l}_+ = \vec{n} + \vec{m}, \tag{1}$$

$$\vec{l}_- = \vec{n} - \vec{m}. \tag{2}$$

In this basis the mean curvature vector $\vec{H}$ decomposes as

$$\vec{H} = -\frac{1}{2} \left(\theta^- \vec{l}_+ + \theta^+ \vec{l}_- \right), \tag{3}$$

[d]Our sign convention for extrinsic curvatures is that the second fundamental form vector of a 2-sphere in the Euclidean 3-space points outwards.

where $\theta^+ \stackrel{\text{def}}{=} l_+{}^\mu H_\mu$ and $\theta^- \stackrel{\text{def}}{=} l_-{}^\mu H_\mu$ are the null expansions of S along $\vec{l}_+$ and $\vec{l}_-$, respectively. It is straightforward to show that

$$\theta^+ = p + q, \qquad \theta^- = -p + q, \tag{4}$$

where

$$p = m^i p_i, \qquad q = \gamma^{AB} K_{AB},$$

and K_{AB} is the pull-back of the second fundamental form K_{ij} onto S. Trapped surfaces are defined in term of the signs of the null expansions $\theta^\pm$. The definitions used in this contribution are the following.

Definition 3. *A closed (i.e. compact and without boundary) surface is*

- **Weakly outer trapped,** *if* $\theta^+ \leq 0$.
- **Marginally outer trapped (MOTS),** *if* $\theta^+ = 0$.
- **Outer trapped,** *if* $\theta^+ < 0$.
- **Outer untrapped,** *if* $\theta^+ > 0$.
- **Past weakly outer trapped,** *if* $\theta^- \geq 0$.
- **Past marginally outer trapped (past MOTS),** *if* $\theta^- = 0$.

An important concept in what follows is the stability of MOTS, defined in terms of the sign of the principal eigenvalue of its stability operator.[4]

Definition 4. *The* **stability operator** $L_{\vec{m}}$ *for a connected MOTS* $S \subset \Sigma$ *is defined by*

$$L_{\vec{m}}\psi \stackrel{\text{def}}{=} \delta_{\psi\vec{m}}\theta^+ = -\Delta_S\psi + 2s^A\nabla^S_A\psi + \left(\frac{1}{2}R_S - Y - s_A s^A + \nabla^S_A s^A\right)\psi, \tag{5}$$

where Δ_S *and* ∇^S_A *denote the Laplacian and the covariant derivative on* S, R_S *is the scalar curvature of* S, $s_A \stackrel{\text{def}}{=} m^i e^j_A K_{ij}$ *and*

$$Y \stackrel{\text{def}}{=} \frac{1}{2}\Pi^\mu_{AB}\Pi^{\nu AB} l_{+\mu} l_{+\nu} + G_{\mu\nu} l^\mu_+ n^\nu. \tag{6}$$

This operator gives the first variation of θ^+ along the vector $\psi\vec{m}$. A similar operator can be defined for past MOTS by replacing θ^+ by $-\theta^-$ above.

Let μ be the principal eigenvalue of $L_{\vec{m}}$, defined as the eigenvalue with smallest real part and which turns out to be always real. Stability properties of MOTS are related to the sign of the principal eigenvalue.[4]

Definition 5. *A MOTS* $S \subset \Sigma$ *is* **stable in** Σ *(or, simply, stable) if the principal eigenvalue* μ *of the stability operator* $L_{\vec{m}}$ *of each connected component of* S *is non-negative.* S *is* **strictly stable in** Σ *if* $\mu > 0$.

It turns out[4] that a MOTS S is (strictly) stable if and only if there exists an outer variation with not decreasing (strictly increasing) θ^+. This suggests that the presence of surfaces with negative θ^+ outside S may be related with the stability property of S. This can be made precise by introducing the following notion.

Definition 6. *A MOTS $S \subset \Sigma$ is* **locally outermost** *if there exists a two-sided neighbourhood of S on Σ whose exterior part does not contain any weakly outer trapped surface.*

The relationship between this concept and stability of MOTS is given by the following result.[3]

Proposition 1. *(1) A strictly stable MOTS (or past MOTS) is necessarily locally outermost.*
(2) A locally outermost MOTS (or past MOTS) is necessarily stable.
(3) None of the converses is true in general.

2.2. *The trapped region*

It is natural to ask whether the locally outermost properties of stable MOTS can be extended to global notions. This is the content of a fundamental theorem by Andersson and Metzger[5] (extended by Eichmair[33] to higher dimensions) on the existence, uniqueness and regularity of the outermost MOTS on a spacelike 3-dimensional manifold. As we will see, this existence theorem plays an essential role in the proof of the results presented in this work. In order to state this theorem, we need to define the concepts of barrier, bounding surface and trapped region.

Definition 7. *Let Σ be a spacelike 3-dimensional manifold possibly with boundary. A closed surface $S_b \subset \Sigma$ is a* **barrier with interior** Ω_b *if there exists a manifold with boundary Ω_b which is topologically closed and such that $\partial \Omega_b = S_b \bigcup \underset{a}{\cup} (\partial \Sigma)_a$, where $\underset{a}{\cup}(\partial\Sigma)_a$ is a union (possibly empty) of connected components of $\partial \Sigma$.*

Remark. If no confusion arises we will refer to a barrier S_b with interior Ω_b simply as a *barrier.* □

The concept of a barrier will give us a criterion to define the exterior and the interior of a special type of surfaces called *bounding.* More precisely,

Definition 8. *Let Σ be a spacelike 3-dimensional manifold possibly with boundary with a barrier S_b with interior Ω_b. A surface $S \subset \Omega_b \setminus S_b$ is* **bounding with respect to the barrier** S_b *if there exists a compact manifold $\Omega \subset \Omega_b$ with boundary such that $\partial \Omega = S \cup S_b$. The set $\Omega \setminus S$ will be called the* **exterior** *of S in Ω_b and $(\Omega_b \setminus \Omega) \cup S$ the* **interior** *of S in Ω_b.*

Remark. Again, for simplicity and when no confusion arises, we will often refer to a surface which is bounding with respect a barrier simply as a *bounding surface.* For graphic examples of surfaces which are bounding with respect to a barrier see Figures 1 and 2. □

The concept of bounding surface allows for a meaningful definition of *outer direction.* The outer normal vector $\vec{m}$ is the unit normal to S in Σ which points into the exterior of S in Ω_b. For S_b itself, we define $\vec{m}$ as the unit normal pointing outside Ω_b. In either case, we define the outer and the inner null vectors, $\vec{l}_+$ and $\vec{l}_-$, by equations (1) and (2).

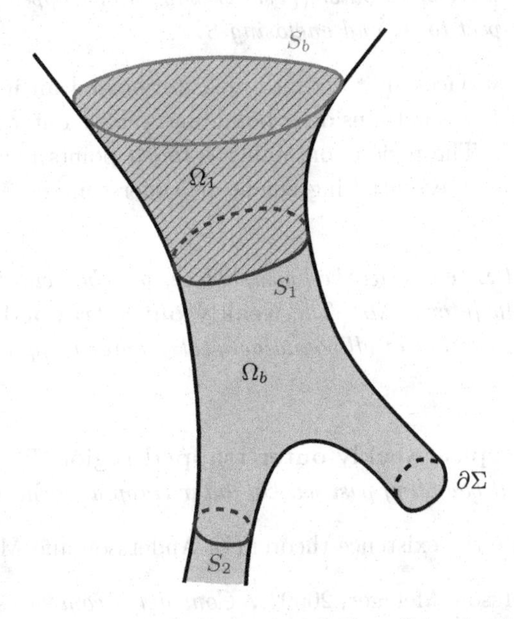

Figure 1. (Color online) In this example, the surface S_b (in red) is a barrier with interior Ω_b (in grey). The surface S_1 is bounding with respect to S_b with Ω_1 (the stripped area) being its exterior in Ω_b. The surface S_2 fails to be bounding with respect to S_b because its "exterior" would contain $\partial\Sigma$.

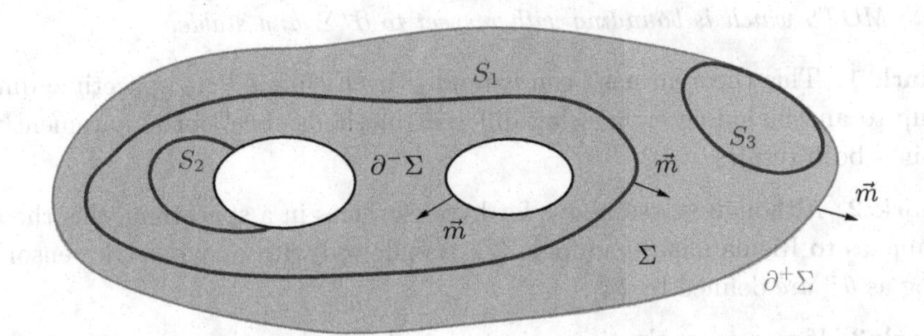

Figure 2. A manifold Σ with boundary $\partial\Sigma = \partial^-\Sigma \cup \partial^+\Sigma$. The boundary $\partial^+\Sigma$ is a barrier whose interior coincides with Σ. The surface S_1 is bounding with respect to $\partial^+\Sigma$, while S_2 and S_3 fail to be bounding. The figure also shows the outer normal $\vec{m}$ as defined in the text.

Definition 9. *Given two surfaces S_1 and S_2 which are bounding with respect to a barrier S_b, we will say that S_1 encloses S_2 if the exterior of S_2 contains the exterior of S_1.*

Definition 10. *A (past) MOTS $S \subset \Sigma$ which is bounding with respect to a barrier S_b is* **outermost** *if there is no other (past) weakly outer trapped surface in Σ which is bounding with respect to S_b and enclosing S.*

Since bounding surfaces split Ω_b into an exterior and an interior region, it is natural to consider the points inside a bounding weakly outer trapped surface S as "trapped points". The region containing trapped points is called *weakly outer trapped region* and is an essential ingredient of Andersson and Metzger's theorem. More precisely,

Definition 11. *Let Σ be a spacelike hypersurface, possibly with boundary, containing a barrier S_b with interior Ω_b. The* **weakly outer trapped region** *T^+ of Ω_b is the union of the interiors of all bounding weakly outer trapped surfaces in Ω_b.*

Analogously,

Definition 12. *The* **past weakly outer trapped region** *T^- of Ω_b is the union of the interiors of all bounding past weakly outer trapped surfaces in Ω_b.*

We can now state the existence theorem of Andersson and Metzger.[5]

Theorem 3 (Andersson, Metzger, 2009[5]). *Consider a compact spacelike hypersurface $\tilde{\Sigma}$ with boundary $\partial\tilde{\Sigma}$ in a spacetime $(M, g^{(4)})$. Assume that the boundary can be split in two non-empty disjoint components $\partial\tilde{\Sigma} = \partial^-\tilde{\Sigma} \cup \partial^+\tilde{\Sigma}$ (neither of which are necessarily connected) and take $\partial^+\tilde{\Sigma}$ as a barrier with interior $\tilde{\Sigma}$. Suppose that $\theta^+[\partial^-\tilde{\Sigma}] \leq 0$ and $\theta^+[\partial^+\tilde{\Sigma}] > 0$ (with respect to the outer normals defined above). Then the topological boundary $\partial^{top}T^+$ of the weakly outer trapped region of $\tilde{\Sigma}$ is a smooth MOTS which is bounding with respect to $\partial^+\tilde{\Sigma}$ and stable.*

Remark 1. This theorem has been extended by Eichmair[33] to spacetime dimension up to and including eight using different methods. See[6] for a statement that combines both results. □

Remark 2. Although stated above for hypersurfaces in a spacetime, this theorem also applies to Riemannian manifolds $(\tilde{\Sigma}, g)$ endowed with a symmetric tensor K_{ij} as long as $\theta^{\pm}$ are defined by (4). □

Remark 3. If we reverse the time orientation of the spacetime, an analogous result for the topological boundary of the past weakly outer trapped region T^- follows. Indeed, if the hypotheses on the sign of the outer null expansion of the components of $\partial\tilde{\Sigma}$ are replaced by $\theta^-[\partial^-\tilde{\Sigma}] \geq 0$ and $\theta^-[\partial^+\tilde{\Sigma}] < 0$ then the conclusion is that $\partial^{top}T^-$ is a smooth *past* MOTS which is bounding with respect to $\partial^+\tilde{\Sigma}$ and stable. □

Since, by construction, no bounding MOTS can penetrate into the exterior of $\partial^{top}T^+$ this theorem shows the existence, uniqueness and smoothness of the outermost bounding MOTS in a compact manifold. Note also that another consequence of this result is the fact that the set T^+ is topologically closed (because it is the interior of the bounding surface $\partial^{top}T^+$).

2.3. *Killing Initial Data (KID)*

We start with the standard definition of initial data set.

Definition 13. *An* ***initial data set*** $(\Sigma, g, K; \rho, \mathbf{J})$ *is a 3-dimensional connected manifold* Σ*, possibly with boundary, endowed with a Riemannian metric* g*, a symmetric, rank-two tensor* K*, a scalar* ρ *and a one-form* $\mathbf{J}$ *satisfying the so-called constraint equations,*

$$2\rho = R^{\Sigma} + (tr_{\Sigma}K)^2 - K_{ij}K^{ij},$$
$$-J_i = \nabla^{\Sigma}{}_j(K_i{}^j - tr_{\Sigma}K\delta_i^j),$$

where R^{Σ} *and* ∇^{Σ} *are respectively the scalar curvature and the covariant derivative of* (Σ, g) *and* $tr_{\Sigma}K = g^{ij}K_{ij}$*.*

For simplicity, we will often write (Σ, g, K) instead of $(\Sigma, g, K; \rho, \mathbf{J})$ when no confusion arises.

In the framework of the Cauchy problem for the Einstein field equations, Σ is a spacelike hypersurface of a spacetime $(M, g^{(4)})$, g is the induced metric and K is the second fundamental form. The initial data **energy density** ρ and **energy flux** $\mathbf{J}$ are defined by $\rho \stackrel{\mathbf{def}}{=} G^{(4)}_{\mu\nu}n^{\mu}n^{\nu}$, $J_i \stackrel{\mathbf{def}}{=} -G^{(4)}_{\mu\nu}n^{\mu}e_i^{\nu}$, where $G^{(4)}_{\mu\nu}$ is the Einstein tensor of $g^{(4)}$, $\vec{n}$ is the unit future directed vector normal to Σ and $\{\vec{e}_i\}$ is a local basis for $\mathfrak{X}(\Sigma)$. When $\rho = 0$ and $\mathbf{J} = 0$, the initial data set is said to be **vacuum**. As remarked in Section 1, we will regard initial data sets as abstract objects on their own, independently of the existence of a spacetime where they may be embedded, unless explicitly stated. The definition of the null energy condition (NEC) can be easily adapted to the initial data level as follows.

Definition 14. *An initial data set* (Σ, g, K) *satisfies the* **null energy condition** *(NEC) if for all* $\mathfrak{p} \in \Sigma$ *the tensor*[e] $G_{\mu\nu} \stackrel{\mathbf{def}}{=} \rho n_{\mu}n_{\nu} + J_{\mu}n_{\nu} + n_{\mu}J_{\nu} + \tau_{\mu\nu}$ *defined on* $T_{\mathfrak{p}}\Sigma \times \mathbb{R}$ *satisfies that* $G_{\mu\nu}k^{\mu}k^{\nu}|_{\mathfrak{p}} \geq 0$ *for any null vector* $\vec{k} \in T_{\mathfrak{p}}\Sigma \oplus \mathbb{R}$*, where this vector space is endowed with the product metric of* $g|_{\mathfrak{p}}$ *and minus the canonical metric on* $\mathbb{R}$*, and* n_{μ} *is unit and tangent to the* $\mathbb{R}$ *factor.*

Consider for a moment a spacetime $(M, g^{(4)})$ possessing a Killing vector field $\vec{\xi}$ and let (Σ, g, K) be an initial data set in this spacetime. We can decompose $\vec{\xi}$ along Σ into a normal and a tangential component as

$$\vec{\xi} = N\vec{n} + \vec{Y} \tag{7}$$

(see Figure 3), where $N = -\xi_{\mu}n^{\mu}$. Note that with this decomposition the squared norm of the Killing vector reads $\xi_{\mu}\xi^{\mu} = -N^2 + \vec{Y}^2$. Inserting (7) into the Killing

[e] Note that, if Σ is embedded in a spacetime, the tensor $G_{\mu\nu}$ is just the Einstein tensor $G^{(4)}_{\mu\nu}$.

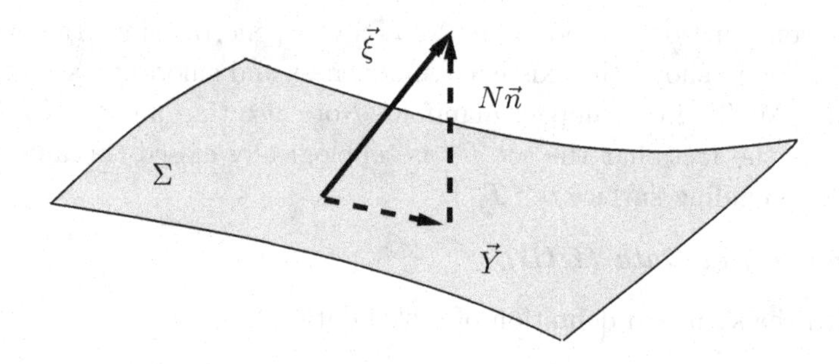

Figure 3. The vector $\vec{\xi}$ decomposed into normal $N\vec{n}$ and tangential $\vec{Y}$ components.

equations and performing a 3+1 splitting on (Σ, g, K) it follows (see,[31, 11]),

$$2NK_{ij} + 2\nabla^{\Sigma}_{(i}Y_{j)} = 0, \tag{8}$$

$$\mathcal{L}_{\vec{Y}}K_{ij} + \nabla^{\Sigma}_i\nabla^{\Sigma}_j N = N\left(R^{\Sigma}{}_{ij} + \mathrm{tr}_{\Sigma}K K_{ij} - 2K_{il}K^l_j - \tau_{ij} + \frac{1}{2}g_{ij}(\mathrm{tr}_{\Sigma}\tau - \rho)\right), \tag{9}$$

where the parentheses in (8) denotes symmetrization, $\tau_{ij} \stackrel{\mathbf{def}}{=} G^{(4)}_{\mu\nu}e^{\mu}_i e^{\nu}_j$ are the remaining components of the Einstein tensor and $\mathrm{tr}_{\Sigma}\tau \stackrel{\mathbf{def}}{=} g^{ij}\tau_{ij}$. Thus, the following definition of Killing initial data becomes natural.[11]

Definition 15. *An initial data set $(\Sigma, g, K; \rho, \mathbf{J})$ endowed with a scalar N, a vector $\vec{Y}$ and a symmetric tensor τ_{ij} satisfying equations (8) and (9) is called a* ***Killing initial data*** *(KID). On a KID we define*

$$\lambda \stackrel{\mathbf{def}}{=} -\xi_{\mu}\xi^{\mu} = N^2 - \vec{Y}^2.$$

In particular, if a KID has $\rho = 0$, $\mathbf{J} = 0$ and $\tau = 0$ then it is said to be a **vacuum KID**.

A point $\mathfrak{p} \in \Sigma$ where $N = 0$ and $\vec{Y} = 0$ is a **fixed point**. This name is motivated by the fact that when the KID is embedded into a spacetime with a local isometry, the corresponding Killing vector $\vec{\xi}$ vanishes at $\mathfrak{p}$ and the isometry has a fixed point there. A fixed point will be called **transverse** if and only if it satisfies $\nabla^{\sigma}_{[i}Y_{j]}|_{\mathfrak{p}} \neq 0$ and **non-transverse** if and only if $\nabla^{\sigma}_{[i}Y_{j]}|_{\mathfrak{p}} = 0$. Here the brackets denote antisymmetrization.

A natural question regarding KID is whether they can be embedded into a spacetime $(M, g^{(4)})$ such that N and $\vec{Y}$ correspond to a Killing vector $\vec{\xi}$. A KID which is embedded in a spacetime will be called **embedded KID**. The simplest case where the spacetime is guaranteed to exist is when $N \neq 0$ everywhere. In this case, the action of the isometry can be used to construct the so-called Killing development (see[11]).

We will finish this subsection by giving the definition of an asymptotically flat end of a KID.

Definition 16. *An* **asymptotically flat end** *of a KID $(\Sigma, g, K; N, \vec{Y})$ is a subset $\Sigma_0^\infty \subset \Sigma$ which is diffeomorphic to $\mathbb{R}^3 \setminus \overline{B_R}$, where B_R is an open ball of radius R. Moreover, in the Cartesian coordinates $\{x^i\}$ induced by the diffeomorphism, the following decay holds*

$$N - A = O^{(2)}(1/r), \qquad g_{ij} - \delta_{ij} = O^{(2)}(1/r),$$
$$Y^i - C^i = O^{(2)}(1/r), \qquad K_{ij} = O^{(2)}(1/r^2).$$

where A and $\{C^i\}_{i=1,2,3}$ are constants such that $A^2 - \delta_{ij}C^iC^j > 0$, and $r = (x^i x^j \delta_{ij})^{1/2}$. Here, a function $f(x^i)$ is said to be $O^{(k)}(r^n), k \in \mathbb{N} \cup \{0\}$ if $f(x^i) = O(r^n)$, $\partial_j f(x^i) = O(r^{n-1})$ and so on for all derivatives up to and including the k-th ones.

Remark. The condition on the constants A, C^i is imposed to ensure that the KID is timelike near infinity. □

The uniqueness theorems of black holes involve asymptotically flat spacetimes. On the other hand, the existence of an outermost MOTS, as presented in Theorem 3, applies to compact hypersurfaces. However, this discrepancy is only apparent since one can always select a sufficiently large outer untrapped surface in Σ_0^∞ and work with its interior. More precisely, the following definition specifies a way of choosing a barrier in manifolds with a selected asymptotically flat end (see Figure 4).

Definition 17. *Let (Σ, g, K) be a spacelike hypersurface, possibly with boundary, with a selected asymptotically flat end Σ_0^∞. Take a sphere $S_b \subset \Sigma_0^\infty$ with $r = r_0$(constant) large enough so that the spheres with $r \geq r_0$ are outer untrapped with respect to the direction pointing into the asymptotic region in Σ_0^∞. Define $\Omega_b = \Sigma \setminus \{r > r_0\}$, which is obviously topologically closed and satisfies $S_b \subset \partial\Omega_b$. Then S_b is a barrier with interior Ω_b. A surface $S \subset \Sigma$ will be called* **bounding** *if it is bounding with respect to S_b.*

Remark 1. Obviously, the definitions of exterior and interior of a bounding surface (Definition 8), enclosing (Definition 9), outermost (Definition 10) and $T^\pm$ (Definitions 11 and 12), given in the previous section, are applicable in the asymptotically flat setting. Moreover, since r_0 can be taken as large as desired, the specific choice of S_b and Ω_b is not relevant for the definition of bounding (once the asymptotically flat end has been selected). Because of that we will refer to the exterior of S in Ω_b simply as the *exterior of S in Σ*. □

Remark 2. Given a bounding weakly outer trapped surface surface S in an initial data set with an asymptotically flat end, Theorem 3 can be applied to the compact manifold with boundary $\tilde{\Sigma}$ defined as the topological closure of the exterior of S in Ω_b. □

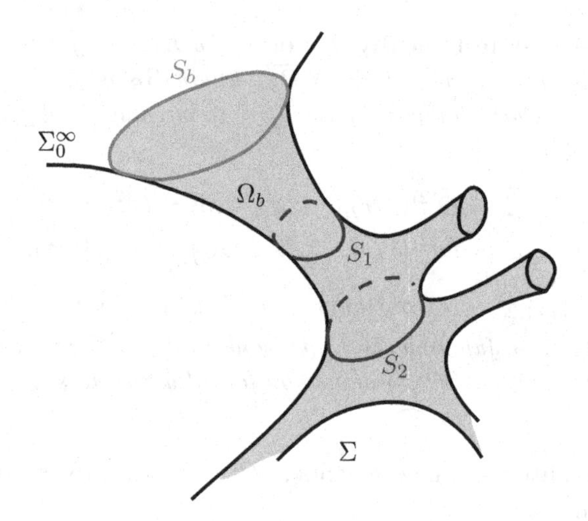

Figure 4. The hypersurface Σ possesses an asymptotically flat end Σ_0^∞ but also other types of ends and boundaries. The surface S_b, which represents a large sphere in Σ_0^∞ and is outer untrapped, is a barrier with interior Ω_b (in grey). The surface S_1 is bounding with respect to S_b (c.f. Definition 8) and therefore is bounding. The surface S_2 fails to be bounding (c.f. Figure 1).

3. Stability of MOTS and Symmetries

In this Section we will present some results concerning the causal character of Killing vectors on stable, strictly stable and locally outermost MOTS in static spacetimes. These results, which appeared in,[19] were obtained for arbitrary vectors and then particularized to concrete symmetries, in particular Killing vectors.

Our idea to study the relation between MOTS and symmetries in[19] starts from the following simple geometric argument: Consider the case of a stationary spacetime and let Σ be a spacelike hypersurface and S the outermost bounding MOTS in Σ, which exists and is unique under suitable assumptions (see Theorem 3). Let us suppose that the stationary Killing vector field $\vec{\xi}$ is timelike on S. Then, move S to the past along the integral lines of $\vec{\xi}$ a small parametric amount τ to obtain a new bounding MOTS S'_τ (note that when moving along an isometry the geometrical properties remain unchanged and therefore this new surface is also a MOTS) which lies in the past of Σ. Later, move S'_τ back to Σ along the outer normal null geodesics. In this way, a new bounding surface $S_\tau \subset \Sigma$ is generated (see Figure 5). Since the parametric amount τ can be varied, we obtain a one-parameter family of surfaces $\{S_\tau\}_\tau$ in Σ. Let us denote by $\vec{\nu}$ the infinitesimal generator of the family $\{S_\tau\}_\tau$ which is clearly tangent to Σ. A simple computation shows that on S we have $\vec{\nu} = \vec{\xi} - N\vec{l}_+ = Q\vec{m} + \vec{Y}^{\parallel}$, where $\vec{m}$ is the unit normal to S in Σ which points into the exterior of S, $\vec{Y}^{\parallel}$ is the component of $\vec{Y}$ tangent to S and the function $Q \stackrel{\text{def}}{=} \xi_\mu l_+^\mu$ determines, to first order, the amount and orientation to which a point $\mathfrak{p} \in S$ moves along the normal direction. Due to the fact that $\vec{\xi}$ was supposed to be timelike on S, the resulting surface S_τ turns out to lie completely in the exterior of S in Σ. Furthermore, if we assume that the NEC holds in the spacetime, then the Raychaudhuri equation implies that the outer null expansion does not increase when passing from S'_τ to S_τ, for τ small enough, which implies that S_τ is weakly outer trapped. Thus, we have constructed a bounding weakly outer trapped surface S_τ which lies in the exterior of S in Σ which contradicts the initial hypothesis that S was the outermost bounding MOTS in Σ. The conclusion is clear: A Killing vector cannot be timelike on an outermost bounding MOTS.

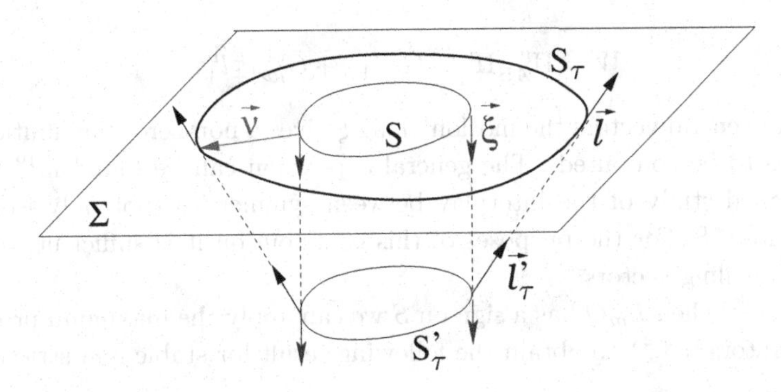

Figure 5. The figure represents how the new surface S_τ is constructed from the original surface S. The intermediate surface S'_τ is obtained from S by dragging along $\vec{\xi}$ a parametric amount τ. Although $\vec{\xi}$ has been depicted as timelike here, this vector can be in fact of any causal character.

Independently of whether $\vec{\xi}$ is a Killing vector or not, the more favorable case to obtain restrictions on the generator $\vec{\xi}$ of a certain symmetry on a given outermost bounding MOTS is when the newly constructed surface S_τ is bounding and weakly outer trapped. This is guaranteed for small enough τ when the variation $\delta_{\vec{\nu}}\theta^+$ of θ^+ along $\vec{\nu}$ is strictly negative everywhere because then first order terms become dominant. Due to the fact that the tangential part of $\vec{\nu}$ does not affect the variation $\delta_{\vec{\nu}}\theta^+$ for a MOTS (see e.g.[4]), it follows that, on any connected component, $\delta_{\vec{\nu}}\theta^+ = L_{\vec{m}}Q$, where $L_{\vec{m}}$ is the stability operator defined above. Since the vector $\vec{\nu} = Q\vec{m} + \vec{Y}^{\parallel}$ determines to first order the direction to which a point $\mathfrak{p} \in S$ moves, it is clear that $L_{\vec{m}}Q < 0$ everywhere and $Q > 0$ somewhere is impossible on any connected component of an outermost bounding MOTS. This is precisely the argument we have used above and is intuitively very clear. However, this geometric method does not provide the most powerful way of finding this type of restriction. Indeed, when the first order term $L_{\vec{m}}Q$ vanishes at some points, then higher order coefficients come necessarily into play, which makes the geometric argument of little use. It is remarkable that using elliptic results most of these situations can be treated in a satisfactory way. Furthermore, since the elliptic methods only use infinitesimal information, there is no need to restrict oneself to outermost bounding MOTS, and the more general case of stable or strictly stable MOTS (not necessarily bounding) can be considered.

The general expression of $L_{\vec{m}}Q$ given by the stability operator (5) is not directly linked to the vector $\vec{\xi}$. In the case of Killing vectors, the point of view of moving S along $\vec{\xi}$ and then back to Σ gives a simple method of calculating $L_{\vec{m}}Q$. Indeed, the only contribution comes from the second step, which can be evaluated directly with the Raychaudhuri equation. The result is simply

$$L_{\vec{m}}Q = NW, \tag{10}$$

where

$$W = \Pi^{\mu}_{AB}\Pi^{\nu AB}l_{+\mu}l_{+\nu} + G^{(4)}_{\mu\nu}l^{\mu}_{+}l^{\nu}_{+}. \tag{11}$$

For more general vectors the motion along $\vec{\xi}$ gives a non-zero contribution to θ^{+} which needs to be computed. The general expression can be found in[19] together with a detailed study of the interplay between symmetries (not only isometries) and stable MOTS. For the purposes of this contribution it is sufficient to restrict ourselves to Killing vectors.

For the case when $L_m Q$ has a sign on S we can apply the maximum principle of elliptic operators (c.f.[4]) to obtain the following result for stable and strictly stable MOTS.

Theorem 4 (Carrasco, Mars, 2009[19]). *Let S be a stable MOTS in a hypersurface Σ of a spacetime $(M, g^{(4)})$ which admits a Killing vector $\vec{\xi}$.*

(i) If $NW|_S \leq 0$ and not identically zero, then $\xi_\mu l^{\mu}_{+}|_S < 0$.
(ii) If S is strictly stable and $NW|_S \leq 0$ then $\xi_\mu l^{\mu}_{+}|_S \leq 0$ and vanishes at one point only if it vanishes everywhere.

Remark: The theorem also holds if all the inequalities are reversed.

Under the NEC, the condition $NW \leq 0$ holds provided $N \leq 0$, i.e. when $\vec{\xi}$ points below Σ everywhere on S (where the term "below" includes also the tangential directions). For strictly stable S, the conclusion of the theorem is that the Killing vector must lie above the null hyperplane defined by the tangent space of S and the outer null normal $\vec{l}_{+}$ at each point $\mathfrak{p} \in S$. If the MOTS is only assumed to be stable, then the theorem requires the extra condition that $\vec{\xi}$ points strictly below Σ at some point with $W \neq 0$. In this case, the conclusion is stronger and forces $\vec{\xi}$ to lie strictly above the null hyperplane everywhere. By changing the orientation of $\vec{\xi}$, it is clear that similar restrictions arise when $\vec{\xi}$ is assumed to point *above* Σ. Figure 6 summarizes the allowed and forbidden regions for $\vec{\xi}$ in this case.

This theorem has an interesting consequence if the Killing vector $\vec{\xi}$ is causal everywhere.

Corollary 1. *Let a spacetime $(M, g^{(4)})$ satisfying the NEC admit a causal Killing vector $\vec{\xi}$ which is future (or past) directed everywhere on a stable MOTS $S \subset \Sigma$. Then,*

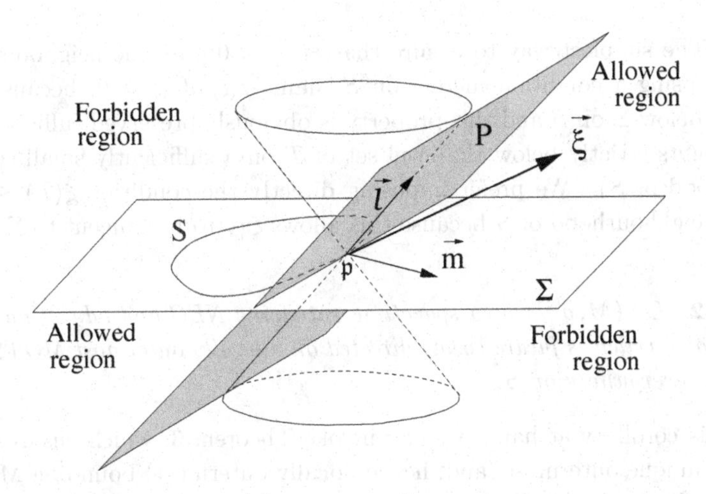

Figure 6. The planes $T_{\mathfrak{p}}\Sigma$ and $P \stackrel{\text{def}}{=} T_{\mathfrak{p}}S \oplus \text{span}\{\vec{l}_+|_{\mathfrak{p}}\}$ divide the tangent space $T_{\mathfrak{p}}M$ in four regions. By Theorem 4, if S is strictly stable and the spacetime satisfies the NEC, then a Killing field $\vec{\xi}$ which points above Σ everywhere on S cannot enter into the forbidden region at any point (and similarly, if $\vec{\xi}$ points below Σ everywhere). The allowed region includes the plane P. However, if there is a point with $W \neq 0$ where $\vec{\xi}$ is not tangent to Σ, then the result is also valid for stable MOTS with P belonging to the forbidden region.

(i) The second fundamental form Π^+_{AB} along $\vec{l}_+$ (i.e. $\Pi^+_{AB} \stackrel{\text{def}}{=} \Pi^\mu_{AB} l_{+\mu}$) and $G^{(4)}_{\mu\nu} l^\mu_+ l^\nu_+$ vanish identically on every point $\mathfrak{p} \in S$ where $\vec{\xi}|_{\mathfrak{p}} \neq 0$.

(ii) If S is strictly stable, then $\vec{\xi} \propto \vec{l}_+$ everywhere.

As it was mentioned above, elliptic methods usually give sharper results than the finite geometric construction of S_τ. However, when $W \equiv 0$ the elliptic methods do not imply anything in the marginally stable (i.e. stable but not strictly stable) case. Nevertheless, the geometric construction does produce restrictions on the sign on $\xi_\mu l^\mu_+$ in this case. It is worth remarking that there is as least one case where marginally stable MOTS play an important role, namely after a jump in the outermost MOTS in a $(3+1)$ foliation of the spacetime (see[2] for details). These results, which provide generalizations of Theorem 4 and point (ii) of Corollary 1 to locally outermost MOTS, read as follows.

Theorem 5 (Carrasco, Mars, 2009[19]). *Let $(M, g^{(4)})$ be a spacetime with a Killing vector $\vec{\xi}$ and satisfying the NEC. Suppose M contains a compact spacelike hypersurface $\tilde{\Sigma}$ with boundary consisting in the disjoint union of a weakly outer trapped surface $\partial^-\tilde{\Sigma}$ and an outer untrapped surface $\partial^+\tilde{\Sigma}$ (neither of which are necessarily connected) and take $\partial^+\tilde{\Sigma}$ as a barrier with interior $\tilde{\Sigma}$. Without loss of generality, assume that $\tilde{\Sigma}$ is defined locally by a level function $T = 0$ with $T > 0$ to the future of $\tilde{\Sigma}$ and let S be the outermost MOTS which is bounding with respect to $\partial^+\tilde{\Sigma}$. If $\vec{\xi}(T) \leq 0$ ($\vec{\xi}(T) \geq 0$) on some spacetime neighbourhood of S, then $\xi^\mu l^+_\mu \leq 0$ ($\xi^\mu l^+_\mu \geq 0$) everywhere on S.*

Remark. The simplest way to ensure that $\vec{\xi}(T) \leq 0$ on some neighbourhood of S is by imposing a condition merely on S, namely $\xi_\mu n^\mu|_S > 0$, because then $\vec{\xi}$ lies strictly below $\tilde{\Sigma}$ on S and this property is obviously preserved sufficiently near S (i.e. $\vec{\xi}$ points strictly below the level set of T on a sufficiently small spacetime neighbourhood of S). We prefer imposing directly the condition $\vec{\xi}(T) \leq 0$ on a spacetime neighbourhood of S because this allows $\vec{\xi}|_S$ to be tangent to Σ. □

Corollary 2. *Let $(M, g^{(4)})$ be a spacetime satisfying NEC and admitting a causal Killing vector $\vec{\xi}$ which is future (past) directed on a locally outermost MOTS $S \subset \Sigma$. Then $\vec{\xi} \propto \vec{l}_+$ everywhere on S.*

With this corollary at hand, we can invoke Theorem 3, which ensures the existence of a unique outermost (and, hence, locally outermost) bounding MOTS, to obtain the following result for stationary spacetimes (c.f. Fig. 7).

Theorem 6 (Carrasco, Mars, 2008[18,19,17]). *Consider a spacelike hypersurface (Σ, g, K) possibly with boundary in a spacetime satisfying the NEC and possessing a Killing vector $\vec{\xi}$ with squared norm $\xi_\mu \xi^\mu = -\lambda$. Assume that Σ possesses a barrier S_b with interior Ω_b which is outer untrapped with respect to the direction pointing outside of Ω_b.*

Consider any surface S which is bounding with respect to S_b. Let us denote by Ω the exterior of S in Ω_b. If S is weakly outer trapped and $\Omega \subset \{\lambda > 0\}$, then λ cannot be strictly positive on any point $\mathfrak{p} \in S$.

Remark. When *weakly outer trapped surface* is replaced by the stronger condition of being a *weakly trapped surface with non-vanishing mean curvature*, then this theorem can be proved by a simple argument based on the first variation of area.[43] In that case, the assumption of S being bounding becomes unnecessary. It would be interesting to know if Theorem 6 holds for arbitrary weakly outer trapped surfaces, not necessarily bounding. □

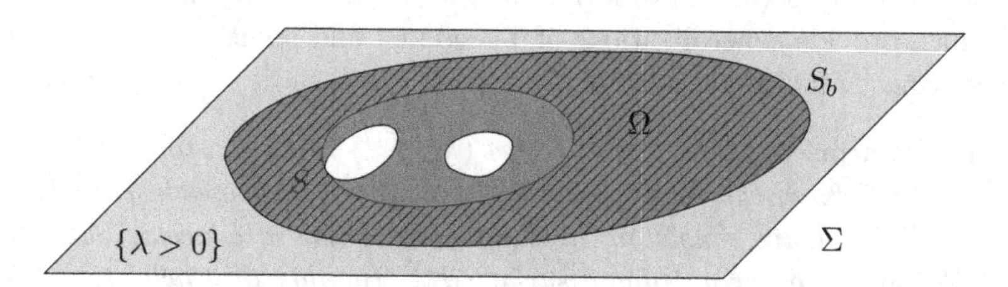

Figure 7. (Color online) Theorems 6 and 7 exclude the possibility pictured in this figure, where S (in blue) is a weakly outer trapped surface which is bounding with respect to the outer trapped barrier S_b. The grey (both light and dark) regions represent the region where $\lambda > 0$. The dark grey region represents the interior of S_b, while the striped area corresponds to Ω, which is the exterior of S in Ω_b.

Finally, the following result is a particularization of Theorem 6 to the case when the hypersurface Σ possesses an asymptotically flat end.

Theorem 7 (Carrasco, Mars, 2008[18,19,17]). *Let (Σ, g, K) be a spacelike hypersurface, possibly with boundary, in a spacetime satisfying the NEC and possessing a Killing vector $\vec{\xi}$. Suppose that Σ possesses an asymptotically flat end Σ_0^∞.*

Consider any bounding surface S (in the sense of Definition 17). Let us denote by Ω the exterior of S in Σ. If S is weakly outer trapped and $\Omega \subset \{\lambda > 0\}$, then λ cannot be strictly positive on any point $\mathfrak{p} \in S$.

We would like to finish this section by pointing out that all the results above imposed the condition that $L_{\vec{m}}Q$ has a sign, as required by the elliptic methods. In[19] we have also shown that the finite geometric construction, together with a smoothing argument by Kriele and Hayward[37] (which, in fact, is used for the proof of several results below), can be exploited to obtain interesting (but weaker) restrictions also in the case when $L_{\vec{m}}Q$ changes sign along S.

4. Confinement of MOTS in Static KID

The results of the previous section involved arbitrary Killing vectors, not necessarily hypersurface orthogonal. In this section we particularize further to the case of *static* spacetimes. The main result of this section (Theorem 8) extends Miao's theorem (Theorem 2) as a confinement result by asserting that, under suitable hypotheses, no bounding MOTS can penetrate into the exterior region where the static Killing is timelike in a static KID. By "exterior region" we mean the connected component of the set $\{\lambda > 0\}$ containing the large sphere used in Definition 17 (we take this sphere sufficiently large so that the Killing vector is timelike everywhere in the asymptotic region outside the sphere). We denote this exterior region by $\{\lambda > 0\}^{ext}$.

Notice that Theorem 7 already forbids the existence of weakly outer trapped surfaces whose exterior lies in $\{\lambda > 0\}^{ext}$ and which penetrate into the timelike region. However, this result does not exclude the existence of a weakly outer trapped surface penetrating into the timelike region but not lying entirely in the causal region. This is the situation we exclude in Theorem 8. We also emphasize that all the results in this section are stated and proved directly at the initial data level, without assuming a spacetime containing such data. Some of these results generalize known properties of static spacetimes to the initial data setting and may be of independent interest.

Thus, our aim is to prove that no bounding weakly outer trapped surface can penetrate into the exterior region $\{\lambda > 0\}^{ext}$. The proof is by contradiction and, in summary, goes as follows. Let us suppose a bounding weakly outer trapped surface S penetrating into $\{\lambda > 0\}^{ext}$. Theorem 3 ensures the existence of an outermost bounding MOTS $\partial^{top}T^+$ which, by construction, necessarily penetrates into $\{\lambda > 0\}^{ext}$. Now we have two possibilities: (i) $\partial^{top}T^+ \subset \overline{\{\lambda > 0\}^{ext}}$, or, (ii)

$\partial^{top}T^+ \not\subset \overline{\{\lambda > 0\}^{ext}}$. Possibility (i) is excluded[f] by Theorem 7. Excluding case (ii) is a harder problem. The key step consists in proving that $\partial^{top}\{\lambda > 0\}^{ext}$ is a bounding MOTS. Once this is proved, the rest is rather easy: construct a new weakly outer trapped surface outside both $\partial^{top}T^+$ and $\partial^{top}\{\lambda > 0\}^{ext}$ by smoothing outwards the corner where they intersect as described by Kriele and Hayward[37] (see Figure 8). This gives the desired contradiction.

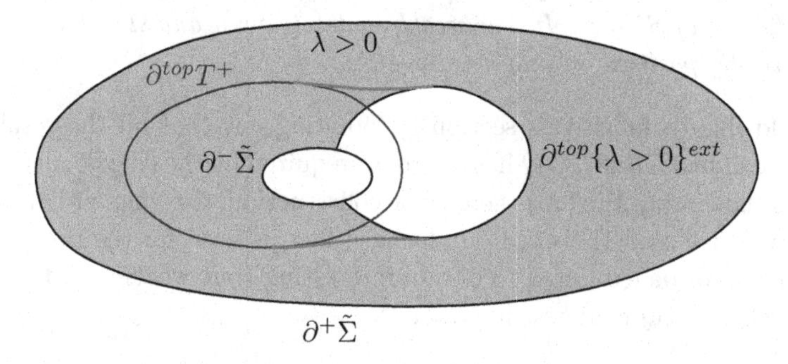

Figure 8. (Color online) The figure illustrates the situation when $\partial^{top}T^+ \not\subset \overline{\{\lambda > 0\}^{ext}}$. The grey region represents the region with $\lambda > 0$ in $\tilde{\Sigma}$. In this case we use the smoothing procedure of Kriele and Hayward to construct a smooth surface S_1 from $\partial^{top}\{\lambda > 0\}^{ext}$ and $\partial^{top}T^+$ (in blue). The red lines represent precisely the part of S_1 which comes from this smoothing.

Thus, the more complicated part is to prove that $\partial^{top}\{\lambda > 0\}^{ext}$ is a bounding MOTS, in particular to show that $\partial^{top}\{\lambda > 0\}^{ext}$ is smooth and embedded. The intuition that $\partial^{top}\{\lambda > 0\}^{ext}$ has to be a MOTS comes from the fact that, in a static spacetime, the arc-connected components of the boundary of the region where the Killing vector is timelike are known to be Killing horizons (this is the content of the Vishveshwara–Carter Lemma[60,21]), except in two cases:

a) When this boundary has fixed points. Indeed, by a result of Boyer,[14] fixed points lie in the closure of Killing horizons and define the so-called bifurcation surface. The spacelike hypersurface Σ may intersect more than one Killing horizon (e.g. the black hole and the white hole Killing horizon) near the fixed point which implies that the corresponding arc-connected component of $\partial^{top}\{\lambda > 0\}^{ext}$ containing the fixed point may fail to be smooth there (see e.g. Figure 9).
b) When the arc-connected component of $\partial^{top}\{\lambda > 0\}^{ext}$ is a spacelike section of a *non-embedded degenerate Killing prehorizon.* Killing prehorizons are null, injectively immersed, hypersurfaces with the property that the Killing vector is non-zero, null and tangent to them. If the gradient of λ vanishes there (degenerate case), these hypersurfaces may *a priori* be non-embedded, which in essence

[f]In fact, Theorem 7 does not immediately exclude possibility (i) because there the KID was assumed to be embedded in a spacetime, which is not the situation here. However, the ideas behind the proof of Theorem 7 can still be applied and the result follows, see[18,17] for details.

means that they may accumulate onto themselves. The occurrence of these non-embedded objects has remained largely overlooked in the black hole uniqueness theory until recently[26] and poses additional complications for proving uniqueness of black holes. For our purposes, the existence of these objects may have the consequence that some arc-connected components of $\partial^{top}\{\lambda > 0\}^{ext}$ may fail to be compact (see e.g. Figure 10), as required by MOTS.

It is clear that these two obstructions to smoothness or embeddedness will show up in any attempt to prove that $\partial^{top}\{\lambda > 0\}^{ext}$ is a MOTS and that restrictions dealing with these difficulties will be necessary.

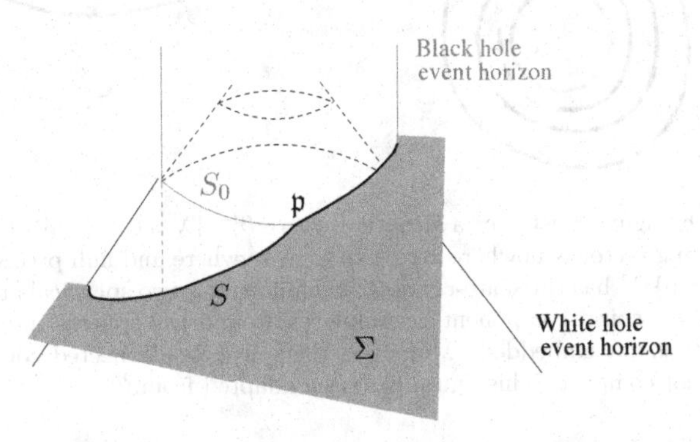

Figure 9. (Color online) An example of non-smooth boundary $S = \partial^{top}\{\lambda > 0\}^{ext}$ in an initial data set Σ of Kruskal spacetime with one dimension suppressed. The region outside the cylinder and the cone corresponds to the domain of outer communications of the Kruskal spacetime. The initial data set Σ intersects the bifurcation surface S_0 (in red). The shaded region corresponds to the intersection of Σ with the asymptotic region, and is in fact a connected component of the subset $\{\lambda > 0\} \subset \Sigma$. Its boundary is non-smooth at the point $\mathfrak{p}$ lying on the bifurcation surface.

In order to describe the results leading to the conclusion that $\partial^{top}\{\lambda > 0\}^{ext}$ is a MOTS (under suitable circumstances) it is convenient to introduce the scalar quantity

$$I_1 = -\frac{1}{2\lambda}\left(g^{ij} - \frac{Y^iY^j}{N^2}\right)\nabla_i^{\Sigma}\lambda\nabla_j^{\Sigma}\lambda, \tag{12}$$

which corresponds to the translation on the KID of the invariant ${\nabla^{(4)}}_{\mu}\xi_{\nu}{\nabla^{(4)}}^{\mu}\xi^{\nu}$ of the Killing form ${\nabla^{(4)}}_{\mu}\xi_{\nu}$. In a spacetime framework this invariant is intimately related to the *surface gravity* of a Killing horizon, so it is not surprising that it plays a relevant role here. Indeed, likewise the surface gravity on each arc-connected component of a Killing horizon, I_1 has the property of being constant on each arc-connected component of $\partial^{top}\{\lambda > 0\}$ (see[18] and Lemma 4.3.10 in[17]).

In order to show that $\partial^{top}\{\lambda > 0\}^{ext}$ is a MOTS, the first step is to investigate its smoothness properties. Smoothness of the open sets of non-fixed points of $\partial^{top}\{\lambda > 0\}^{ext}$ follows easily by using the Killing development of the KID and applying the

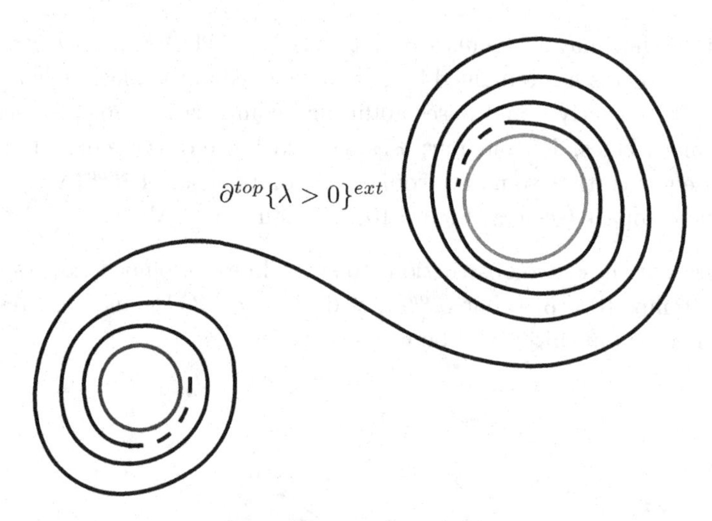

Figure 10. The figure illustrates a situation where $\partial^{top}\{\lambda > 0\}^{ext}$ fails to be embedded. In this figure, the Killing vector is nowhere zero, causal everywhere and null precisely on the plotted line. Here, $\partial^{top}\{\lambda > 0\}^{ext}$ has three arc-connected components: two spherical and one with spiral form. The fact that the spiral component accumulates around the spheres implies that the whole set $\partial^{top}\{\lambda > 0\}^{ext}$ is not embedded. Moreover, the spiral arc-connected component, which is itself embedded, is not compact. This figure has been adapted from.[26]

Vishveshwara–Carter Lemma.[60,21] Open sets of fixed points are easy to deal with once the following lemma is proved (see,[18] Lemma 4.3.5 in[17]).

Lemma 1. *Let* $\mathfrak{p} \in \overline{\{\lambda > 0\}}$ *be a fixed point of a static KID, then* $I_1|_{\mathfrak{p}} < 0$.

Indeed, this lemma implies immediately that $\nabla_i^{\Sigma} N \neq 0$ in any fixed point of $\partial^{top}\{\lambda > 0\}^{ext}$ (Lemma 8 in[18]). Moreover, it is easy to show that the open sets of fixed points of $\partial^{top}\{\lambda > 0\}^{ext}$ coincide with open subsets of $\{N = 0\}$ (see Proposition 4.3.10 in[17]) and, as a consequence, that open sets of fixed points are smooth and embedded and totally geodesic surfaces in Σ.

As expected, the most difficult case involves fixed points which are limits of non-fixed points in $\partial^{top}\{\lambda > 0\}^{ext}$. Here is where the obstruction mentioned in case a) appears and where additional hypotheses need to be made. It turns out that the sign condition

$$NY^i \nabla_i^{\Sigma} \lambda \geq 0 \tag{13}$$

on every connected component of $\partial^{top}\{\lambda > 0\}^{ext}$ is sufficient. More precisely,

Proposition 2. *Let* $(\Sigma, g, K; N, \vec{Y})$ *be a static KID and consider a connected component* $\{\lambda > 0\}_0$ *of* $\{\lambda > 0\}$. *If either* $Y^i \nabla_i^{\Sigma} \lambda \geq 0$ *or* $Y^i \nabla_i^{\Sigma} \lambda \leq 0$ *everywhere on an arc-connected component* $\mathcal{S}$ *of* $\partial^{top}\{\lambda > 0\}_0$, *then* $\mathcal{S}$ *is a smooth submanifold (i.e. injectively immersed) of* Σ.

The idea of the proof of Proposition 2 is the following: First, we prove that the condition on the sign of $Y^i \nabla_i^{\Sigma} \lambda$ implies that every fixed point $\mathfrak{p} \in \partial^{top}\{\lambda > 0\}$ is non-transverse. Second, we show that for every non-tranverse fixed point $\mathfrak{p} \in \partial^{top}\{\lambda > 0\}^{ext}$ there exists a neighbourhood with coordinates $\{x, z^A\}$ such that $\lambda = x^2 - \zeta(z^A)$, with ζ smooth and non-negative. The final step is to prove that $+\sqrt{\zeta}$ is a smooth function (see Proposition 2 in[20]).

Once $\partial^{top}\{\lambda > 0\}^{ext}$ is known to be a smooth submanifold, the next step is to show that under suitable circumstances it has vanishing outer null expansion. For this result, it is important to notice that $\vec{Y}$ is always normal to $\partial^{top}\{\lambda > 0\}^{ext}$. For fixed points, this statement is trivial. For non-fixed points the statement is a consequence of the fact that a Killing vector is orthogonal to its Killing horizon (notice that one can pass to the Killing development near non-fixed points and apply the Vishveshwara-Carter Lemma,[60,21] mentioned above). Then, it is immediate to recover the MOTS condition $p + q = 0$ from equation (8) provided $NY^i m_i \geq 0$.

Proposition 3. *Let* $(\Sigma, g, K; N, \vec{Y})$ *be a static KID and consider a connected component* $\{\lambda > 0\}_0$ *of* $\{\lambda > 0\}$ *with non-empty topological boundary. Let* S *be an arc-connected component* $\partial^{top}\{\lambda > 0\}_0$ *and assume*

(i) If S *contains at least one fixed point:* $NY^i \nabla_i^{\Sigma} \lambda|_S \geq 0$.
(ii) If S *contains no fixed points:* $NY^i m_i|_S \geq 0$, *where* $\vec{m}$ *is the unit normal pointing towards* $\{\lambda > 0\}_0$.

Then S *is a smooth submanifold (i.e. injectively immersed) with* $\theta^+ = 0$ *provided the outer direction is defined as the one pointing towards* $\{\lambda > 0\}_0$. *Moreover, if* $I_1 \neq 0$ *in* S, *then* S *is embedded.*

This proposition ensures (under suitable conditions) that $\partial^{top}\{\lambda > 0\}^{ext}$ has $\theta^+ = 0$. However, it does not show embeddedness except on arc-connected components with $I_1 \neq 0$. Since, for our contradiction argument, we need to construct a bounding weakly outer trapped surface, and these are necessarily embedded, we still need to study under which conditions the injective immersion is an embedding. The case $I_1 \neq 0$ is already covered by Proposition 3. For so-called *degenerate* components, i.e. components with $I_1 = 0$, the potential lack of embeddedness is very closely related to the potential existence of non-embedded Killing prehorizons in a static spacetime (as mentioned in case b) above). Studying whether non-embedded Killing prehorizons can exist is a very interesting problem. In the context of black hole spacetimes (i.e. under suitable global-in-time assumptions on the spacetime), these objects have been excluded to occur in the domain of outer communications in.[28] For our purposes we simply assume an extra condition on degenerate components of $\partial^{top}\{\lambda > 0\}^{ext}$ which implies rather easily that they are embedded submanifolds. This condition is that every arc-connected component of $\partial^{top}\{\lambda > 0\}^{ext}$ with $I_1 = 0$ is topologically closed (for the proof that this condition implies embeddedness see[20]). In all the results below we include this topological condition on degenerate components. If non-existence of non-embedded Killing prehorizons can be shown, then this condition can simply be dropped from every statement[g].

Once sufficient conditions guaranteeing that $\partial^{top}\{\lambda > 0\}^{ext}$ is a MOTS have been found, we are ready to state our confinement result (c.f. Fig. 11). For technical reasons in the proof, we formulate this result for outer trapped surfaces instead of weakly outer trapped surfaces (see[18] for more details). However, as discussed in the Remark below, this theorem can be extended to weakly outer trapped surfaces in many circumstances.

Theorem 8 (Carrasco, Mars, 2011[20]). *Consider a static KID $(\Sigma, g, K; N, \vec{Y})$ satisfying the NEC and possessing a barrier S_b with interior Ω_b (see Definition 7) which is outer untrapped and such that such that $\lambda\big|_{S_b} > 0$. Let $\{\lambda > 0\}^{ext}$ be the connected component of $\{\lambda > 0\}$ containing S_b. Assume that every arc-connected component of $\partial^{top}\{\lambda > 0\}^{ext}$ with $I_1 = 0$ is topologically closed and*

(1) $NY^i\nabla^\Sigma_i\lambda \geq 0$ in each arc-connected component of $\partial^{top}\{\lambda > 0\}^{ext}$ containing at least one fixed point.
(2) $NY^i m_i \geq 0$ in each arc-connected component of $\partial^{top}\{\lambda > 0\}^{ext}$ which contains no fixed points, where $\vec{m}$ is the unit normal pointing towards $\{\lambda > 0\}^{ext}$.

Consider any surface S which is bounding with respect to S_b. If S is outer trapped then it does not intersect $\{\lambda > 0\}^{ext}$.

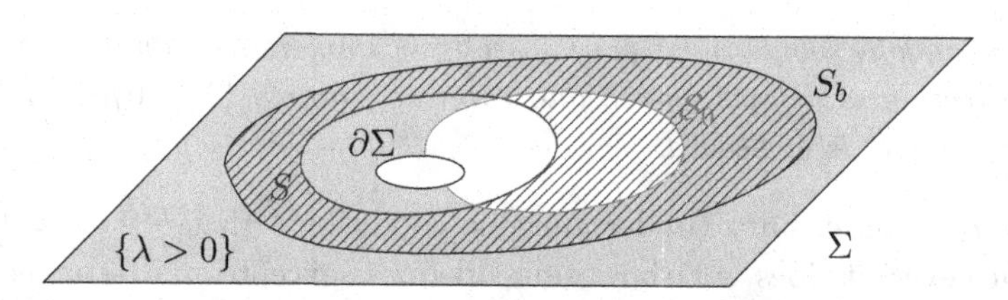

Figure 11. (Color online) Theorem 8 forbids the existence of an outer trapped surface S like the one in the figure (in blue). The striped area corresponds to the exterior of S in Ω_b and the shaded area corresponds to the set $\{\lambda > 0\}^{\text{ext}}$ whose boundary is $\mathcal{S}_0$ (in red). Note that $\mathcal{S}_0$ may intersect $\partial\Sigma$.

Remark. The hypothesis of the theorem can be relaxed to $\theta^+ \leq 0$ if one of the following conditions holds:

(1) S is not the outermost MOTS.
(2) $S \cap \partial\Sigma = \emptyset$.

[g]Recent work by one of us and M. Reiris[42] has shown that Killing prehorizons cannot exist in any static (not necessarily time-symmetric) and asymptotically flat Killing initial data set with a boundary which does not touch the closure of the exterior region where the Killing vector is timelike. The only requirement on the matter model is that the null energy condition is satisfied. In turn, the condition on the boundary is shown to hold as soon as the matter model is well-posed and the inner boundary is enclosed by a marginally outer trapped surface (in particular, if it has non-positive and not identically zero outer null expansion) see[42] for details.

(3) The KID $(\Sigma, g, K; N, \vec{Y})$ can be isometrically embedded into another KID $(\hat{\Sigma}, \hat{g}, \hat{K}, \hat{N}, \hat{\vec{Y}})$ with $\partial\Sigma \subset \text{int}(\hat{\Sigma})$

In this case, Theorem 8 includes Miao's theorem in the particular case of asymptotically flat time-symmetric vacuum static KID with minimal compact boundary. This is because in the time-symmetric case all points with $\lambda = 0$ are fixed points and hence there are no arc-connected components of $\partial^{top}\{\lambda > 0\}$ with $I_1 = 0$ (see Lemma 1) and $Y^i \nabla_i^\Sigma \lambda$ is identically zero on $\partial^{top}\{\lambda > 0\}^{ext}$. □

For the particular case of KID possessing an asymptotically flat end we have the following immediate corollary.

Corollary 3. *Consider a static KID $(\Sigma, g, K; N, \vec{Y})$ with a selected asymptotically flat end Σ_0^∞ and satisfying the NEC. Denote by $\{\lambda > 0\}^{ext}$ the connected component of $\{\lambda > 0\}$ which contains the asymptotically flat end Σ_0^∞. Assume that every arc-connected component of $\partial^{top}\{\lambda > 0\}^{ext}$ with $I_1 = 0$ is closed and*

(1) $NY^i \nabla_i^\Sigma \lambda \geq 0$ in each arc-connected component of $\partial^{top}\{\lambda > 0\}^{ext}$ containing at least one fixed point.

(2) $NY^i m_i \geq 0$ in each arc-connected component of $\partial^{top}\{\lambda > 0\}^{ext}$ which contains no fixed points, where $\vec{m}$ is the unit normal pointing towards $\{\lambda > 0\}^{ext}$.

Then, any bounding (in the sense of Definition 17) outer trapped surface S in Σ cannot intersect $\{\lambda > 0\}^{ext}$.

5. Uniqueness of Static Spacetimes with MOTS

In this section we will present our results on the extension of the static black hole uniqueness theorems to static, asymptotically flat KID containing weakly outer trapped surfaces. As already mentioned, the most powerful method known so far to prove uniqueness of static black holes is the doubling method of Bunting and Masood-ul-Alam,[16] which has been successfully applied to a number of different matter models. As already mentioned in the Introduction, our strategy is trying to recover the framework of the doubling method from a generic static KID containing a weakly outer trapped surface. This framework consists of an asymptotically flat spacelike hypersurface Σ with topological boundary $\partial^{top}\Sigma$ which is a closed (i.e. compact and without boundary) embedded topological manifold and such that the static Killing field is causal on Σ and null only on $\partial^{top}\Sigma$. For black hole uniqueness, the existence of this topological manifold $\partial^{top}\Sigma$ is ensured precisely by the presence of a black hole[h].

Consequently, our strategy consists in showing that, under suitable conditions, the topological boundary $\partial^{top}\{\lambda > 0\}^{ext}$ is a closed embedded topological submanifold.

[h]Note that $\partial^{top}\Sigma$ is not required to be smooth.

Since *a priori* MOTS have nothing to do with black holes, $\partial^{top}\{\lambda > 0\}^{ext}$ may fail to be closed as required in the doubling method. The main difficulty comes from the fact that, since we are considering manifolds Σ with boundary $\partial\Sigma$, the topological boundary of a set, say $\partial^{top}\{\lambda > 0\}^{ext}$, can very well have non-empty boundary (necessarily lying in $\partial\Sigma$). A schematic example showing this behaviour is given in Figure 12. Consequently, it becomes necessary to study under which conditions we can guarantee that $\partial^{top}\{\lambda > 0\}^{ext}$ is compact and without boundary.

It turns out that, under the hypotheses of Theorem 8 (and its Corollary 3) the conclusion that $\partial^{top}\{\lambda > 0\}^{ext}$ is a closed surface follows without much difficulty because the non-penetration property of $\partial^{top}T^+$ inside $\{\lambda > 0\}^{ext}$ prevents $\partial^{top}\{\lambda > 0\}^{ext}$ from reaching the manifold boundary $\partial\Sigma$. This leads to our first uniqueness result.

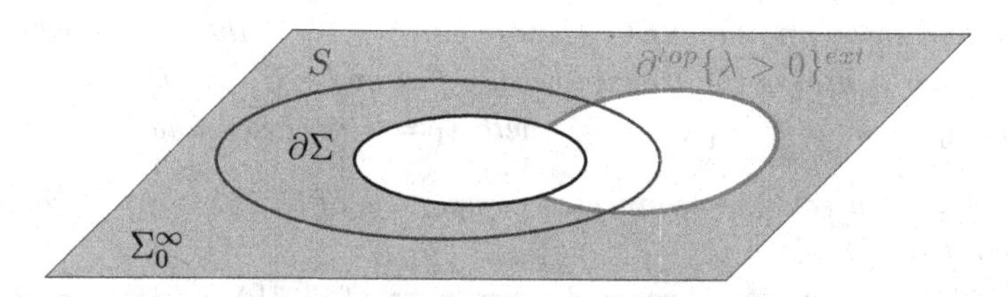

Figure 12. (Color online) The figure illustrates a situation where $\partial^{top}\{\lambda > 0\}^{ext}$ (in red) has non-empty manifold boundary (which lies in $\partial\Sigma$) and, therefore, is not closed. Here, S (in blue) represents a bounding MOTS and the grey region corresponds to $\{\lambda > 0\}^{ext}$. In a situation like this the doubling method cannot be applied.

Theorem 9 (Carrasco, Mars, 2011[20]). *Consider a static KID $(\Sigma, g, K; N, \vec{Y})$ with a selected asymptotically flat end Σ_0^∞ and satisfying the NEC. Assume that Σ possesses an outer trapped surface S which is bounding (Definition 17). Denote by $\{\lambda > 0\}^{ext}$ the connected component of $\{\lambda > 0\}$ which contains the asymptotically flat end Σ_0^∞. If*

(1) Every arc-connected component of $\partial^{top}\{\lambda > 0\}^{ext}$ with $I_1 = 0$ is topologically closed.
(2) $NY^i\nabla_i^\Sigma\lambda \geq 0$ in each arc-connected component of $\partial^{top}\{\lambda > 0\}^{ext}$ containing at least one fixed point.
(3) $NY^i m_i \geq 0$ in each arc-connected component of $\partial^{top}\{\lambda > 0\}^{ext}$ which contains no fixed points, where $\vec{m}$ is the unit normal pointing towards $\{\lambda > 0\}^{ext}$.
(4) The matter model is such that Bunting and Masood-ul-Alam doubling method gives uniqueness of black holes.

Then, $(\{\lambda > 0\}^{ext}, g, K)$ is a slice of such a unique spacetime.

Theorem 9 has been formulated for outer trapped surfaces instead of weakly outer trapped surfaces for the same reason as in Theorem 8. Consequently, the hypotheses of this theorem can also be relaxed to $\theta^+ \leq 0$ if one of the following conditions hold: S is not the outermost MOTS, $S \cap \partial\Sigma = \emptyset$, or the KID can be extended. Under these circumstances, this result already extends Miao's theorem as a uniqueness result.

This theorem, although of interest, is not fully satisfactory in the sense that it imposes conditions directly on the boundary $\partial^{top}\{\lambda > 0\}^{ext}$, which is a fundamental object in the doubling procedure. In the remainder of this section we will discuss a uniqueness result which does not involve any *a priori* restriction on $\partial^{top}\{\lambda > 0\}^{ext}$.

We already know that, without suitable conditions, $\partial^{top}\{\lambda > 0\}^{ext}$ is not a smooth submanifold in general (see Figure 9). As already mentioned, the key difficulty for uniqueness lies in proving that $\partial^{top}\{\lambda > 0\}^{ext}$ is a manifold without boundary. In the previous theorem, the non-penetration property of $\partial^{top}T^{+}$ into $\{\lambda > 0\}^{ext}$ is the essential ingredient that allows us to conclude that $\partial^{top}\{\lambda > 0\}^{ext}$ must lie in the exterior of the bounding outer trapped surface S and hence it cannot intersect $\partial\Sigma$. In turn, this non-penetration property is strongly based on the smoothness of $\partial^{top}\{\lambda > 0\}^{ext}$, which we do not have in general. The main problem is therefore: How can we exclude the possibility (as in Figure 12) that $\partial^{top}\{\lambda > 0\}^{ext}$ reaches $\partial\Sigma$ in the general case?

To address this issue we need to understand better the structure of $\partial^{top}\{\lambda > 0\}^{ext}$ (and, more generally, of $\partial^{top}\{\lambda > 0\}$) when conditions 2 and 3 of Theorem 9 are not satisfied. This forces us to view KID as hypersurfaces embedded in a spacetime, instead of as abstract objects on their own, as we were able to do in the results of the previous section. Let us explain briefly the reason for this.

As mentioned above, the smoothness of any arc-connected component of $\partial^{top}\{\lambda > 0\}$ consisting exclusively of fixed points or exclusively of non-fixed points holds without any additional hypotheses. However, for arc-connected components having both types of points, condition (13) was required to conclude smoothness. As briefly explained there, this hypothesis was imposed in order to avoid the existence of *transverse* fixed points in $\partial^{top}\{\lambda > 0\}$. Actually, the existence of transverse points is, by itself, not very problematic because the structure of $\partial^{top}\{\lambda > 0\}$ in a neighbourhood of a transverse fixed points is well understood and consists of two intersecting branches (see Lemma 4.3.13 in[20]). The problematic situation happens when a sequence of transverse fixed points tends to a non-transverse point $\mathfrak{p}$. In this case the intersecting branches can have a very complicated limiting behavior at $\mathfrak{p}$. As we pointed out before, locally near any non-transverse point $\mathfrak{p}$ there exists coordinates such that $\lambda = x^2 - \zeta(z^A)$, with ζ a non-negative smooth function. In order to understand the behavior of $\partial^{top}\{\lambda > 0\}$ we need to take the square root of ζ. Under the assumptions of Proposition 2 it follows that the *positive* square root is C^∞. For general non-transverse points, this positive square root is in general not even C^1. Moreover, it is not clear at all whether any C^1 square root exists, even allowing such square root to change sign (see[20] for further details). It is plausible that the equations that are satisfied in a static KID forbid the existence of ζ functions with no C^1 square root. This is, however, a difficult issue and we have not been able to resolve it. This is the reason why we need to restrict ourselves to embedded static KID in this section. Assuming the existence of a static spacetime where the KID is embedded, it follows that, irrespective of the structure of fixed points in Σ, a suitable square root of ζ always exists.

In fact, assuming a KID embedded in a spacetime we can use the Rácz–Wald–Walker[52] construction in a suitable neighbourhood of any fixed point and prove the following result[20] on the structure of $\partial^{top}\{\lambda > 0\}$ (note that this is the boundary of the full set of points where the Killing vector is timelike and not of one of its connected component as before).

Proposition 4. *Consider an embedded static KID $(\tilde{\Sigma}, g, K; N, \vec{Y})$, compact and possibly with boundary $\partial\tilde{\Sigma}$. Assume that every arc-connected component of $\partial^{top}\{\lambda > 0\}$ with $I_1 = 0$ is topologically closed. Then*

$$\partial^{top}\{\lambda > 0\} = \underset{a}{\cup} S_a, \tag{14}$$

where each S_a is a smooth, compact, connected and orientable surface, possibly with boundary (and then $\partial S_a \subset \partial\tilde{\Sigma}$). Moreover, at least one of the two null expansions of S_a vanishes everywhere.

With this proposition at hand we are in a situation where can prove that $\partial^{top}\{\lambda > 0\}^{ext} = \partial^{top}T^+$, which is the crucial ingredient for our uniqueness result later. The strategy of the proof is again by contradiction. We assume $\partial^{top}\{\lambda > 0\}^{ext} \neq \partial^{top}T^+$, and construct a bounding weakly outer trapped surface outside $\partial^{top}T^+$. From $\partial^{top}\{\lambda > 0\}^{ext} \neq \partial^{top}T^+$ and the fact that $\partial^{top}T^+$ cannot be fully contained in the closure of the exterior region $\{\lambda > 0\}^{ext}$ (by Theorem 7) it follows that at least one of the surfaces S_a (say S_0) penetrates both the exterior of T^+ and the interior of T^+ and, at the same time, contains points lying on $\partial^{top}\{\lambda > 0\}^{ext}$. According to Proposition 4, this surface satisfies that either θ^+ vanishes everywhere or θ^- vanishes everywhere. If the second possibility occurs (for any possible choice of S_0) then our method would fail to give a contradiction because the smoothing between S_0 and $\partial^{top}T^+$ would not give a MOTS. In order to deal with this difficulty we impose the additional condition $T^- \subset T^+$ (under this assumption, we can perform the Kriele–Hayward smoothing procedure between S_0 and $\partial^{top}T^-$ and still obtain a contradiction). If this condition is not imposed, then hypothetical situations like the one illustrated in Figure 13 can *a priori* happen.

A second difficulty comes from the fact that the surfaces S_a (in particular S_0) are not guaranteed to be bounding. Hence, the smoothing between S_0 and $\partial^{top}T^+$ gives a closed, orientable surface which *a priori* could fail to be bounding. Again, this would spoil the method of proof because no bounding MOTS penetrating the exterior of the trapped region would be constructed. However, this difficulty is much milder than before and can be dealt with simply by imposing a topological condition on $\text{int}(\tilde{\Sigma})$ which forces that all closed and orientable surfaces separate the manifold into disconnected subsets. This topological condition is rather weak and only requires that the first homology group $H_1(\mathrm{int}(\tilde{\Sigma}), \mathbb{Z}_2)$ with coefficients in $\mathbb{Z}_2$ vanishes.

Putting things together (see[20] for details), the following theorem is proved.

Theorem 10 (Carrasco, Mars, 2011[20]). *Consider an embedded static KID $(\tilde{\Sigma}, g, K; N, \vec{Y})$ compact, with boundary $\partial\tilde{\Sigma}$ and satisfying the NEC. Suppose that the boundary can be split into two non-empty disjoint components $\partial\tilde{\Sigma} = \partial^-\tilde{\Sigma} \cup \partial^+\tilde{\Sigma}$ (neither of which are necessarily connected). Take $\partial^+\tilde{\Sigma}$ as a barrier with interior $\tilde{\Sigma}$ and assume $\theta^+[\partial^-\tilde{\Sigma}] \leq 0$ and $\theta^+[\partial^+\tilde{\Sigma}] > 0$. Let T^+, T^- be, respectively, the weakly outer trapped and the past weakly outer trapped regions of $\tilde{\Sigma}$. Assume also the following hypotheses:*

(1) Every arc-connected component of $\partial^{top}\{\lambda > 0\}^{ext}$ with $I_1 = 0$ is topologically closed.
(2) $\lambda|_{\partial^+\tilde{\Sigma}} > 0$.
(3) $T^- \subset T^+$.
(4) $H_1\left(int(\tilde{\Sigma}), \mathbb{Z}_2\right) = 0$.

Denote by $\{\lambda > 0\}^{ext}$ the connected component of $\{\lambda > 0\}$ which contains $\partial^+\tilde{\Sigma}$. Then

$$\partial^{top}\{\lambda > 0\}^{ext} = \partial^{top}T^+.$$

Therefore, $\partial^{top}\{\lambda > 0\}^{ext}$ is a non-empty stable MOTS which is bounding with respect to $\partial^+\tilde{\Sigma}$ and, moreover, it is the outermost bounding MOTS.

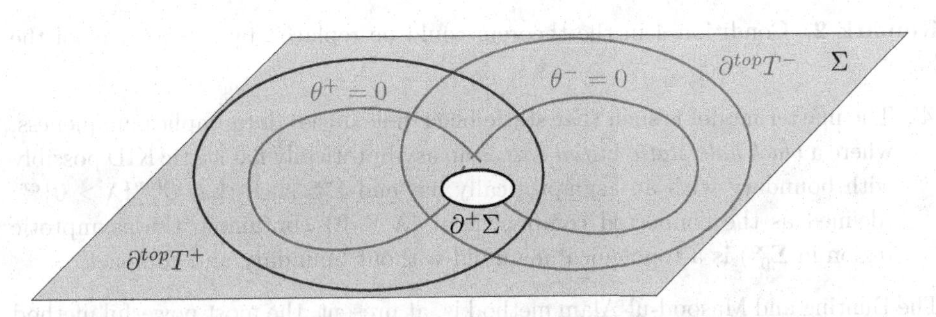

Figure 13. (Color online) The figure illustrates a hypothetical situation where $T^- \subset T^+$ does not hold and the conclusions of the Theorem 10 would not be true. The red continuous line represents the set $\partial^{top}\{\lambda > 0\}^{ext}$ which is composed by a smooth surface with $\theta^+ = 0$, lying inside of $\partial^{top}T^+$ (in blue) and partly outside of $\partial^{top}T^-$ (in green), and a smooth surface with $\theta^- = 0$, which lies partly outside of $\partial^{top}T^+$ and inside of $\partial^{top}T^-$.

Uniqueness of static KID with MOTS follows as an immediate consequence of this theorem.

Theorem 11 (Carrasco, Mars, 2011[20]). *Let $(\Sigma, g, K; N, \vec{Y})$ be an embedded static KID with a selected asymptotically flat end Σ_0^∞ and satisfying the NEC. Assume* *that Σ possesses a weakly outer trapped surface S which is bounding (in the sense of Definition 17). Assume the following:*

(1) Every arc-connected component of $\partial^{top}\{\lambda > 0\}^{ext}$ with $I_1 = 0$ is topologically closed.
(2) $T^- \subset T^+$.
(3) $H_1(\Sigma, \mathbb{Z}_2) = 0$.
(4) The matter model is such that Bunting and Masood-ul-Alam doubling method for time-symmetric initial data sets gives uniqueness of black holes.

Then $(\Sigma \setminus T^+, g, K)$ is a slice of such a unique spacetime.

Remark 1. Except for the subtlety that Theorem 11 applies to embedded static KID instead of abstract static KID, this theorem includes Miao's Theorem 2 as a particular case. This is because all conditions 1 to 4 of the theorem are fulfilled for asymptotically flat, time-symmetric vacuum KID with a compact minimal boundary. Indeed, condition 4 is obviously satisfied for vacuum. Moreover, the property of time-symmetry implies that all points with $\lambda = 0$ are fixed points and hence no arc-connected component of $\partial^{top}\{\lambda > 0\}$ with $I_1 = 0$ exists. Thus, condition 1 is automatically satisfied. Time-symmetry also implies $T^- = T^+$ and condition 2 is trivial. Finally, the region outside the outermost minimal surface in a Riemannian manifold with non-negative Ricci scalar is $\mathbb{R}^3$ minus a finite number of closed balls (see e.g.[36]). This manifold is simply connected and hence satisfies condition 3. □

Remark 2. Condition 4 in the theorem could be replaced by a statement of the form

4'. The matter model is such that static black hole initial data implies uniqueness, where a *black hole static initial data* is an asymptotically flat static KID possibly with boundary with an asymptotically flat end Σ_0^∞ such that $\partial^{top}\{\lambda > 0\}^{ext}$ (defined as the connected component of $\{\lambda > 0\}$ containing the asymptotic region in Σ_0^∞) is a topological manifold without boundary and compact.

The Bunting and Masood-ul-Alam method is, at present, the most powerful method to prove uniqueness under the circumstances of 4'. However, if a new method is invented, Theorem 11 would still give uniqueness. □

Remark 3. A comment on the condition $T^- \subset T^+$ is in order. First of all, in the static regime, T^+ and T^- are expected to be the intersections of both the black and the white hole with Σ. Therefore, the hypothesis $T^- \subset T^+$ could be understood as the requirement that the first intersection, as coming from Σ_0^∞, of Σ with an event horizon occurs with the black hole event horizon. Therefore, this hypothesis is in some sense similar to the hypotheses on $\partial^{top}\{\lambda > 0\}^{ext}$ made in Theorem 3. However, there is a fundamental difference between them: The hypothesis $T^- \subset T^+$

is a hypothesis on the weakly outer trapped regions which, *a priori*, have nothing to do with the location and properties of $\partial^{top}\{\lambda > 0\}^{ext}$. In a physical sense, the existence of past weakly outer trapped surfaces in the spacetime reveals the presence of a white hole region. Moreover, given a (3+1) decomposition of a spacetime satisfying the NEC, the Raychaudhuri equation implies that T^- shrinks to the future while T^+ grows to the future (see[2]) ("grow" and "shrink" is with respect to any timelike congruence in the spacetime). It is plausible that by letting the initial data evolve sufficiently long, only the black hole event horizon is intersected by Σ. The uniqueness Theorem 11 could be applied to this evolved initial data. Although this requires much less global assumptions than for the theorem that ensures that no MOTS can penetrate into the domain of outer communications, it still requires some control on the evolution of the initial data. □

We conclude this section with a simple corollary of Theorem 11, which is nevertheless interesting.[20]

Corollary 4. *Let $(\Sigma, g, K = 0; N, \vec{Y} = 0; \vec{E})$ be a time-symmetric electro-vacuum embedded static KID, i.e a static KID with an electric field $\vec{E}$ satisfying*

$$\nabla_i^{\Sigma} E^i = 0, \quad \rho = |\vec{E}|^2_g, \quad \tau_{ij} = |\vec{E}|^2 g_{ij} - 2E_iE_j.$$

Let $\Sigma = \mathcal{K} \cup \Sigma_0^{\infty}$ where $\mathcal{K}$ is compact and Σ_0^{∞} is an asymptotically flat end and assume that $\partial\Sigma$ is non-empty and has mean curvature $p \leq 0$ (with respect to the normal pointing inside Σ).

Then $(\Sigma \setminus T^+, g, K = 0; N, \vec{Y} = 0; \vec{E})$ can be isometrically embedded in the Reissner–Nordström spacetime with $M > |Q|$, where M is the ADM mass of (Σ, g) and Q is the total electric charge of $\vec{E}$, defined as $Q = \frac{1}{4\pi}\int_{S_{r_0}} E^i m_i \eta_{S_{r_0}}$ where $S_{r_0} \subset \Sigma_0^{\infty}$ is the coordinate sphere $\{r = r_0\}$ and $\vec{m}$ its unit normal pointing towards infinity.

6. Conclusions and Open Problems

The results summarized in this contribution constitute an attempt to perform a systematic study of MOTS in static spacetimes. As a consequence, several properties of MOTS and weakly outer trapped surfaces in static spacetimes which were well-known under global assumptions have been extended to the initial data level (or to slabs of spacetimes around the initial data hypersurface). The main result is a uniqueness theorem for embedded static KID containing an asymptotically flat end, satisfying the NEC and containing a bounding weakly outer trapped surface. The matter model is arbitrary as long as it admits a static black hole uniqueness proof with the Bunting and Masood-ul-Alam doubling method. This result extends a previous theorem by Miao valid on vacuum and time-symmetric slices, and allows to conclude that, under suitable conditions and at least regarding uniqueness of black holes, event horizons and MOTS do coincide in static spacetimes.

To that aim we have exploited a method which is strongly based in the existence, uniqueness and regularity of the outermost bounding MOTS which in 4-dimensions is guaranteed by the Andersson–Metzger Theorem 3. This theorem is known to hold in dimensions to up and including eight by the results of Eichmair[33] (c.f.[6]). Although we have used dimensional dependent arguments in a few places, it seems reasonable that the arguments above can be extended to these dimensions without much effort.

Our methodology has forced us to impose a number of conditions along the way in order to prove the results. A major assumption is that all the trapped surfaces we have used are *bounding*. It would be very interesting to extend these results to general surfaces not necessarily bounding (in particular the uniqueness theorem). This would certainly need completely different methods not based on the existence theorem of Andersson and Metzger.

In addition, we have been forced to make a few other technical hypotheses. Obviously, it would be of interest to drop as many of those conditions as possible. Perhaps the most remarkable one is the topological condition on arc-components of $\partial^{top}\{\lambda > 0\}^{ext}$ with $I_1 = 0$, directly related to the possible existence of non-embedded Killing prehorizons. The second condition that one would certainly like to relax (or drop altogether) is the hypothesis that $\emptyset \neq T^- \subset T^+$.

After this work was completed, an interesting development on uniqueness of static initial data sets with outer trapped boundary has taken place. One of the authors together with M. Reiris, from the Albert Einstein Institute, have proven[42] a uniqueness statement that drops several of the technical issues mentioned above. More precisely, under the assumptions that the initial data set boundary is outer trapped and compact and assuming that the initial data is stationary and asymptotically flat with a matter model satisfying the null energy condition and well-posed field equations, it proven that the future Cauchy development is a black hole. More precisely, it is shown that the future Cauchy development is large enough so as to contain the future Killing development of the exterior of a large coordinate sphere in the asymptotically flat region. This result, together with the non-existence of Killing prehorizons mentioned before (see footnote g) allowed for a general proof of uniqueness for static, asymptotically flat KID with outer trapped boundary provided the matter model is well-posed, satisfies the null energy condition and admits a static black hole uniqueness theorem.

A very interesting problem related to the uniqueness issue discussed in this work, would be to analyze the stability of the uniqueness result, in the sense of studying whether such uniqueness results also hold in an approximate sense, i.e. if a spacetime is *nearly* static and contains a MOTS, then the spacetime is *nearly* unique. This problem is, of course, very difficult because it needs a suitable concept of "being close to". In the particular case of the Kerr metric (and hence also of Schwarzschild), there exists a notion of an initial data being close to Kerr[7] which is based on a suitable characterization of this spacetime.[41] This closeness notion is defined for initial data sets without boundary and has been extended to manifolds with boundary under certain circumstances.[8] It would be of interest to extend it to the case with a non-empty boundary which is a MOTS.

In any case, it is clear that the static uniqueness result for MOTS is only a first step in understanding the relation between trapped surfaces and black holes. Future work should try to extend this result to the stationary setting. The problem is, however, considerably more difficult because the techniques known at present to prove uniqueness of stationary black holes are much less developed than those for proving uniqueness of static black holes. Assuming however, that the spacetime is axially symmetric (besides being stationary) simplifies the black hole uniqueness proof considerably (the problem becomes essentially a uniqueness proof for a boundary value problem of a non-linear elliptic system on a domain in the Euclidean plane, see[35]). The next natural step would be to try and extend this uniqueness result to a setting where the black hole is replaced by a MOTS.

Acknowledgements

Financial support under the projects FIS2009-07238 (Spanish MEC) and P09-FQM-4496 (Junta de Andalucía and FEDER funds) are gratefully acknowledged.

References

1. L. Andersson, "The global existence problem in General Relativity" in *The Einstein equations and the large scale behaviour of gravitational fields*, eds. P.T. Chruściel and H. Friedrich, Birkhäuser (2004).
2. L. Andersson, M. Mars, J. Metzger and W. Simon, "The time evolution of marginally trapped surfaces", *Class. Quantum Grav.* **26**, 085018 (14pp) (2009).
3. L. Andersson, M. Mars and W. Simon, "Local existence of dynamical and trapping horizons", *Phys. Rev. Lett.* **95**, 111102 (4pp) (2005).
4. L. Andersson, M. Mars and W. Simon, "Stability of marginally outer trapped surfaces and existence of marginally outer trapped tubes", *Adv. Theor. Math. Phys.* **12**, 853–888 (2008).
5. L. Andersson and J. Metzger, "The area of horizons and the trapped region", *Commun. Math. Phys.* **290**, 941–972 (2009).
6. l. Andersson, M. Eichmair and J. Metzger, "Jang's equation and its applications to marginally trapped surfaces", *Proceedings of the Complex Analysis & Dynamical Systems IV Conference (Nahariya, Israel, May 2009), arXiv:1006.4601*
7. T. Bäckdahl and J.A. Valiente-Kroom, "On the construction of a geometric invariant measuring the deviation from Kerr data", *Annales Henri Poincaré* **11**, 1225–1271 (2010).
8. T. Bäckdahl and J.A. Valiente-Kroom, "The 'non-Kerrness' of domains of outer communication of black holes and exteriors of stars", *Proc. Roy. Soc. Lond. A* **467**, 1701–1718 (2011).
9. R. Bartnik and P.T. Chruściel, "Boundary value problems for Dirac-type equations, with applications", *Journal für die reine und angewandte Mathematik (Crelle's Journal)* **579**, 13–73 (2005).

10. R. Bartnik and J. McKinnon, "Particle solutions of the Einstein–Yang–Mills equations", *Phys. Rev. Lett.* **61**, 141–144 (1988).
11. R. Beig and P.T. Chruściel, "Killing initial data", *Class. Quantum Grav.* **14**, A83–A92 (1997).
12. I. Ben-Dov, "Outer trapped surfaces in Vaidya spacetimes", *Phys. Rev. D* **75**, 064007 (33pp) (2007).
13. I. Bengtsson and J.M.M. Senovilla, "A Note on trapped Surfaces in the Vaidya Solution", *Phys. Rev. D* **79**, 024027 (6pp) (2009).
14. R.H. Boyer, "Geodesic Killing orbits and bifurcate Killing horizons", *Proc. Roy. Soc. A* **311**, 245–253 (1969).
15. P. Breitenlohner, D. Maison and G. Gibbons, "4-dimensional black holes from Kaluza–Klein", *Commun. Math. Phys.* **120**, 295–333 (1988).
16. G. Bunting and A.K.M. Masood-ul-Alam, "Nonexistence of multiple black holes in asymptotically euclidean static vacuum space-time", *Gen. Rel. Grav.* **19**, 147–154 (1987).
17. A. Carrasco, "Trapped surfaces in spacetimes with symmetries and applications to uniqueness theorems", Ph.D. Thesis, Universidad de Salamanca (2011), arXiv:1201.1640
18. A. Carrasco and M. Mars, "On marginally outer trapped surfaces in stationary and static spacetimes", *Class. Quantum Grav.* **25**, 055011 (19pp) (2008).
19. A. Carrasco and M. Mars, "Stability in marginally outer trapped surfaces in spacetimes with symmetries", *Class. Quantum Grav.* **26**, 175002 (19pp) (2009).
20. A. Carrasco and M. Mars, "Uniqueness theorem for static spacetimes containing marginally outer trapped surfaces", *Class. Quantum Grav.* **28**, 175018 (30pp) (2011).
21. B. Carter, "Killing horizons and orthogonally transitive groups in space-time", *J. Math. Phys.* **10**, 70–81 (1969).
22. B. Carter, "Axisymmetric black holes has only two degrees of freedom", *Phys. Rev. Lett.* **26**, 331–332 (1971).
23. B. Carter, "Black holes equilibrium states", in *Black Holes*, Eds. C. DeWitt and B.S. DeWitt, Gordon & Breach (1973).
24. P.T. Chruściel, "The classification of static vacuum space-times containing an asymptotically flat spacelike hypersurface with compact interior", *Class. Quantum Grav.* **16**, 661–687 (1999).
25. P.T. Chruściel, "Towards a classification of static electro-vacuum space-times containing an asymptotically flat spacelike hypersurface with compact interior", *Class. Quantum Grav.* **16**, 689–704 (1999).
26. P.T. Chruściel, "The classification of static vacuum space-times containing an asymptotically flat spacelike hypersurface with compact interior", *arXiv:gr-qc/9809088v2 (corrigendum to arXiv:gr-qc/9809088 submitted in 2010).*
27. P.T. Chruściel, E. Delay, G.J. Galloway and R. Howard, "Regularity of horizons and the area theorem", *Annales Henri Poincaré* **2**, 1779–1817 (2001).
28. P.T. Chruściel and G.J. Galloway, "Uniqueness of static black holes without analyticity", *Class. Quantum Grav.* **27** 152001 (6pp) (2010).
29. P. Chruściel and J. Lopes Costa, "On uniqueness of stationary vacuum black holes" in *Géométrie Différentielle, Physique Mathématique, Mathématique et Société, Volume en l' honneur de Jean Pierre Bourguignon*, Ed. O. Hijazi, *Astérisque* **321**, pp. 195–265 (2008).
30. P.T. Chruściel and P. Tod, "The classification of static electro-vacuum space-times containing an asymptotically flat spacelike hypersurface with a compact interior", *Commun. Math. Phys.* **271**, 577–589 (2007).

31. B. Coll, "On the evolution equations for Killing fields", *J. Math. Phys.* **18**, 1918–1922 (1997).
32. D.M. Eardley, "Black hole boundary conditions and coordinate conditions", *Phys. Rev. D* **57**, 2299 (6pp) (1998).
33. M. Eichmair "Existence, regularity and properties of generalized apparent horizons", *Commun. Math. Phys.* **294**, 745–760 (2009).
34. S.W. Hawking and G.F.R. Ellis, *The large scale structure of space-time* (Cambridge monographs on mathematical physics), Cambridge University Press (1973).
35. M. Heusler, *Black hole uniqueness theorems* (Cambridge Lecture Notes in Physics **6**), Cambridge University Press (2006).
36. G. Huisken and T. Ilmanen, "The inverse mean curvature flow and the Riemannian Penrose inequality", *J. Diff. Geom.* **59**, 353–437 (2001).
37. M. Kriele and S.A. Hayward, "Outer trapped surfaces and their apparent horizon", *J. Math. Phys.* **38**, 1593–1604 (1997).
38. W. Israel, "Event horizons in static vacuum space-times", *Phys. Rev* **164**, 397–399 (1967).
39. W. Israel, "Event horizons in static electrovac space-times", *Commun. Math. Phys.* **8**, 245–260 (1968).
40. J. Lopes Costa, "On black hole uniqueness theorems", Ph. D. Thesis (2010).
41. M. Mars, "Uniqueness properties of the Kerr metric", *Class. Quantum Grav.* **17**, 3353–3374 (2000).
42. M. Mars, M. Reiris, "Global and uniqueness properties of stationary and static spacetimes with enter trapped surfaces; *to appear in comm. Math. Physics.*, arXiv:1206.0671v1(2012).
43. M. Mars and J.M.M. Senovilla, "Trapped surfaces and symmetries", *Class. Quantum Grav.* **20**, L293–L300 (2003).
44. M. Mars and W. Simon, "On uniqueness of static Einstein-Maxwell-Dilaton black holes", *Adv. Theor. Math. Phys.* **6**, 279–306 (2002).
45. A.K.M. Masood-ul-Alam, "Uniqueness proof of static black holes revisited", *Class. Quantum Grav.* **9**, L53–L55 (1992).
46. A.K.M. Masood-ul-Alam, "Uniqueness of a static charged dilaton black hole", *Class. Quantum Grav.* **10**, 2649–2656 (1993).
47. P.O. Mazur, "Proof of uniqueness of the Kerr-Newman black hole solution", *J. Phys. A* **15**, 3173–3180 (1982).
48. P. Miao, "A remark on boundary effects in static vacuum initial data sets", *Class. Quantum Grav.* **22**, L53–L59 (2005).
49. H. Müller zum Hagen, D.C. Robinson and H.J. Seifert, "Black holes in static vacuum space-times", *Gen. Rel. Grav.* **4**, 53–78 (1973).
50. H. Müller zum Hagen, D.C. Robinson and H.J. Seifert, "Black holes in static electrovac space-times", *Gen. Rel. Grav.* **5**, 61–72 (1974).
51. R. Penrose, "Gravitational collapse — the role of general relativity", *Nuovo Cimiento* **1**, 252–276 (1965).
52. I. Rácz and R.M. Wald, "Extensions of spacetimes with Killing horizons", *Class. Quantum Grav.* **9**, 2643–2656 (1992).
53. D.C. Robinson, "Uniqueness of the Kerr black hole", *Phys. Rev. Lett.* **34**, 905–906 (1975).
54. D.C. Robinson, "A simple proof of the generalization of the Israel's theorem", *Gen. Rel. Grav.* **8**, 695–698 (1977).
55. P. Ruback, "A new uniqueness theorem for charged black holes", *Class. Quantum Grav.* **5**, L155–L159 (1988).

56. R. Schoen and S.-T. Yau, "Proof of the positive mass theorem II", *Commun, Math. Phys.* **79**, 231–260 (1981).
57. J.M.M. Senovilla, "Singularity theorems and their consequences", *Gen. Rel. Grav.* **30**, 701–848 (1998).
58. W. Simon, "A simple proof of the generalized Israel theorem", *Gen. Rel. Grav.* **17**, 761–768 (1985).
59. W. Simon, "Radiative Einstein–Maxwell spacetimes and 'no-hair' theorems", *Class. Quantum Grav.* **9**, 241–256 (1992).
60. C.V. Vishveshwara, "Generalization of the 'Schwarzschild surface' to arbitrary static and stationary metrics", *J. Math. Phys.* **9**, 1319–1322 (1968).
61. R.M. Wald, *General Relativity*, The University of Chicago Press (1984).
62. R.M. Wald, "Gravitational collapse and cosmic censorship", in *Black Holes, Gravitational Radiation and the Universe*, Eds. B.R. Iyer and B. Bhawal (Fundamental Theories of Physics **100**), pp. 69–85, Kluwer Academic (1999).

Chapter 4

Horizons in the Near-Equilibrium Regime

Ivan Booth

Departament de Física Fonamental, Universitat de Barcelona,
Marti i Franquès 1, E-08028 Barcelona, Spain
on sabbatical leave from:
Department of Mathematics and Statistics, Memorial University of Newfoundland
St. John's, Newfoundland and Labrador, A1C 5S7, Canada
ibooth@mun.ca

1. Introduction

When trying to understand a physical system, one often starts with its equilibrium states. For example, introductions to mechanics generally begin with statics while the zeroth law of thermodynamics defines a system in thermal equilibrium. After the time-independent physics is understood, one turns to dynamics. However this generally entails a significant leap in complexity: thermodynamics is over 150 years old but non-equilibrium thermodynamics is still an active area of research. Thus as an intermediate step in the approach to full dynamics it is often useful to start with the near-equilibrium regime where many of the lessons from equilibrium physics still apply. For thermodynamics, this is the *quasi-static* regime where a system smoothly (and generally slowly) transitions between equilibrium states.

The study of black holes follows the usual pattern. In four-dimensional asymptotically flat space-times, the equilibrium states are identified with the Kerr-Newman family of exact solutions. Well-known theorems identify them as the unique stationary, axisymmetric and asymptotically flat vacuum black hole solutions. As such they have been intensively studied since their discovery almost 50 years ago. Much of what we know about black holes comes from a study of these equilibrium solutions and their near-equilbrium perturbations.[1]

That said, as has been repeatedly emphasized elsewhere in this volume, this family of solutions is not the end of the story. Real black holes don't sit alone in an otherwise empty universe and so there is great interest in developing a more localized theory which characterizes black holes by their physical and geometric properties rather than as any particular exact solution. Hence the closely related programmes studying apparent, trapping, isolated and dynamical horizons (review articles include[3,4,2,5,6]). In this context, isolated trapping horizons are the equilibrium states of interest and dynamical trapping horizons are the corresponding time-dependent states.

In this chapter we will review the geometric characterization and physical properties of near-equilibirum black hole horizons. These are the "almost-isolated" *slowly evolving horizons* first defined by Booth and Fairhurst.[7,8] In section 2 we consider the geometrical background common to all types of horizons. With this material in hand, section 3 reviews the various definitions of horizons, focussing on slowly evolving horizons. Section 4 discusses two of the key properties of these horizons: they obey dynamical laws of black hole mechanics and are accompanied by a candidate event horizon. Section 5 considers a couple of examples of the slowly evolving horizons: a simple Vaidya spacetime and some much more interesting five-dimensional black brane spacetimes that manifest the fluid–gravity duality. Finally section 6 closes this chapter with a summary of our main points.

2. Background

We begin with a brief review of the geometry of two- and three-dimensional surfaces in four-dimensional spacetime. Much of this material is repeated in other chapters of this volume, however this section will serve to establish notation and perhaps present things from a slightly different perspective. For those who would like more details, this is essentially an abbreviated version of the extended discussion in.[8]

2.1. *Intrinsic and extrinsic horizon geometry*

Let (M, g_{ab}, ∇_a) be a four-dimensional, time-orientable spacetime and (H, q_{ab}, D_a) be a three-dimensional hypersurface which can be foliated into two-dimensional spacelike surfaces $(S_v, \tilde{q}_{ab}, d_a)$. The normal space to each of the S_v can be spanned by a pair of future-oriented null vectors ℓ^a (outward-pointing) and n^a (inward-pointing). The direction of these vectors is fixed, but their scaling isn't. One degree of freedom is usually removed by requiring that $\ell \cdot n = -1$ which leaves a single degree of rescaling freedom:

$$\ell \to e^f \ell \quad \text{and} \quad n \to e^{-f} n \,, \tag{1}$$

for an arbitrary function f. Independent of the choice of scaling, we can write the two-metric as

$$\tilde{q}_{ab} = g_{ab} + \ell_a n_b + n_a \ell_b \tag{2}$$

and the four-dimensional volume element ϵ as

$$\boldsymbol{\epsilon} = \boldsymbol{\ell} \wedge \mathbf{n} \wedge \widetilde{\boldsymbol{\epsilon}} \tag{3}$$

where $\boldsymbol{\ell}$ and $\mathbf{n}$ are the one-form versions of the corresponding vectors and $\widetilde{\boldsymbol{\epsilon}}$ is the induced volume element on the S_v.

The intrinsic geometry of the S_v is determined by the induced metric $\tilde{q}_{ab}$ but the extrinsic geometry is defined by how the null normals vary along the surface. We have the extrinsic curvature analogues:

$$k_{ab}^{(\ell)} = \tilde{q}_a^c \tilde{q}_b^d \nabla_c \ell_d \text{ and } k_{ab}^{(n)} = \tilde{q}_a^c \tilde{q}_b^d \nabla_c n_d \tag{4}$$

as well as the connection on the normal bundles:

$$\tilde{\omega}_a = -\tilde{q}_a^b n_c \nabla_b \ell^c . \tag{5}$$

It is useful to decompose the extrinsic curvatures into their trace and trace-free parts:

$$k_{ab}^{(\ell)} = \frac{1}{2}\theta_{(\ell)}\tilde{q}_{ab} + \sigma_{ab}^{(\ell)} \text{ and } k_{ab}^{(n)} = \frac{1}{2}\theta_{(n)}\tilde{q}_{ab} + \sigma_{ab}^{(n)} , \tag{6}$$

where the traces are called *expansions* and the trace-free parts are *shears.* These all depend the scaling of the null vectors and under the rescalings defined by (1):

$$k_{ab}^{(\ell)} \to e^f k_{ab}^{(\ell)} , \; k_{ab}^{(n)} \to e^{-f} k_{ab}^{(\ell)} \text{ and } \tilde{\omega}_a \to \tilde{\omega}_a + d_a f . \tag{7}$$

Returning to H, one can further reduce the freedom allowed to the null vectors by tying their scaling to the foliation. Given a coordinate labelling v of the foliating two-surfaces, a unique future-oriented vector field $\mathcal{V}^a$ on H satisfies the following conditions: 1) it is normal to the S_v, 2) it is tangent to H and 3) it satisfies $\mathcal{L}_{\mathcal{V}} v = 1$. This is the *evolution vector field* for that foliation labelling: it evolves leaves of the foliation into each other. In coordinate language $\mathcal{V} = \partial / \partial v$ for any parameterization in which $\mathcal{V}^a$ Lie-drags the spatial-coordinates between S_v. Independently of whether or not such a system is constructed, the scaling of the null vectors can be fixed by the requirement that

$$\mathcal{V}^a = \ell^a - C n^a , \tag{8}$$

for some function C which we call the *expansion parameter.* Note that if $C > 0$, H is spacelike, while if $C < 0$ it is timelike and if $C = 0$ it is null. We will mainly be interested in the cases where $C \geq 0$.

Given this construction, the scaling freedom of the null vectors is restricted to the freedom to reparameterize the foliation labelling. For an alternative labelling $\tilde{v} = \tilde{v}(v)$ we have

$$[d\tilde{v}]_a = \frac{1}{\alpha(v)}[dv]_a \Rightarrow \tilde{\mathcal{V}}^a = \alpha(v)\mathcal{V}^a \tag{9}$$

where $\alpha(v) = \frac{dv}{d\tilde{v}}$ is constant over each S_v. Then

$$\tilde{\ell}^a = \alpha \ell^a , \; \tilde{n}^a = \frac{1}{\alpha} n^a \text{ and } \tilde{C} = \alpha^2 C . \tag{10}$$

Under this restricted class of rescalings with α constant over each individual S_v, $\tilde{\omega}_a$ is invariant.

The intrinsic and extrinsic geometry of the full H can be expressed mainly in terms of the two-surface quantities. First, the induced metric is

$$q_{ab} = 2C[dv]_a[dv]_b + \tilde{q}_{ab} \, , \tag{11}$$

which explicitly demonstrates how the sign of C determines the signature of the metric. Next, defining

$$\tau_a = \ell_a + C n_a \tag{12}$$

as a future-oriented normal to the surface, the associated extrinsic curvature is

$$K^{(\tau)}_{ab} \equiv q_a^c q_b^d \nabla_c \tau_d = 2C\bar{\kappa}_{\mathcal{V}}[dv]_a[dv]_b + 2C\left(\bar{\omega}_a[dv]_b + [dv]_a\bar{\omega}_b\right) + \left(k^{(\ell)}_{ab} + Ck^{(n)}_{ab}\right) , \tag{13}$$

where

$$\bar{\kappa}_{\mathcal{V}} \equiv \kappa_{\mathcal{V}} - \frac{1}{2}\mathcal{L}_{\mathcal{V}} \ln C \quad \text{and} \tag{14}$$

$$\bar{\omega}_a \equiv \tilde{\omega}_a - \frac{1}{2} d_a \ln C \, .$$

with $\kappa_{\mathcal{V}} = -\mathcal{V}^a n_b \nabla_a \ell^b$. Note that we have not unit-normalized τ_a so as to allow for all values of C. If H is spacelike, the usual extrinsic curvature defined in terms of the timelike unit normal $\hat{\tau}_a$ is

$$K^{(\hat{\tau})}_{ab} = \frac{1}{\sqrt{2C}} K^{(\tau)}_{ab} \, . \tag{15}$$

2.2. *Evolutions and deformations of two-surfaces*

Next, consider how these geometric quantities change if the surfaces are deformed. We start with the geometry of H for which the formalism will probably be more familiar as it identical to that used in the $(3+1)$-decomposition of general relativity.[1,9] Restricting our attention to normal deformations and a spacelike H, potential evolutions are generated by vector fields of the form $T^a = N\hat{\tau}^a$, where N is a function. Then

$$\delta_T q_{ab} = 2NK^{(\hat{\tau})}_{ab} \tag{16}$$

while

$$\delta_T K^{(\hat{\tau})}_{ab} = D_a D_b N - N\left(R_{ab} + K^{(\hat{\tau})} K^{(\hat{\tau})}_{ab} - 2K^{(\hat{\tau})c}_a K^{(\hat{\tau})}_{cb} + q_a{}^c q_b{}^d \mathcal{R}_{cd}\right) , \tag{17}$$

where as specified earlier D_a is the induced covariant derivative on the surface, R_{ab} is the (intrinsic three-dimensional) Ricci tensor and $\mathcal{R}_{ab}$ is the Ricci tensor for the full spacetime. In $(3+1)$-general relativity these are the time-evolution equations for evolutions with lapse function N but vanishing shift vector.

Similar equations can be developed for deformations of the individual two-surfaces. It is probably easiest to visualize this with the help of a coordinate system. Relative to a full coordinate system $\{x^\alpha\}$ on some region of M, we can parameterize any S_v by coordinates θ^A so that S_v is embedded in M by the relation $x^\alpha = \mathcal{X}^\alpha(\theta^A)$ for four functions $\mathcal{X}^\alpha$. Again we restrict our attention to normal deformations, noting that any normal vector field $X^\alpha(\theta)$ defined over S_v can be written as

$$X^\alpha = A\ell^\alpha - Bn^\alpha \tag{18}$$

for some functions A and B. Then

$$\mathcal{X}^\alpha(\theta^A) \rightarrow \mathcal{X}^\alpha(\theta^A) + \epsilon X^\alpha(\theta^A) \tag{19}$$

defines a new surface S'_v by deforming S_v a coordinate distance ϵ in the direction X^α. Further it identifies points on the two-surfaces (essentially by Lie-dragging coordinates between surfaces) as shown in Figure 1. Then, for example, the *deformation* of the two-metric $\tilde{q}_{ab}$ in the direction X^α is defined by

$$\delta_X \tilde{q}_{AB} = \lim_{\epsilon \to 0} \frac{\tilde{q}_{AB}|_{\mathcal{X}+\epsilon X} - \tilde{q}_{AB}|_{\mathcal{X}}}{\epsilon}, \tag{20}$$

where we have written the components of the metric in terms of the surface coordinates.

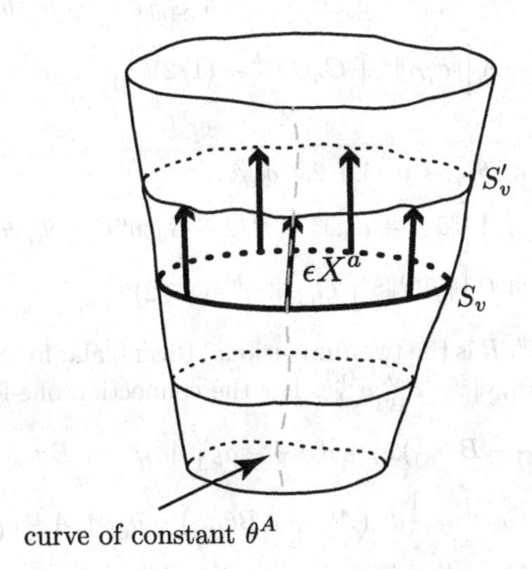

Figure 1. This diagram can be interpreted in two ways. First letting $X^a = \mathcal{V}^a$ it may be thought of as a full horizon H with the S_v and S'_v as foliation surfaces. Alternatively it depicts a more general deformation of a S_v into S'_v in which case X^a is the deforming vector field. Of course the dual interpretation arises because the evolution of the horizon is just a special case of more general deformations.

As for the $(3+1)$-decomposition of general relativity, computationally this operation amounts to taking Lie derivatives with some extra conditions imposed to ensure that the geometric quantities are calculated on the correct surface and that all quantities are appropriately pulled-back into those surfaces (full details may be found in[8]). Then it is straightforward to show that

$$\delta_\ell \tilde{q}_{ab} = 2k^{(\ell)}_{ab} \text{ and } \delta_n \tilde{q}_{ab} = 2k^{(n)}_{ab} \tag{21}$$

while

$$\delta_\ell \tilde{\epsilon} = \tilde{\epsilon}\theta_{(\ell)} \text{ and } \delta_n \tilde{\epsilon} = \tilde{\epsilon}\theta_{(n)} \,. \tag{22}$$

Now it is clear why the traces of the extrinsic curvatures are called expansions: they tell us how the area element changes when the surface is evolved in the given direction. The shears are the part of the evolution that deforms the surface (but does not change its area).

Normal deformations are linear for the first derivatives of the metric. For example:

$$\delta_X \tilde{\epsilon} = A\delta_\ell \tilde{\epsilon} - B\delta_n \tilde{\epsilon} = \tilde{\epsilon}(A\theta_{(\ell)} - B\theta_{(n)}) \tag{23}$$

for arbitrary functions A and B. However, as in the three-dimensional case, things become significantly more complicated for second derivatives. For example

$$\begin{aligned} \delta_X \theta_{(\ell)} = \;& \kappa_X \theta_{(\ell)} - d^2 B + 2\tilde{\omega}^a d_a B \\ & -B\left[\|\tilde{\omega}\|^2 - d_a\tilde{\omega}^a - \tilde{R}/2 + G_{ab}\ell^a n^b - \theta_{(\ell)}\theta_{(n)}\right] \\ & -A\left[\|\sigma_{(\ell)}\|^2 + G_{ab}\ell^a\ell^b + (1/2)\theta^2_{(\ell)}\right] \,, \end{aligned} \tag{24}$$

and

$$\begin{aligned} \delta_X \theta_{(n)} = \;& -\kappa_X \theta_{(n)} + d^2 A + 2\tilde{\omega}^a d_a A \\ & +A\left[\|\tilde{\omega}\|^2 + d_a\tilde{\omega}^a - \tilde{R}/2 + G_{ab}n^a\ell^b - \theta_{(\ell)}\theta_{(n)}\right] \\ & +B\left[\|\sigma^{(n)}\|^2 + G_{ab}n^a n^b + (1/2)\theta^2_{(n)}\right] \,, \end{aligned} \tag{25}$$

where $\kappa_X = -X^a n_b \nabla_a \ell^b$, $\tilde{R}$ is the two-dimensional Ricci scalar for S_v, $\|\tilde{\omega}\|^2 = \tilde{\omega}^a\tilde{\omega}_a$, $\|\sigma_{(\ell)}\|^2 = \sigma^{ab}_{(\ell)}\sigma^{(\ell)}_{ab}$ and $\|\sigma_{(n)}\|^2 = \sigma^{ab}_{(n)}\sigma^{(n)}_{ab}$. For the connection one-form:

$$\begin{aligned} \delta_X \tilde{\omega}_a = \;& -(A\theta_{(\ell)} - B\theta_{(n)})\tilde{\omega}_a + d_a\kappa_{\mathcal{V}} - d_b\left(A\sigma_{(\ell)}{}^b{}_a + B\sigma_{(n)}{}^b{}_a\right) \\ & -\tilde{q}_a^{\,b} G_{bc}\tau^c + \frac{1}{2} d_a\left(A\theta_{(\ell)} + B\theta_{(n)}\right) - \theta_{(\ell)} d_a A - \theta_{(n)} d_a B \,, \end{aligned} \tag{26}$$

where, as before, $\tau^c = A\ell^c + Bn^c$ is normal to X^c. All of these second derivatives depend on how A and B vary over the S_v.

Probably the most intuitive application of these deformations is in understanding the time-evolution of a horizon. Then X^a is identified with the $\mathcal{V}^a$ evolution vector field and, for example,

$$\begin{aligned} \delta_{\mathcal{V}} \theta_{(\ell)} = \;& \kappa_{\mathcal{V}} \theta_{(\ell)} - d^2 C + 2\tilde{\omega}^a d_a C \\ & -C\left[\|\tilde{\omega}\|^2 - d_a\tilde{\omega}^a - \tilde{R}/2 + G_{ab}\ell^a n^b - \theta_{(\ell)}\theta_{(n)}\right] \\ & -\left[\|\sigma_{(\ell)}\|^2 + G_{ab}\ell^a\ell^b + (1/2)\theta^2_{(\ell)}\right] \,. \end{aligned} \tag{27}$$

In particular for a null surface $\mathcal{V}^a = \ell^a$ and we recover the Raychaudhuri equation

$$\delta_\ell \theta_{(\ell)} = \kappa_\ell \theta_{(\ell)} - \left[\|\sigma_{(\ell)}\|^2 + G_{ab}\ell^a \ell^b + (1/2)\theta_{(\ell)}^2\right] , \tag{28}$$

which is key to many of the calculations involving isolated[2] or event[1,9] horizons.

3. Geometric Horizons

We are now ready to define slowly evolving horizons. In this section we will use geometric arguments, leaving physical properties for the next section. We begin with a review of the definitions of isolated and dynamical trapping horizons.

3.1. *General quasilocal horizons*

In a four-dimensional spacetime (M, g_{ab}, ∇_a) a *future outer trapping horizon (or FOTH)* is a three-surface H that is foliated by spacelike two-surfaces $(S_v, \tilde{q}_{ab}, d_a)$ such that on each surface: i) $\theta_{(\ell)} = 0$, ii) $\theta_{(n)} < 0$ and iii) there is a positive function β such that $\delta_{\beta n}\theta_{(\ell)} < 0$.[10] These conditions are intended to (locally) mimic those used to define apparent horizons:[11] each slice of a FOTH is marginally outer trapped $(\theta_{(\ell)} = 0)$ but the other two conditions guarantee that it is possible to deform the S_v inwards so that they become fully trapped.

That the inward expansion be negative is straightforward to check and independent of the scaling of the null vectors; however the last condition is not quite so simple. From the general deformation equation for $\theta_{(\ell)}$ we see that with $\theta_{(\ell)} = 0$ on the horizon:

$$\delta_{\beta n}\theta_{(\ell)} = d^2\beta - 2\tilde{\omega}^a d_a \beta - (\tilde{R}/2 + \|\tilde{\omega}\|^2 - d_a\tilde{\omega}^a + G_{ab}\ell^a n^b)\,\beta\,. \tag{29}$$

Now, for a given scaling of the null vectors, deciding whether condition iii) holds will amount to a study of the possible behaviours of this second order elliptical differential operator. Even for fairly simple cases, a non-trivial β can be required. For example, rapidly rotating Kerr solutions and the standard scaling of the null vectors obtained from Boyer–Lindquist co-ordinates has $\delta_n\theta_{(\ell)} \nless 0$ (appendix C of[8]). However, the horizons certainly are FOTHs and this can be demonstrated with a suitable choice of β. Similarly for general FOTHs away from spherical symmetry, it will often be the case that the scaling determined by fixing $\mathcal{V}^a = \ell^a - Cn^a$, is not a scaling for which $\delta_n\theta_{(\ell)} < 0$. Again a non-trivial β is required.

Since $\theta_{(\ell)} = 0$ on a FOTH we must also have $\mathcal{L}_{\mathcal{V}}\theta_{(\ell)} = 0$. Then applying a maximum principle argument to (24) along with the fact that $\delta_{\beta n}\theta_{(\ell)} < 0$ (for some β) one can show that the null energy condition implies that $C \geq 0$ which in turn means that

$$\mathcal{L}_{\mathcal{V}}\tilde{\epsilon} = -C\theta_{(n)}\tilde{\epsilon} \geq 0\,, \tag{30}$$

since $\theta_{(n)} < 0$. Thus, there is a second law of FOTH mechanics: their area is non-decreasing.

Next, we summarize isolated horizons (see Engle and Liko[2] in this volume for a more information or original papers[12] for the full details). A three-dimensional submanifold in that same spacetime is a *non-expanding horizon* if: i) it is null and topologically $S \times \mathbb{R}$ for some closed two-manifold S, ii) $\theta_{(\ell)} = 0$ and iii) $-T^{ab}\ell_b$ is future directed and causal. As usual ℓ^a is a future oriented, outward-pointing normal; since the horizon is null, in this case no foliation is needed for its construction. That said a foliation is certainly no hindrance to a three-surface being a non-expanding horizon and any null FOTH satisfying the null energy condition will certainly be a non-expanding horizon.

Non-expanding horizons are the simplest objects in the isolated horizon family. Any non-expanding horizon can be turned into a *weakly isolated* horizon if the scaling of the null vectors is chosen so that $\mathcal{L}_\ell \omega_a = 0$ where

$$\omega_a \equiv -n_b \nabla_{\underleftarrow{a}} \ell^b = -\kappa_\ell n_a + \tilde{\omega}_a \,, \tag{31}$$

and the arrow indicates a pull-back into the cotangent bundle of the non-expanding horizon. With this scaling, zeroth and first laws of isolated horizon mechanics may be established. Finally, if there exists a scaling of the null vectors for which the entire extrinsic geometry of the horizon is invariant in time then the weakly isolated horizon is an *isolated horizon*. Thus in that case not only do $\mathcal{L}_\ell \kappa_\ell$ and $\mathcal{L}_\ell \tilde{\omega}_a$ vanish, but so do $\mathcal{L}_\ell \theta_{(n)}$ and $\mathcal{L}_\ell \sigma^{(n)}_{ab}$. All Killing horizons are isolated in this sense.

Our final object is a *dynamical horizon*.[13] A three-dimensional sub-manifold of a spacetime (M, g_{ab}) is a dynamical horizon if: i) it is spacelike and ii) can be foliated by spacelike two-surfaces such that the null normals to those surfaces satisfy $\theta_{(\ell)} = 0$ and $\theta_{(n)} < 0$. Perhaps the key property of dynamical horizons is that they always expand: the spacelike assumption means that $C > 0$ and so by the strictly greater-than version of (30), dynamical horizons always expand. Not surprisingly, if the null energy condition holds and $\delta_\ell \theta_{(\ell)} \neq 0$ at least somewhere on each S_v, then a FOTH is a dynamical horizon. Another important property of dynamical horizons is that their foliation into S_v surfaces with $\theta_{(\ell)} = 0$ is *unique*.[14] This is very convenient as we do not need to worry about whether or not geometric properties are foliation dependent: they are, but since the foliation is unique this is fine!

Note that this foliation rigidity contrasts with the corresponding situation for isolated horizons where any foliation is irrelevant to the intrinsic geometry and so can be freely deformed. However, the situation is very different if we consider the rigidity of the full three-dimensional H. Marginally outer trapped surfaces in an isolated horizon can only be deformed along the horizon itself. By contrast for a dynamical FOTH, the $\theta_{(\ell)} = 0$ surfaces can only be deformed *out of* H (this freedom is equivalent to how the exact location of a dynamical apparent horizon depends on the foliation of the full spacetime). An extended discussion of this point can be found in the references.[8]

We will take FOTHs as our basic objects and classify them by hybridizing the naming systems.[8] Thus a FOTH that also satisfies one of these other sets of properties will be referred to as non-expanding, (weakly) isolated, or dynamical as appropriate. There are also more exotic forms of FOTHs associated with apparent horizon "jumps",[15,16] inner horizons or white holes.[10] However, they can be left aside for the purposes of this article. The horizons that we will be interested in will always be dynamical or weakly isolated FOTHs.

3.2. *Near-equilibrium quasilocal horizons*

Intuitively, a near-equilibrium black hole boundary should be "almost" isolated. Thus, since isolated horizons are non-expanding and null, a near-equilbrium horizon should expand slowly and be almost null. The complication here is quantifying "expand slowly" and "almost null": expanding dynamical horizons are, by their very nature, spacelike so it is not immediately clear what "slowly" means and further it is certainly not obvious how a spacelike surface can be "almost null".

In analogy with isolated horizons, we break the definition up into two parts: *slowly expanding horizons* which focus on restrictions to the intrinsic geometry and *slowly evolving horizons* which add constraints to the extrinsic geometry.

3.2.1. *Slowly expanding horizons*

Definition: Let $\triangle H$ be a section of a future outer trapping horizon foliated by two-surfaces S_v so that $\triangle H = \{\cup_v S_v : v_1 \leq v \leq v_2\}$. Further let $\mathcal{V}^a$ be an evolution vector field that generates the foliation so that $\mathcal{L}_{\mathcal{V}} v = \alpha(v)$ for some positive function v, and scale the null vectors so that $\mathcal{V}^a = \ell^a - Cn^a$. Finally let R_H be the characteristic length scale for the problem. Then $\triangle H$ is a *slowly expanding horizon* if the dominant energy condition holds and

(1) $|\tilde{R}|, \tilde{\omega}_a \tilde{\omega}^a, |d_a \tilde{\omega}^a|$ and $T_{ab}\ell^a n^b \lesssim 1/R_H^2$.
(2) Two-surface derivatives of horizon fields are at most of the same magnitude as the (maximum) of the original fields. For example, $\|d_a C\| \lesssim C_{max}/R_H$, where C_{max} is largest absolute value attained by C on S_v.
(3) $\epsilon \ll 1$ where $\epsilon^2/R_H^2 = \text{Maximum}\left[C\left(\|\sigma^{(n)}\|^2 + T_{ab}n^a n^b + \theta_{(n)}^2/2\right)\right]$.

If the $\alpha(v)$ is chosen so that $C \approx \epsilon^2$ then this scaling of the null normals is said to be compatible with the *evolution parameter* ϵ.

This definition requires some discussion and we analyze it point-by-point. Starting with the preamble, we specify a section $\triangle H$ rather than a full horizon so that the definition will cover sections of horizons that may be slowly expanding only for finite periods of time. For standard four-dimensional black holes the characteristic length scale R_H will be set by the areal radius $R_H = \sqrt{A/(4\pi)}$.

Next, following from the discussion back in section 2.1, choosing $\mathcal{L}_{\mathcal{V}} v = \alpha(v)$ for some positive function $\alpha(v)$ ensures that the flow generated by $\mathcal{V}^a$ evolves leaves of the horizon into each other and so can be thought of as a coordinate vector field compatible with the foliation. The free function is included so that the scaling is not tied to a particular choice of foliation labelling; including it is equivalent to allowing for relabellings of the S_v.

Turning to the main clauses, the notation $X \lesssim Y$ means that $X \leq k_o Y$ for some constant k_o of order one. Then our first condition that

$$|\tilde{R}|\,,\ \tilde{\omega}^a\tilde{\omega}_a\,,\ |d_a\tilde{\omega}^a| \,\text{and}\, |G_{ab}\ell^a n^b| \lesssim \frac{1}{R_H^2}\,, \tag{32}$$

is essentially a restriction on the allowed geometries of the horizons. The first part ensures that the curvature is not too extreme, the second two assumptions place some restriction on the extrinsic curvature while the last bounds the Einstein tensor (or equivalently stress-energy). In particular it has been shown[8] that these restrictions hold for all members of the Kerr family of horizons. An immediate consequence of these assumptions is that by (29),

$$|\delta_n \theta_{(\ell)}| \lesssim \frac{1}{R_H^2} \tag{33}$$

is similarly bounded. Note the use of the absolute value sign. Even though we must have $\delta_{\beta n}\theta_{(\ell)} < 0$ for β scaling, in general this will not be true for $\beta = 1$.

The second condition restricts the size of the derivatives of the horizon fields so that they are at most commensurate with their maximum values. For tensor quantities such as $\sigma^{(n)}_{ab}$ the magnitudes are taken to be judged relative to an orthornormal frame. For example

$$\left\| d_a \sigma^{(n)}_{bc} \right\| \lesssim \frac{1}{R_H} \left\| \sigma^{(n)}_{bc} \right\| \tag{34}$$

indicates that this inequality should hold for all components relative to any orthonormal frame on S_v.

We can then consider the last condition. After the set-up of the earlier conditions, this is the one that captures the essence of what it means to be slowly evolving. Beginning with the evolution of the area-form $\tilde{\epsilon}$ on S_v, by (23) we have

$$\mathcal{L}_{\mathcal{V}}\tilde{\epsilon} = -C\theta_{(n)}\tilde{\epsilon}\,, \tag{35}$$

which is certainly dependent on the scaling of the null vectors. However this dependence may be easily isolated by rewriting

$$\mathcal{L}_{\mathcal{V}}\tilde{\epsilon} = -C\theta_{(n)}\tilde{\epsilon} = ||\mathcal{V}|| \left(-\sqrt{\frac{C}{2}}\theta_{(n)}\tilde{\epsilon} \right)\,, \tag{36}$$

so that the effects of the rescaling freedom are restricted to pre-factor $||\mathcal{V}|| = \sqrt{2C}$. The term in parentheses (which is required to be small by the third condition) then provides an invariant measure of the rate of expansion. Among other properties, it vanishes if the horizon is non-expanding while on a dynamical horizon section it is equal to the rate expansion of $\tilde{\epsilon}$ with respect to the unit-normalized version of the evolution vector field[a] The examples of section 5 provide intuitive support for the identification of small ϵ with a slow expansion.

This notion of a slow area change is the key part of the entire definition. The rest of the third condition forces the surface to be "almost-null" with a slowly changing intrinsic metric. To simplify notation we now adapt a evolution-parameter-compatible scaling of the null vectors so that

$$C \approx \epsilon^2 \tag{37}$$

which also ensures that any transition to equilibrium and so a fully isolated horizon will be smooth. With this scaling the first two conditions along with (24) imply that

$$|\delta_\ell \theta_{(\ell)}| = \sigma^{(\ell)}_{ab}\sigma^{ab}_{(\ell)} + T_{ab}\ell^a\ell^b \lesssim \frac{\epsilon^2}{R_H^2} \, . \tag{38}$$

and so the rate of expansion becomes

$$\mathcal{L}_{\mathcal{V}}\widetilde{\epsilon} = \underbrace{-C\theta_{(n)}\widetilde{\epsilon}}_{O(\epsilon^2)} \, . \tag{39}$$

while the "time"-derivative of the full metric is

$$\mathcal{L}_{\mathcal{V}}\tilde{q}_{ab} = \underbrace{2\sigma^{(\ell)}_{ab}}_{O(\epsilon)} + \underbrace{\left(C\theta_{(n)}\tilde{q}_{ab} - 2C\sigma^{(n)}_{ab}\right)}_{O(\epsilon^2)} \, . \tag{40}$$

The leading terms are the same as they would be for a truly null surface. Further the leading terms of the energy flux across the horizon

$$\mathcal{V}^a T^b_a \tau_b \approx \underbrace{T_{ab}\ell^a\ell^b}_{O(\epsilon^2)} - \underbrace{C^2 T_{ab}n^a n^b}_{O(\epsilon^4)} \, , \tag{41}$$

(where $\tau_a = \ell_a + Cn_a$ is the usual timelike normal) also match those of a truly null surface.

We refer the reader to the references[8] for further discussion of these and other consequences of the definition.

3.2.2. *Slowly evolving horizons*

For slowly expanding horizons we imposed conditions to ensure that $\triangle H$ was almost null and that the two-geometry changed only slowly. We now move on to slowly evolving horizons where additional conditions are imposed to also restrict the evolution of much of the rest of the horizon geometry. This can be motivated by the analogous development of the isolated horizon formalism from non-expanding horizons (restricting intrinsic geometry) to fully isolated horizons (also restricting extrinsic geometry).

[a] One might be concerned as to whether this quantity vanishes in the limit $C \to 0$ (in which case $\widehat{\mathcal{V}}^a$ is no longer defined). It does. For an explicit demonstration see the discussion.[17]

The two-metric $\tilde{q}_{ab}$, extrinsic curvatures $(k_{ab}^{(\ell)}, k_{ab}^{(n)})$ and normal connection $\tilde{\omega}_a$ fully specify the geometry of the S_v. The rest of the geometry of H is then fixed by $\kappa_{\mathcal{V}}$ and C. From the previous section, for a slowly expanding horizon with an adapted evolution parameter, we have a good notion of a "time"-derivative. Thus, for these remaining quantities we can directly bind their rates of change with respect to $\mathcal{V}^a$, demanding that in each case their rate of change be of a lower order than the original quantity.

Definition: Let $\triangle H$ be a slowly expanding section of a FOTH with a compatible scaling of the null normals. Then it is said to be a *slowly evolving horizon* (SEH) if in addition

(1) $\|\mathcal{L}_{\mathcal{V}}\tilde{\omega}_a\|$, $|\mathcal{L}_{\mathcal{V}}\kappa_{\mathcal{V}}|$ and $|\mathcal{L}_{\mathcal{V}}\theta_{(n)}| \lesssim \epsilon/R_H^2$ and
(2) $|\mathcal{L}_{\mathcal{V}}C| \lesssim \epsilon^3/R_H$.

Note that the inclusion the $\mathcal{L}_{\mathcal{V}}C$ is new in this article as compared to earlier definitions of slowly evolving horizons. It is not required for the first law however, and as we shall see, it is required to show that there is always an event horizon candidate in close proximity to any SEH.

4. Key Properties

In this section we review two of the key properties of slowly evolving horizons.

4.1. *Laws of Mechanics*

Slowly evolving horizons are intended to be the black hole analogues of quasi-static systems in thermodynamics. As such we would expect them to (approximately) obey the laws of black hole mechanics (a nice discussion of the equilibrium forms of these laws can be found in the isolated horizon chapter of this book[2]).

Starting with the zeroth law, we expect that surface gravity will be approximately constant on each surface S_v. This is the case. Applying the various defining conditions to the equation for the variation of $\tilde{\omega}_a$ (26), it is immediate that

$$\|d_a\kappa_{\mathcal{V}}\| \lesssim \frac{\epsilon}{R_H^2} \tag{42}$$

(the analogous equation for null surfaces is used to show that the surface gravity is constant for a (weakly) isolated horizon). Given that the surface gravity has already been required to be slowly changing up the horizon, it follows that over a foliation parameter range on $\triangle H$ that is small relative to $1/\epsilon$,

$$\kappa_{\mathcal{V}} = \kappa_o + O(\epsilon) \tag{43}$$

for some constant κ_o. Note however that, as for temperature in a standard quasi-static process, the surface gravity can accumulate larger changes over sufficiently long periods of time.

We have already seen that FOTHs that satisfy the null energy condition are non-decreasing in area. SEHs inherit this property. This is one form of the second law, however given that SEHs represent quasi-static black holes one might also expect them to dynamically satisfy the second law in its Clausius form. For reversible thermodynamic processes, entropy is defined by

$$\delta S = \frac{\delta Q}{T} . \tag{44}$$

Switching to black hole mechanics we expect the analogous law to link area increase, energy flux and surface gravity.

Slowly evolving horizons do obey a law of this form.[7,8] First linearly combining (24) and (25) one can show that

$$\kappa_{\mathcal{V}} \theta_{\mathcal{V}} = \delta_{\mathcal{V}} \theta_{(\ell)} + C \delta_{\mathcal{V}} \theta_{(n)} + d_a (d^a C - 2C \tilde{\omega}^a) + \sigma^{(\tau)}_{ab} \sigma^{ab}_{(\mathcal{V})} + G_{ab} \mathcal{V}^a \tau^b + \frac{1}{2} \theta_{(\mathcal{V})} \theta_{(\tau)} ,$$

where the shears associated with $\mathcal{V}^a$ and τ^a are defined in the obvious way. Integrating over S_v eliminates the total derivative term and then applying slowly expanding/evolving conditions (including the zeroth law) we find that to order $O(\epsilon^2)$:

$$\frac{\dot{a}}{4\pi G} \approx \frac{2}{\kappa_o} \int_{S_v} \tilde{\epsilon} \left\{ \frac{\|\sigma^{(\ell)}\|^2}{8\pi G} + T_{ab} \ell^a \ell^b \right\} . \tag{45}$$

Inside the integral on the right-hand side, the terms can respectively be identified as fluxes of gravitational radiation and stress-energy and, as would be expected, they are equivalent to the terms that one finds for truly null surfaces.[18,9]

Note, though we have called this the Clausius form of the second law, it is more traditional to refer to it as the first law of black hole mechanics.[19] For stationary black holes and isolated horizons this first law takes the form

$$\delta M = \frac{1}{8\pi G} \kappa \delta A + \Omega \delta J + \Phi \delta Q , \tag{46}$$

where the M is the black hole mass (taken as the internal energy), J is angular momentum and Q is the charge while Ω and Φ are their associated potentials. However, it has been pointed out[20] that while there are energy flow terms in (45), there is no explicit splitting of those terms into work terms versus those involving changes in the internal energy of the system. The notion of internal energy is key to the thermodynamic interpretations of the first law and as such it probably makes more sense to view the energy flow terms as more analogous to the heat flow in (44). Hence we refer to it as a form of the second law.

4.2. *Spacetime near a SEH*

In this section we introduce a new property: there is an actual null surface close to each slowly evolving FOTH. This null surface is a candidate event horizon and is, in fact, the actual event horizon if the FOTH remains a SEH forever. Further, despite the teleological definition of event horizons the behaviour of this event horizon candidate is locally determined and its evolution closely mirrors that of the SEH. Full details of this result can be found in the references.[21]

4.2.1. *Event horizons*

An event horizon is the boundary of a *causal black hole*: a region of spacetime from which no causal signal can escape. Such a surface is necessarily null and, for outside observers, it is the boundary between the unobservable events inside the black hole and those outside that can be seen. This definition is teleological: one determines the extent (or existence) of a black hole by tracing all causal paths "until the end of time" and then retroactively identifying any black hole region. The most obvious manifestation of this type of definition is that, as shown in Figure 2, infalling matter curtails rather than causes the expansion of an event horizon.

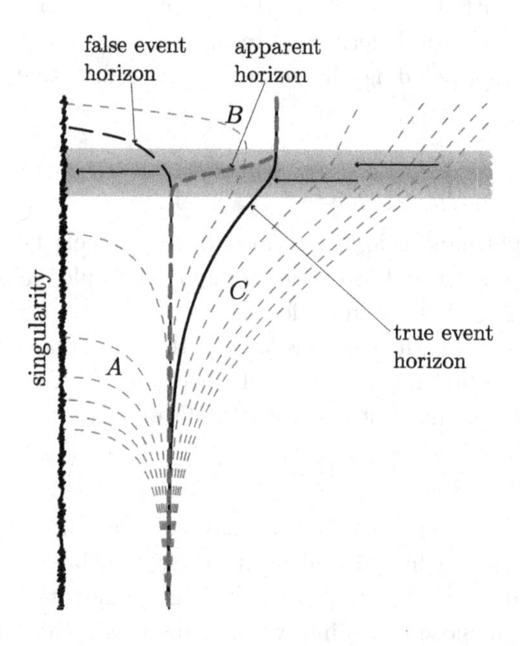

Figure 2. A schematic that plots both the quasilocal and event horizons for a typical Vaidya spacetime in which a shell of dust (the shaded gray region) falls into a pre-existing black hole. In this figure, horizontal location measures the areal radius of the associated spherical shell while the direction of increasing time is roughly vertical outside the event horizon but tipping horizontal-and-to-the-left inside. On both sides, inward-moving null geodesics are horizontal while "outward-pointing" null geodesics are represented by gray dashed lines. The false event horizon is a null surface that would appear to be the event horizon if one is unaware that the matter shell will fall into the black hole in the future and so capture it.

The reason for this behaviour is mathematically easy to understand. As a null surface, the event horizon is necessarily ruled by null geodesics. Thus, in the language of earlier sections, it is a three-surface with $C = 0$ and so the Raychaudhuri equation holds:

$$\mathcal{L}_{\mathcal{V}}\theta_{(\ell)} = \kappa_\ell \theta_{(\ell)} - \left(\theta_{(\ell)}^2/2 + \|\sigma_{(\ell)}\|^2 + G_{ab}\ell^a\ell^b\right) . \tag{47}$$

Now, given the Einstein equations and null energy condition, if we temporarily adopt an affine scaling of the null vectors ($\kappa_\ell = 0$), it follows that

$$\mathcal{L}_\ell \theta_{(\ell)} \leq -\frac{1}{2}\theta_{(\ell)}^2 . \tag{48}$$

Thus the expansion naturally decreases with time. There is nothing profound about this statement. Consider for example a sphere of light expanding outwards in flat space so that $R(t) = R_o + ct$. Then

$$\theta_{(\ell)} = \frac{2\dot{R}}{R} = \frac{2c}{R(t)} \tag{49}$$

and so the rate of expansion decreases as time (and therefore) R increases: the rate of expansion scales with the (inverse) radius of the horizon. Returning to general relativity a non-zero flux of gravitational energy or positive-energy density matter will cause the rate to decrease even more. Again this is to be expected: the gravitational influence of more mass inside the shell should decrease the rate of expansion.

The apparently counterintuitive nature of event horizon evolution is then seen to be not so mysterious. Event horizons are families of null geodesics evolving exactly as null geodesics should. The particular set of geodesics has been chosen based on future boundary conditions (not escaping to infinity and not falling into the singularity) but at any given moment they evolve causally just like any other null congruence.

The real mystery is then: why is it that numerical[22] and perturbative calculations[23] often show event horizons evolving in the apparently intuitive way, expanding in response to infalling radiation and matter? Rewriting the Raychaudhuri equation as

$$\kappa_\ell \theta_{(\ell)} - \mathcal{L}_\ell \theta_{(\ell)} = \|\sigma_{(\ell)}\|^2 + G_{ab}\ell^a\ell^b + \frac{1}{2}\theta_{(\ell)}^2 \tag{50}$$

it is straightforward to see that just a couple of slowly-evolving-type conditions will be sufficient to enforce a Clausius-type second law (a.k.a the first law of black hole mechanics). For a small $\theta_{(\ell)}$ one would have $\theta_{(\ell)}^2 \ll \theta_{(\ell)}$, while one might also expect the rate of change of the expansion to be smaller than the expansion itself $\mathcal{L}_\ell \theta_{(\ell)} \ll \theta_{(\ell)}$ (as in a standard derivative-expansion like those considered in the fluid–gravity correspondence[24]).

We now consider how to express this intuition in geometrically invariant way. First note that on right-hand side of (50) the $\theta_{(\ell)}^2$ term can be neglected if

$$\frac{1}{2}\theta_{(\ell)}^2 \lesssim \epsilon \left(\|\sigma_\ell\|^2 + G_{ab}\ell^a\ell^b\right) , \tag{51}$$

for some $\epsilon \ll 1$. Both sides of this inequality change in the same way with respect to scalings of the null vectors so this condition is geometrically meaningful and can be used to determine the scale of "slowness". Next, if we can scale the null vectors so that on the left-hand side

$$\mathcal{L}_{\ell}\theta_{(\ell)} \lesssim \left(\frac{\epsilon}{R_H}\right)\theta_{(\ell)}, \tag{52}$$

and κ_ℓ is of order $1/R_H$, then the Clausius-type law is recovered for this null surface:

$$\kappa_\ell \theta_{(\ell)} \approx \|\sigma_\ell\|^2 + G_{ab}\ell^a\ell^b. \tag{53}$$

Thus, in this limit fluxes will appear to drive the expansion of these null surfaces in the intuitive way. Further discussion of these slowly evolving null surfaces can be found in the literature.[25]

4.2.2. *Event horizon candidate near a SEH*

Intuitively one might expect a truly null surface to lie close to any slowly evolving horizon. In fact, this is the case and that null surface is also slowly evolving in the sense discussed above. We now motivate this result.

As noted earlier, the extrinsic curvature of a hypersurface is the rate of change of the induced metric under normal deformations. If we assume that the induced metric on surfaces near the slowly evolving horizon may be expanded as a Taylor series in the proper time measured along timelike normal geodesics, then we have:

$$q_{ab}^{(\Delta\tau)} \approx q_{ab}^{(0)} + (\Delta s)\left[\mathcal{L}_{\hat{\tau}} q_{ab}\right]^{(0)} = q_{ab}^{(0)} + 2(\Delta s)K_{ab}^{(0)}, \tag{54}$$

where $\hat{\tau}$ is the future-oriented timelike normal to $\triangle H$ and Δs measures proper time. Substituting in (11) and (13) this becomes

$$\begin{aligned} q_{ab}^{(\Delta\tau)} \approx & \left(2C + (\Delta s)\sqrt{2C}\bar{\kappa}_{\mathcal{V}}\right)[dv]_a[dv]_b \\ & + \sqrt{2C}(\Delta s)\left(\bar{\omega}_a[dv]_b + [dv]_a\bar{\omega}_b\right) + \left(\tilde{q}_{ab} + \frac{\Delta s}{\sqrt{2C}}\left(k_{ab}^{(\ell)} + 2Ck_{ab}^{(n)}\right)\right). \end{aligned} \tag{55}$$

Thus, the surface at proper time

$$\Delta s \approx -\sqrt{2C}/\kappa_{\mathcal{V}} \tag{56}$$

into the past of the SEH is null to leading order[b]. It can be shown that this surface is also slowly evolving in the sense considered in the last section and so evolves in a local way which closely mirrors that of the SEH. If C remains small and ultimately asymptotes to zero then this null surface also asymptotes to the SEH and truly is

[b]For purposes of this preview, this calculation nicely demonstrates the general principles involved in finding the event horizon candidate. That said, q_{ab} cannot always be expanded in this way: in general moving small proper distances along the timelike geodesics will correspond to large changes in, for example, K_{ab}. In such cases it is not valid to work with a Taylor expansion in Δs and a more involved analysis is required to demonstrate the existence of this null surface.[21]

the event horizon. As noted earlier, full details of these results can be found in reference.[21]

It is important to keep in mind that even in this case, the event horizon is still teleologically defined. We identify it based on the assumption that it will remain slowly evolving for all eternity. If, at some point in the future the FOTH is no longer slowly evolving, then the construction breaks down and this set of null geodesics will dive into the singularity as did the false horizon in Figure 2.

5. Examples and Applications

In order to build some intuition about these horizons, we now turn to a couple of concrete examples. The first is very simple, a SEH in a Vaidya spacetime, and it mainly serves to help to establish the range of astrophysical processes that we might expect to be slowly evolving. The second example is a SEH in a five-dimensonal black brane spacetime. This is technically much more complicated and also has the very interesting property that it is dual to a four-dimensional conformal fluid through the fluid–gravity duality. As such we can use it transfer lessons from slowly evolving horizons to fluid dynamics and vice versa.

5.1. *Vaidya black holes*

We begin with a spherically symmetric example in four-dimensions: slowly evolving Vaidya black holes. This is taken from reference[26] which discusses both this example and some other, slightly more realistic, ones. As in that paper we are primarily interested in using the example to build physical intuition and so take one solar mass as our standard unit:

$$M_\odot = 1.9 \times 10^{30}\,\text{kg}\,. \tag{57}$$

Hence, with $G = 6.7 \times 10^{-11}\,\text{m}^3\,\text{kg}^{-1}\,\text{s}^{-2}$ and $c = 3.0 \times 10^{8}\,\text{m}\,\text{s}^{-1}$, distances and times that come out of our equations will be measured in units of

$$R_\odot = GM_\odot/c^2 \approx 1.4 \times 10^3\,\text{m} \text{ and} \tag{58}$$

$$T_\odot = R_\odot/c \approx 4.7 \times 10^{-6}\,\text{s} \tag{59}$$

respectively. Relative to everyday experience, our unit of mass is huge while our unit of time is very small. As such we might expect to find that some fairly dramatic events lie in the slowly evolving regime. We will see that this is the case.

5.1.1. *The spacetime*

The Vaidya solution[9] is the simplest example of a dynamical black hole spacetime. It models the collapse of null dust and is described by the metric

$$ds^2 = -\left(1 - \frac{2m(v)}{r}\right) dv^2 + 2dvdr + r^2 d\Omega^2\,, \tag{60}$$

where v is an advanced time coordinate, $m(v)$ measures the mass of the black hole on a hypersurface of constant v and the infalling null dust has stress-energy tensor

$$T_{ab} = \frac{dm/dv}{4\pi r^2}[dv]_a[dv]_b\,. \tag{61}$$

Figure 2 shows one such a spacetime in which a shell of null dust accretes onto a pre-existing black hole.

To calculate the various quantities of interest we fix a scaling of the null vectors to surfaces of constant r and v:

$$\ell^a = \left[1, 1 - \frac{2m(v)}{r}, 0, 0\right] \text{ and } n^a = [0, -1, 0, 0]\,. \tag{62}$$

Then straightforward calculations show that for these spherically symmetric two-surfaces:

$$\theta_{(\ell)} = \frac{1}{r}\left(1 - \frac{2m(v)}{r}\right) \text{ and } \theta_{(n)} = -\frac{2}{r} \tag{63}$$

while

$$\kappa_\ell = -n_b \ell^a \nabla_a \ell^b = \frac{m}{r^2} \text{ and } G_{ab}\ell^a\ell^b = \frac{2\dot{m}}{r^2}\,. \tag{64}$$

Overdots indicate derivatives with respect to v. Thanks to both the symmetry and the special nature of null dust, most other geometrical quantities (including the shear and all other components of the stress-energy tensor) vanish.

5.1.2. *The slowly evolving horizon*

Now, there is clearly a FOTH located at $r = 2m(v)$ with characteristic length $R_H = 2m$. For this surface

$$C = 2\dot{m} \tag{65}$$

so

$$\epsilon^2 \equiv \frac{1}{2} C \theta_{(n)}^2 (R_H)^2 = 4\dot{m} \tag{66}$$

and it is straightforward to see that if $\dot{m} \ll 1$, then this is a slowly expanding horizon. Next,

$$\kappa_{\mathcal{V}} = \frac{1}{4m} \text{ and } \mathcal{L}_{\mathcal{V}} C = 4\ddot{m}\,. \tag{67}$$

and so, if $\ddot{m} \ll \dot{m}$ the horizon is also slowly evolving.

Before checking implications, we restrict to a specific example to get a feeling for the kind of situations under which Vaidya is slowly evolving. Consider a piecewise linear mass function :

$$m(v) = \begin{cases} 1 & v \leq 0, \\ 1 + \alpha v & 0 \leq v \leq v_o, \\ 1 + \alpha v_o & v > v_o. \end{cases} \tag{68}$$

That is, we irradiate an initially solar mass black hole with null dust for a finite period of time.

To interpret this physically, consider a fleet of observers who use rockets to hold themselves some distance from the horizon at constant areal radius r_o. These observers will see a total mass of αv_o fall past them during time

$$T = \int_0^{v_o} \sqrt{1 - \frac{2m(v)}{r}} dv < \alpha v_o \, , \tag{69}$$

with the inequality coming close to saturation for very large r_o. Of course, these observers can't actually see the horizon (since it is spacelike there aren't any observers who can see the horizon from outside the black hole).

Adding in some numbers, consider $\alpha = 1/40000$ and $v_o \approx 2.1 \times 10^5 M_\odot$ (one second in standard units). Then, our observers would see $\alpha v_o \approx 5.3$ solar masses of material sweep past them in less then a second. However, at the horizon a quick calculation shows that $\epsilon^2 = 4\alpha = 10^{-4}$. Thus, even in this very dramatic situation the horizon would be slowly evolving to order $\epsilon \approx 0.01$.

There are many possible objections to this example, from the fact that null dust doesn't exist to the assumption of spherical symmetry. However, it is in line with our expectations based on the units $M_\odot$ and $T_\odot$ and broadly comparable results include timelike-dust examples and also Schwarzschild as perturbed by quite strong incoming gravitational waves.[26] Details vary, but the general message is that black holes probably evolve slowly in almost all astrophysical situations, with one notable exception being black hole mergers.

5.1.3. *Mechanics*

For spherically symmetric Vaidya, the zeroth law holds trivially (and exactly) while the FOTH at $r = 2m(v)$ is clearly increasing in area for $\dot{m} > 0$. Thus the only part of the mechanics that needs checking is the Clausius form of the second law (45). From the preceding section

$$\frac{\kappa_o \dot{a}}{8\pi G} \approx \frac{dm}{dv} \tag{70}$$

while the right-hand side is also

$$\int_{S_v} \widetilde{\epsilon} \left\{ \frac{\|\sigma^{(\ell)}\|^2}{8\pi G} + T_{ab}\ell^a \ell^b \right\} \approx \frac{dm}{dv} \, . \tag{71}$$

In fact[26] the Vaidya spacetime is, in many ways, not the best spacetime to demonstrate that this law holds: thanks to the symmetries and special nature of null dust, (71) actually holds exactly whether or not the horizon is slowly evolving. This however is an exceptional circumstance and for all other examples (including Tolman–Bondi) the slowly-evolving condition is necessary.

5.1.4. *The event horizon candidate*

For the Vaidya spacetime one can directly search for null surfaces just outside the event horizon by trying to solve

$$\frac{dr}{dv} = \frac{1}{2}\left(1 - \frac{2m}{r}\right) \tag{72}$$

for a solution of the form $r = 2m(1+\rho)$, with $\rho \ll 1$. If one assumes a hierarchy of derivatives (so that $m \gg m\dot{m} \gg m^2\ddot{m}$), one can iteratively solve this equation order-by-order to any desired accuracy. In particular to second order one finds that:[24]

$$r_{null} \approx 2m + 8m\dot{m} + \left(64m\dot{m}^2 + 32m^2\ddot{m}\right) . \tag{73}$$

If this coordinate-dependent distance is converted into a proper-time measurement,[25] it matches the first order prediction of Eq. (55).

5.2. *Black branes in the fluid–gravity correspondence*

A much more interesting example of near-equilibrium horizons can be found in the context of the fluid–gravity correspondence. This is not the place to go into a full discussion of this version of the AdS-CFT correspondence, but very briefly it states that a near-equilbrium asymptotically AdS black hole in N-dimensions is dual to a near-perfect conformal fluid in $(N-1)$-dimensions. From the CFT perspective, these near-equilbrium black holes correspond to the long-wavelength regime where the field theory reduces to fluid mechanics. That said, from a purely gravitational perspective, there is no need to reference the full correspondence. A direct examination of the long-wavelength perturbations of AdS black branes demonstrates that the gravitational equations of motion governing those perturbations are equivalent to the Navier–Stokes equations governing a near-perfect conformal fluid.[24]

Thermodynamically the temperature of this fluid should correspond to that of the black brane while the entropy flow should map onto the expansion of its horizon, and indeed this is the case to leading order. However, at higher order things become more interesting and several ambiguities arise in the exact meaning of the phrase "the entropy flow should map onto the expansion of its horizon".

In this section we briefly review the role of slowly evolving horizons in the fluid–gravity correspondence. Further details can be found in the literature.[27–29] Up until now we have restricted our attention to four-dimensional spacetimes with spatially compact black holes. Switching to five-dimensions for this example, the formalism that we have developed remains essentially unchanged. The only changes are that we relax the compactness condition (with the scale now being set by the surface gravity rather than the areal radius) and have to replace some factors of 1/2 with 1/3 in equations such as (6), (24)–(27) and (28).

5.2.1. *The spacetime*

Starting with a five-dimensional AdS black-brane metric, the fluid–gravity correspondence may be derived in the following way.[24] This solution is static and on inducing a compatible time-foliation, each instant may be thought of as a stack of three-dimensional intrinsically flat planes that stretch from the spacetime singularity to infinity with a similarly planar event horizon in between. In Eddington–Finkelstein-like form the metric is

$$ds^2 = -r^2\left(1 - \frac{1}{(rb_o)^4}\right)dv^2 + 2dvdr + r^2(dx^2 + dy^2 + dz^2)\,, \tag{74}$$

where for computational convenience the cosmological constant has been scaled to take the value -6 and the event horizon at $r = 1/b_o$ has temperature

$$T = \frac{1}{\pi b_o}\,. \tag{75}$$

An alternate form of this metric arises if one performs a Lorentz boost within the planes so that we switch to a coordinate frame which moves with four-velocity u_o^α relative to the static system:

$$ds^2 = -r^2\left(1 - \frac{1}{(rb_o)^4}\right)u^o_\mu u^o_\nu dx^\mu dx^\nu + 2u^o_\mu dx^\mu dr + r^2\left(\eta_{\mu\nu} + u^o_\mu u^o_\nu\right)dx^\mu dx^\nu\,. \tag{76}$$

In this case the greek indices run over $\{v, x, y, z\}$ and $\eta_{\mu\nu}$ is the standard four-dimensional Minkowski metric.

The fluid–gravity correspondence comes from considering long-wavelength perturbations of this boosted black brane metric: one replaces the constant b_o, u_o^α and $\eta_{\mu\nu}$ with b, u^α and $g_{\mu\nu}$ which agree with their unperturbed counterparts at leading order but beyond that have a coordinate dependence. In the planar directions the scale of variation of these perturbations must be much longer than the length scale set by b (they are long-wavelength perturbations). Thus, for example,

$$\|d_\alpha u^\beta\| \ll \left\|\frac{u^\beta}{b}\right\|. \tag{77}$$

Further, one assumes that this pattern continues for the entire hierarchy of derivatives in the planes. Continuing with the u^β example:

$$\|u^\delta\| \gg b\|d_\alpha u^\delta\| \gg b^2\|d_\alpha d_\beta u^\delta\| \gg b^3\|d_\alpha d_\beta d_\gamma u^\delta\| \ldots \tag{78}$$

It is then possible to separate and solve the Einstein equations for these perturbations order-by-order. As might be expected, though straightforward in principle, these computations rapidly become very intricate.[24,33]

For the quasi-local horizon enthusiast, these approximate solutions are of interest in their own right. At second order in this perturbation theory we can track the evolution of both event and apparent horizons and they are distinct from the unperturbed results. By contrast, for Schwarzschild spacetimes current state-of-the-art perturbation theory cannot distinguish these horizons and places them both at

$r \approx 2m$; corrections to the position are at a higher order than can be tracked.[30–32] However, before considering those horizons in more detail we complete our discussion of the fluid–gravity correspondence as it turns out that the conformal fluid interpretation eases the calculations on the gravity side.

The perturbed solution is entirely described by the quantities $\{b, u_\mu, g_{\mu\nu}\}$ with all tensor-indices being in the cotangent space to the surfaces of constant r. Each of these surfaces has an intrinsic geometry defined by $g_{\mu\nu}$ (it is weakly perturbed from Minkowski) and on each such surface one can construct a four-dimensional stress-energy tensor:

$$T^{\mu\nu} = \epsilon u^\mu u^\nu + p h_{\mu\nu} \,, \tag{79}$$

where

$$h_{\mu\nu} = g_{\mu\nu} + u_\mu u_\nu \tag{80}$$

is the three-metric transverse to the u_μ and the pressure and energy density are defined by b and obey the equation of state:

$$\epsilon = 3p = \frac{3}{16\pi G b^4} \,. \tag{81}$$

This may be regarded as a formal definition (which will be justified by later developments), however it may also be derived systematically as a Brown–York boundary stress-energy tensor.[35] In any case, for the unperturbed metric this is the stress-energy of a conformally invariant perfect fluid. However on implementing the perturbations it is only near-perfect with the corrections depending on gradients of $\{b, u^\mu, g_{\mu\nu}\}$ (and corresponding to non-perfect-fluid corrections such as viscosity).

Implementing the Einstein equations turns out to be equivalent to requiring conservation of the stress-energy tensor on each of these surfaces:

$$\mathcal{D}_\mu T^{\mu\nu} = 0 \,, \tag{82}$$

where $\mathcal{D}_\mu$ is the conformally invariant covariant derivative compatible with $g_{\mu\nu}$ (see the original sources or the summary in reference[29] for more details). This can be observed directly from computation with the components, but it may also be seen as a consequence of the fact that the momentum constraint must hold on each surface.[35]

The dual fluid is generally taken to live on the surface at $r \to \infty$. The conformal invariance can be used to rescale away diverging quantities in this limit and so guarantee that the theory remains well-defined.

5.2.2. *The slowly evolving horizon*

For the Vaidya example, finding a FOTH was straightforward; guided by the spherical symmetry, we simply examined the surfaces of constant r (and ignored all other candidates). Here things are a bit more complicated. The geometric symmetry of the problem is broken and so one cannot use it to select and search for a preferred

FOTH. However, there is still the conformal symmetry and this turns out to be sufficient to select a preferred FOTH (to second order). Again we will be brief, leaving the reader to consult the literature[24, 33, 34, 29] for more details.

The conformal symmetry of the fluid means that solutions are invariant under rescalings of the form

$$g_{\mu\nu} \to e^{-2\phi} g_{\mu\nu}, \quad u^\mu \to e^\phi u^\mu, \quad b \to e^{-\phi} b \text{ and } r \to e^\phi r\,. \tag{83}$$

Physically meaningful quantities should also respect this symmetry and we take this to include the location and evolution vector field of the FOTH. This is a strong restriction as there are only a limited number of conformally invariant quantities available at each order of the gradient expansion. For example, at first order, there is only the Weyl-invariant shear of u^μ:

$$\sigma^{fluid}_{\mu\nu} = \mathcal{D}_{(\mu} u_{\nu)}\,, \tag{84}$$

where the parentheses indicate the usual symmetrization. At second order in gradients there are ten quantities (three scalars, two vectors, and five transverse traceless tensors). We require that corrections to both the location of the horizon and its evolution vector field $\mathcal{V}^\mu$ be constructible out of these quantities.

After some fairly extensive algebra one can show that to second order there is a unique conformally invariant FOTH at

$$r_H = \frac{1}{b} + \frac{1}{b}\left(h_1^{AH} S_1 - \frac{3}{8} S_2 - \frac{1}{24} S_3\right) \tag{85}$$

where h_1^{AH} is a non-trivial numerical factor, while the conformally invariant factors are built out of the conformally invariant shear and vorticity of u^μ as well as the similarly conformally invariant curvature scalars: $S_1 = b^2 \sigma_{\mu\nu}\sigma^{\mu\nu}$, $S_2 = b^2 \omega_{\mu\nu}\omega^{\mu\nu}$ and $S_3 = b^2 \mathcal{R}$. Note that the corrections are all at second order in gradients. At first order the horizon is still at $r = 1/b$. This FOTH is easily seen to be slowly evolving with characteristic length scale $R_H = 1/b$.

The exact implications of this uniqueness are unclear. It has been conjectured[29] that the uniqueness might extend to higher order corrections and possibly even imply that in this case the conformal invariance selects a unique FOTH just as in the Vaidya case the spherical symmetry selected a preferred horizon. If true this would be a fascinating result. However it is also possible that the uniqueness to second order is simply a consequence of the fact that the horizon is slowly evolving. The resolution of this question will be left to future investigations.

5.2.3. *Mechanics: horizon and fluid*

On the gravity side it is easy to check that both the (approximate) zeroth law holds ($\kappa_{\mathcal{V}} \approx 1/b$) as does the Clausius form of the second law. However for these solutions there is an additional interest as we can compare the mechanics of the near-equilibrium horizon to the thermodynamics of the similarly near-equilbrium fluid. By the fluid–gravity correspondence we would expect them to match.

Before we can do this, we first need to understand the thermodynamics of the fluid. It is standard to identify the fluid temperature with that of the horizon so that to first order

$$T_{fluid} \approx \frac{1}{b} \,. \tag{86}$$

Then the equation of state (81) defines energy density and pressure as functions of temperature.

The next step is to identify an entropy current J^μ: the divergence of such a current $\mathcal{D}_\mu J^\mu$ determines the rate of entropy production in the fluid. For equilibrium flows this takes the form

$$J^\mu_{eq} = s u^\mu \tag{87}$$

where s is the (equilibrium) entropy density. However, away from equilibrium the correct form of the entropy current is not so clear and it is usually argued that one should allow for correction terms.[24,36,27,28] In our situation we must consider all conformally invariant causal flows with non-negative divergence. As for the horizons, requiring that the flow be conformally invariant significantly restricts the freedom of construction, but in this case it does not completely remove it: there are allowed second order corrections to J^μ_{eq} and these correspond to different values of entropy production.

The rate of expansion of the slowly evolving FOTH defined by (85) corresponds to one of these values of entropy production. Given that FOTHs are non-unique (recall that they are not rigid and may be deformed), it is very tempting to then conjecture that the other J^μ correspond to other (nearby) FOTHs. However, this cannot be the case: even though the conformal symmetry does not select a unique entropy current, it did select a preferred slowly evolving horizon. Thus any gravitational interpretation of this uncertainty must be sought elsewhere.

5.2.4. *The event horizon candidate*

Our FOTH is slowly evolving, and so we would expect to find a (similarly slowly evolving) event horizon candidate in close proximity. This is the case[24] and so this candidate provides another potential gravitational dual to the entropy current. Indeed it is straightforward to check that the rate of expansion of the event horizon candidate is matched by the divergence of another member of the family of allowed entropy currents.

5.2.5. *Implications and complications*

From the horizon perspective, these black brane spacetimes with their (measurably) dynamical horizons are nice examples in their own right providing testing grounds for ideas about slowly evolving horizons. The fluid–gravity duality adds another level of interest to these examples. In the first place we can see that dual to the near-equilibrium horizons are the similarly near-equilibrium fluids and that the mechanics of the horizon is similarly dual to a thermodynamics of the fluid.

Note the phrase "a thermodynamics" rather than "the thermodynamics". The entropy flow in the fluid is not uniquely defined and so therefore neither is its thermodynamics. Both the FOTH and event-horizon candidates have their own (closely-related but distinct) thermodynamic duals. However, these are just two of the potential thermodynamics and so it is natural to consider duals to the other entropy flows. There are at least two possible ways to account for this freedom in a geometric way:

(1) The conformal fluid lives on the AdS boundary of the spacetime and should be dual to the entire spacetime, not just the horizon. It can be argued that the brane dynamics dominate the spacetime; however even in that case one must still define a map between the horizon and the boundary. Given the coordinate system, it is computationally easiest to just match up points with matching x^μ coordinates. However this mapping is not geometrically preferred and there are other acceptable ways to set it up. It turns out that the freedom in constructing the mapping closely matches that in defining the entropy flow.[24] So, the first possible geometric explanation for the entropy flow freedom is that it is a manifestation of the horizon-boundary mapping uncertainty. Either the FOTH or the event horizon may be viewed as dual to each of the entropy flows.

(2) The second possibility is to take the freedom on the fluid side as signalling that a similar freedom should exist in defining entropy and entropy flows in the bulk. References[27,28] suggest one way of doing this in which the SEH and event horizon candidate are taken to be just two examples of a family of horizon-like structures any one of which can be associated with a mechanics (and each of which is dual to a fluid flow). These surfaces can have any signature but each signals the existence of a nearby casual boundary and is (at least approximately) null and slowly expanding with $\theta_{(\ell)} \approx 0$. In any approach to equilibrium they all are required to have the same isolated horizon limit. In particular the timelike ones are similar to those surfaces considered in the membrane paradigm.[37]

Of course these two possibilities are not mutually exclusive. If the conformal fluid reflects the true physics of the system, then the first resolution means that the horizon-candidates proposed by the second resolution are indistinguishable since both have the same range of freedom in their definition. Thus each candidate would furnish an equally acceptable dynamical black hole mechanics. These would differ in detail but remain closely related and in particular all reduce to the same equilibrium limit.

A definitive resolution of these issues will have to wait for future work. However it may be worth pointing out that uncertainties of this type would not be too surprising. While there are several well-defined measures of total energy in general relativity,[1] its localization is famously ill-defined and the study of potential quasi-local definitions of energy remains an area of active research.[38] Thermodynamics and black hole mechanics are intimately tied to energy flows and so if there are uncertainties in amounts of energy transferred it would be consistent if there were similar uncertainties in entropy.

6. Summary

The first law of black hole mechanics as formulated for exact solutions or isolated horizons is a phase space relation: it considers variations through the phase space of equilibrium states rather than a dynamical evolution between those states. The original motivation for the investigation of slowly evolving horizons was to better understand the near-equilibrium regime for dynamical trapping horizons and so reinterpret this first law in a dynamical way.[7] We have seen that this reformulation is possible and the resulting dynamical version arises as a particular limit of the equations governing the geometric deformation of the surfaces.

For null surfaces the deformation equations reduce to the Raychaudhuri equation and so also govern the evolution of event horizons, though the identification of any particular null surface as the event horizon remains a teleological procedure. There is also a slowly evolving limit for null surfaces and, though we only demonstrated this for a particularly simple case, there is a surface of this type in close proximity to each slowly evolving horizon. If the FOTH remains near-equilibrium at all times in the future, that null surface will be the event horizon. Its physical properties and geometry are very similar, though not identical, to those of the slowly evolving horizon.

A wide variety of astrophysical processes are probably in the near-equilibrium regime. One can quickly see this by noting that the characteristic time scale for black holes would be the horizon-crossing time. For a solar mass black hole this is on the order of 10^{-6} seconds. Thus we would expect anything that happens on a time scale much greater than this to be slowly evolving. Particularly dramatic examples were demonstrated in the Vaidya spacetime.

Finally we considered the interesting case of five-dimensional black branes and the fluid–gravity duality. In that case the perturbatively constructed spacetime gives rise to a slowly evolving horizon in the bulk and a corresponding near-equilibrium fluid on the boundary, with the mechanics of the horizon closely matching the thermodynamics of the fluid. However, the thermodynamics is not uniquely defined and can be adjusted to match both the SEH and its attendant event horizon candidate. It can be matched with the evolution of other nearly null, nearly isolated "horizons". This behaviour is reminiscent of the classical membrane paradigm which demonstrated that much black hole physics can be modelled by considering the interaction of near-horizon timelike surfaces with their environment.

Acknowledgements

This work was supported by the Natural Sciences and Engineering Research Council of Canada. The author thanks Aghil Alaee for his comments on the penultimate version of the manuscript.

References

1. R.M. Wald, *General relativity* (University of Chicago Press, 1984).
2. J. Engle and T. Liko, "Isolated horizons in classical and quantum gravity", arXiv:1112.4412.
3. A. Ashtekar and B. Krishnan, "Isolated and dynamical horizons and their applications", *Living Rev. Rel.* **7**, 10 (2004).
4. A. Ashtekar, C. Beetle, O. Dreyer, S. Fairhurst, B. Krishnan, J. Lewandowski and J. Wiśniewski, "Generic isolated horizons and their applications", *Phys. Rev. Lett.* **85** 3564 (2000).
5. E. Gourgoulhon and J. L. Jaramillo, "A 3+1 perspective on null hypersurfaces and isolated horizons", *Physics Reports,* **423** 159 (2006).
6. I. Booth, "Black Hole Boundaries" *Can. J. Phys.* **83** 1073 (2005).
7. I. Booth and S. Fairhurst, "The first law for slowly evolving horizons", arxiv: 1207-6955. *Phys. Rev. Lett.* **92**, 011102 (2004).
8. I. Booth and S. Fairhurst, "Isolated, slowly evolving, and dynamical trapping horizons: Geometry and mechanics from surface deformations", *Phys. Rev,* **D75** 084019 (2007).
9. E. Poisson, *A Relativist's Toolkit: The Mathematics of Black-Hole Mechanics* (Cambridge University Press, 2004).
10. S.A. Hayward, "General laws of black hole dynamics", *Phys. Rev.* **D49** 6467 (1994).
11. S.W. Hawking and G.F.R. Ellis, *The large scale structure of space-time*, (Cambridge University Press, 1973).
12. A. Ashtekar, S. Fairhurst and B. Krishnan, *Phys. Rev.* **D62** 104025 (2000); A. Ashtekar, C. Beetle and J. Lewandowski, *Phys. Rev.* **D64** 044016 (2001).
13. A. Ashtekar and B. Krishnan, "Dynamical horizons: energy, angular momentum, fluxes and balance laws", *Phys. Rev. Lett.* **89** 261101 (2002); "Dynamical horizons and their properties", *Phys. Rev.* **D68** 104030 (2003).
14. A. Ashtekar and G. J. Galloway, *Adv. Theor. Math. Phys.* **9**, 1 (2005) [gr-qc/0503109].
15. I. Bendov, "The Penrose inequality and apparent horizons", *Phys. Rev.* **D70** (2004) 124031.
16. I. Booth, L. Brits, J. Gonzalez, and C. Van Den Broeck, "Marginally trapped tubes and dynamical horizons", *Class. Quant. Grav.* **22**, 4515 (2005).
17. I. Booth, "Two physical characteristics of numerical apparent horizons", arXiv:0709.0934 (2007).
18. S. W. Hawking and J. B. Hartle, *Commun. Math. Phys.* **27** 283 (1972).
19. J. M. Bardeen, B. Carter, S. W. Hawking, "The Four laws of black hole mechanics", *Commun. Math. Phys.* **31** 161 (1973).
20. E. Gourgoulhon and J. L. Jaramillo, "Area evolution, bulk viscosity and entropy principles for dynamical horizons," *Phys. Rev.* **D74**, 087502 (2006).
21. I. Booth, "Spacetime in the vicinity of a slowly evolving horizon" arXiv: 1207.7955.
22. P. Anninos et. al, *Phys. Rev. Lett.* **71** 2851 (1993); P. Anninos et. al., *Phys. Rev. Lett.* **74**, 630 (1995); Baiotti et.al, *Phys. Rev.* **D71** 024035 (2005); Schnetter et. al. *Phys. Rev.* **D74** 024028 (2006)

23. E. Poisson, *Phys. Rev.* **D70** 084044 (2004); E. Poisson, *Phys. Rev. Lett.* **94** 161103 (2005); E. Poisson and I. Vlasov, *Phys. Rev.* **D81** 024029 (2010).
24. S. Bhattacharyya, V. E. Hubeny, S. Minwalla and M. Rangamani, "Nonlinear Fluid Dynamics from Gravity", JHEP **0802**, 045 (2008); S. Bhattacharyya et al., "Local Fluid Dynamical Entropy from Gravity", JHEP **0806**, 055 (2008).
25. I. Booth and J. Martin, "Proximity of black hole horizons: Lessons from Vaidya spacetime", *Phys. Rev.* **D82** 124046 (2010).
26. W. Kavanagh and I. Booth, "Spacetimes containing slowly evolving horizons", *Phys.Rev.* **D74** 044027 (2006).
27. I. Booth, M. Heller and M. Spalinski, "Black brane entropy and hydrodynamics: The boost-invariant case", *Phys. Rev.* **D80** 126013 (2009).
28. I. Booth, M. Heller and M. Spalinski, "Black brane entropy and hydrodynamics", *Phys. Rev.* **D83** 061901 (2011).
29. I. Booth, M. Heller, G. Plewa and M. Spalinski, "On the apparent horizon in fluid–gravity duality", *Phys. Rev.* **D83** 106005 (2011).
30. E. Poisson, "Retarded coordinates based at a world line, and the motion of a small black hole in an external universe", *Phys. Rev.* **D69**, 084007 (2004); "Absorption of mass and angular momentum by a black hole: Time-domain formalisms for gravitational perturbations, and the small-hole/slow-motion approximation", *Phys. Rev.* **D70**, 084044 (2004); "Metric of a tidally distorted, nonrotating hole", *Phys. Rev. Lett.* **94**, 161103 (2005).
31. E. Poisson and I. Vlasov, *Phys. Rev.* **D81**, 024029 (2010)
32. I. Vega, E. Poisson and R. Massey, "Intrinsic and extrinsic geometries of a tidally deformed black hole", *Class. Quant. Grav.* **28**, 175006 (2011)
33. S. Bhattacharyya et. al, "Conformal nonlinear fluid dynamics from gravity in arbitrary dimensions", *J. High Energy Phys.* **12** 116 (2008).
34. R. Loganayagam, "Entropy Current in Conformal Hydrodynamics", JHEP **0805**, 087 (2008).
35. J. D. Brown and J. W. York, Jr., "Quasilocal energy and conserved charges derived from the gravitational action", *Phys. Rev.* **D47**, 1407 (1993).
36. P. Romatschke, "Relativistic Viscous Fluid Dynamics and Non-Equilibrium Entropy", *Class. Quant. Grav.* **27**, 025006 (2010)
37. K. S. Thorne, R. H. Price and D. A. Macdonald, *Black holes: the membrane paradigm*, (Yale University Press,1986).
38. L. B. Szabados, "Quasi-Local Energy-Momentum and Angular Momentum in GR: A Review Article", *Living Rev. Rel.* **7**, 4 (2004).

Chapter 5

Isolated Horizons in Classical and Quantum Gravity

Jonathan Engle and Tomáš Liko

Department of Physics, Florida Atlantic University, Boca Raton, FL 33431, U.S.A.

Department of Mathematical and Statistical Sciences, University of Alberta, Edmonton, AB T6G 2G1, Canada

1. Introduction

The classic, global approach to black holes in terms of event horizons has accomplished much: One has a definition of black hole horizon which is unambiguous and general, depending only on the assumption that space-time is asymptotically flat or asymptotically (anti-)de Sitter. It is a definition valid for fully dynamical black holes that possess no symmetry. For event horizons, one can prove among other things Hawking's very general area theorem, which implies that the areas of event horizons always grow or stay constant in time. If one furthermore restricts consideration to *stationary* black holes, then the ADM angular momentum and ADM energy can be interpreted as the angular momentum and energy of the black hole itself, giving a notion of these quantities for black holes that is well-rooted in their deeper meaning as generators of flows on phase space. By using the global stationary Killing field available for such space-times, one also arrives at well-defined notions of surface gravity and angular velocity of the horizon. These quantities, together with surface area, energy and angular momentum, satisfy the zeroth and first laws of black hole mechanics, part of the evidence suggesting that black holes are thermodynamic objects.

In spite of these successes, both of these notions of black hole are, from a physical perspective, not completely satisfactory:

- The event horizon definition requires knowledge of the entire space-time all the way to future null infinity. However, physically, one can never know the full history of the universe, nor can one measure quantities at infinity.
- The use of stationary space-times to derive black hole thermodynamics is also not ideal: In all other physical situations, in order to derive laws of

equilibrium thermodynamics, it is only necessary to assume equilibrium of the system in question, not the entire universe.

- Furthermore, beyond such physical considerations, the global nature of event horizons and Killing horizons make them difficult to use in practice. In particular, in quantum theory, in order for a definition of black hole to make sense, one needs to be able to formulate it in terms of phase space functions which can be quantized. Event horizons are not amenable to such a characterization in any obvious or simple way. By contrast, the condition defining a Killing horizon can be formulated even in terms of local functions on space (if the Killing field is fixed). However, because the restriction is imposed on fields outside, as well as within, the horizon, the degrees of freedom outside the horizon are reduced and it is again no longer clear if canonical quantum gravity methods, in particular those of loop quantum gravity, can be applied reliably there.
- Additionally, in numerical relativity, the *global* notions of ADM energy and ADM angular momentum are of limited use, because they do not distinguish the mass of black holes from the energy of surrounding gravitational radiation. If a quasi-local framework could enable a clear definition of angular momentum and energy of a black hole using standard canonical notions, this would be useful for interpreting numerical simulations in a gauge-invariant and systematic way.

For all of the above reasons it is both physically desireable and practical to seek a *quasi-local* alternative to the event horizon and Killing horizon frameworks.

One can ask: Is it possible to find a framework for describing black holes which is quasi-local yet *nevertheless retains all of the desireable features of Killing horizons*? It is not *a priori* obvious that this is possible, because part of the reason for the success of the Killing horizon framework is that one can use global stationarity to relate physics at the horizon to physics at infinity, where the asymptotic flat metric is available to aid in the definition of various quantities. In a quasi-local approach, such access to the structures at infinity will not be possible. In spite of this challenge, as we shall see, the *isolated horizon* framework allows one to answer the above question in the affirmative. Moreover, due to the fact that the isolated horizon conditions restrict only the *intrinsic geometric structure of the horizon*, they can be implemented in quantum theory, giving rise to a quantum description of black holes within loop quantum gravity. In this framework of quantum isolated horizons, one can account for the statistical mechanical origin of the Bekenstein–Hawking entropy of black holes in a large number of physically relevant situations.

The isolated horizon framework has been developed by many authors.[1–14] Furthermore, the framework is closely related to, and inspired by, not only the Killing horizons as hinted above, but also the trapping horizons of Hayward.[15–17] Roughly speaking, isolated horizons correspond to portions of trapping horizons that are null. Portions of trapping horizons that are space-like roughly correspond to a complementary notion called 'dynamical horizons'.[18] A broader review of both of these topics and their relation to trapping horizons has been given in.[19]

In this chapter, after reviewing Killing horizons and black hole thermodynamics for motivational purposes, we review the geometrical observations regarding null surfaces which naturally lead to the isolated horizon definition. We then review a covariant canonical framework for space-times with isolated horizons, leading to a canonical, quasi-local notion of angular momentum and mass which then satisfy a *quasi-local* version of the first law of black hole mechanics. Motivated by the expression for the angular momentum, quasi-local angular momentum and mass multipoles for isolated horizons are also reviewed, in passing. Finally, we review the quantization of this framework, using the methods of loop quantum gravity,[20–22] leading to a statistical mechanical explanation of the Bekenstein–Hawking entropy of black holes.

2. Motivation from Killing Horizons and Black Hole Thermodynamics

2.1. *Black hole horizons and the four laws of black hole mechanics*

A black hole is by definition an object with a gravitational field so strong that no radiation can escape from it to the asymptotic region of space-time. Mathematically, a *black hole region* B of a space-time $\mathcal{M}$ is that which excludes all events that belong to the causal past from future null infinity, here denoted $\mathcal{I}^+$: $B = \mathcal{M} \backslash J^-(\mathcal{I}^+)$, with $J^-(N)$ representing the union of the causal pasts of all events contained in N.[23] The *event horizon*, $\mathcal{H}$, is the boundary of the black hole region: $\mathcal{H} = \partial B$. The two-dimensional *cross section* $\mathcal{S}$ of $\mathcal{H}$ is the intersection of $\mathcal{H}$ with a three-dimensional spacelike (partial Cauchy) hypersurface M: $\mathcal{S} = \mathcal{H} \cap M$.

A space-time is said to be *stationary* if the metric admits a Killing field t^a $(a, b, \cdots \in \{0, 1, 2, 3\})$ which approaches a time-translation at spatial infinity. A space-time is said to be *axi-symmetric* if the metric admits a Killing field ϕ^a that generates an $SO(2)$ isometry. Examples of space-times that are both stationary and axisymmetric are the Schwarzschild solution and more generally the Kerr–Newman family of solutions. The stationary Killing field t^a is normalized by requiring that it be unit at spatial infinity, while the axisymmetric Killing field ϕ^a is normalized by requiring that the affine length of its orbits be 2π.

A space-time is said to contain a *Killing horizon* $\mathcal{K}$ if it possesses a Killing vector field ξ^a that becomes null at $\mathcal{K}$ and generates $\mathcal{K}$. That is, $\mathcal{K}$ is a null hypersurface with ξ^a as its null normal.[24,25] For stationary space-times, under very general assumptions, the event horizon is also a Killing horizon [23, p.331]. As ξ^a is by definition hypersurface-orthogonal at $\mathcal{K}$, by the Frobenius theorem it satisfies

$$\xi_{[a} \nabla_b \xi_{c]} \hat{=} 0$$

where $\widehat{=}$ denotes equality on $\mathcal{K}$. From this one can furthermore deduce that ξ is geodesic at $\mathcal{K}$:

$$\xi^a \nabla_a \xi^b \widehat{=} \kappa \xi^b \,. \tag{1}$$

This defines the surface gravity κ of the black hole. From this definition it is clear that the surface gravity rescales as $\kappa \to c\kappa$ under rescalings of the Killing vector field $\xi \to c\xi$, with c a constant. Equation (1) can be rewritten

$$2\kappa \xi_a \widehat{=} \nabla_a(-\xi_b \xi^b) \,.$$

Making use of the Frobenius theorem, geodesic equation and Killing equation for the vector ξ^a, one obtains the following explicit expression for the surface gravity κ:

$$\kappa^2 = -\frac{1}{2}(\nabla_a \xi_b)(\nabla^a \xi^b) \,. \tag{2}$$

If the dominant energy condition is satisfied, then it follows that κ not only is constant along the null generators of $\mathcal{H}$, but also it does not vary from generator to generator. This means that κ remains constant over all of $\mathcal{H}$. This is the zeroth law of black hole mechanics:

Zeroth Law. *If the dominant energy condition is satisfied, then the surface gravity κ is constant over the entire event horizon $\mathcal{H}$.*

The black hole uniqueness theorems[26,27,23,28–30] state that there is a unique three–parameter set of solutions to the Einstein–Maxwell field equations that are stationary and asymptotically flat; these are the Kerr–Newman family of solutions parameterized by the ADM mass M, ADM angular momentum J and electric charge Q of the space-time. For this family, the Killing vector field defining $\mathcal{K}$ is given by $\xi^a = t^a + \Omega\phi^a$, where Ω is called the angular velocity of $\mathcal{K}$. The electric potential at $\mathcal{K}$ is furthermore defined as $\Phi = -\boldsymbol{A}_a \xi^a|_{\mathcal{K}}$ where $\boldsymbol{A}_a$ is the 4-vector potential of the electromagnetic field. If the mass of a solution contained in the Kerr–Newman family is perturbed by an amount δM, then the changes in the surface area a and asymptotic charges (M, Q, J) of $\mathcal{K}$ are governed by the following law.

First Law. *Let a, κ, Ω, Φ, and (M, J, Q) denote the horizon area, surface gravity, angular velocity, electric potential, and asymptotic charges of a stationary black hole. For any variation δ within the space of stationary black holes, one has*

$$\delta M = \frac{\kappa}{8\pi G}\delta a + \Phi \delta Q + \Omega \delta J \,. \tag{3}$$

This form of the first law is known as the "equilibrium" form. That is, the condition (3) describes the changes of the black hole parameters from a solution to *nearby solutions* within the phase space. There is, however, a "physical process" interpretation: if one drops a small mass δM of matter into a black hole, the resulting changes δQ and δJ in the charge and angular momentum of the black hole will be such that the first law equation is satisfied. Such an interpretation is valid if the space-time is quasi-stationary — i.e., approximately stationary at each point in time.

Table 1. A summary of the four laws of black hole mechanics. Here, we make the identifications $U = M$, $T = \kappa/(2\pi)$ and $S = a/4$. [Adapted from [25].]

Law	Thermodynamics	Black holes
Zeroth	T constant throughout body in thermal equilibrium	κ constant over horizon of a stationary black hole
First	$dU = TdS +$ work terms	$dM = \frac{\kappa}{8\pi}da + \Omega_H dJ + \Phi_H dQ$
Second	$\Delta S \geq 0$ in any process	$\Delta a \geq 0$ in any process
Third	Impossible to achieve $T = 0$ by a physical process	Impossible to achieve $\kappa = 0$ by a physical process

If some matter with stress energy T_{ab} is dropped into a black hole, then the mass of the black hole is going to change. From the first law, the surface area a will have to change as well. A remarkable fact is that, if $T_{ab}\xi^a\xi^b \geq 0$, then this change cannot be a decrease. This is a statement of the second law of black hole mechanics:

Second Law *The surface area a can never decrease in a physical process if the stress-energy tensor T_{ab} satisfies the null energy condition $T_{ab}\xi^a\xi^b \geq 0$.*

One final property of quasi-stationary horizons is that it is impossible for one to become extremal within finite advanced time (i.e., finite time as experienced by free-falling observers near the horizon).[31] That is, the surface gravity cannot be reduced to zero in finite advanced time. This is the statement of the third law of black hole mechanics:

Third Law. *The surface gravity κ of a quasi-stationary black hole cannot be reduced to zero by any physical process within finite advanced time.*

2.2. *Black hole thermodynamics*

At this point it is instructive to pause for a moment to summarize the established four laws of black hole mechanics, and compare them to the corresponding four laws of thermodynamics. In the above Table 1, T is the temperature of a system, U its internal energy and S its entropy. As one can see, there is a striking formal similarity between these two sets of laws. Motivated by this similarity, Bekenstein conjectured that for a black hole:[32,33]

$$\kappa \propto T \quad \text{and} \quad a \propto S\,.$$

This identification of the surface area of the horizon with thermal entropy also offered a way to compensate for the apparent violation of the second law of thermodynamics which would seem to occur when matter falls into a black hole. The idea is that, because black holes are now endowed with entropy, a *generalized second law* of thermodynamics would still hold, even during such processes:

$$\delta S_{\text{Universe}} + \delta S_{\text{Blackhole}} \geq 0 \, .$$

However, it turns out that when trying to match a with S, and in order for S to remain dimensionless, we require a combination of the physical constants c, G *and* $\hbar$. By convention we take temperature to be measured in units of energy so that the Boltzmann constant is unity, whence the unique combinations of fundamental constants which will fit into the proportionality relations are

$$S = \text{constant} \times \frac{a}{4l_{\text{P}}^2} \quad \text{and} \quad T = \text{constant} \times \hbar\kappa \, ,$$

with $l_{\text{P}} = \sqrt{G\hbar/c^3}$, the Planck length.

In 1974, Hawking[34] discovered that black holes radiate a blackbody spectrum with a temperature of $T = \hbar\kappa/(2\pi)$. This is known as the Hawking Effect. The result came from considering quantum field theory on a fixed black hole background space-time. Hawking was able to use this result to fix the proportionality constant in Bekenstein's surface-gravity/temperature relation by requiring that

$$T\delta S = \frac{\kappa}{8\pi G}\delta a \, . \tag{4}$$

One finds that

$$S \equiv \frac{a}{4l_{\text{P}}^2} \tag{5}$$

in Planck units with $\ell_{\text{P}} = \sqrt{G\hbar/c^3}$ the Planck length, or

$$S \equiv \frac{ak_{\text{B}}c^3}{4\hbar G} \tag{6}$$

in SI units with k Boltzmann's constant.

The four laws were first formulated for stationary space-times in four-dimensional Einstein–Maxwell theory,[35] but later were extended using covariant phase space methods to include stationary black holes in arbitrary diffeomorphism-invariant theories.[36–39] This work revealed, remarkably, that the zeroth law holds for *any* stationary black hole space-time if the matter fields satisfy an appropriate energy condition. In addition, the surface area term in the first law is modified only in cases when gravity is supplemented with nonminimally coupled matter or higher-curvature interactions. For a general Lagrangian density $L = L(g_{ab}, \nabla_c g_{ab}, R_{abcd}, \nabla_e R_{abcd})$ (where $L(\cdot,\cdot,\cdot,\cdot)$ involves no derivatives of its arguments), the entropy of a stationary black hole space-time is given by

$$S = -2\pi \oint_{\mathcal{S}} \frac{\delta L}{\delta R_{abcd}} n_{ab} n_{cd} \, , \tag{7}$$

with n_{ab} the binormal to the cross-section $\mathcal{S}$ of the horizon, with normalization $n_{ab}n^{ab} = -2$. The fact that higher-order terms in the curvature affect the entropy of black hole space-times is a consequence of the fact that such terms modify the gravitational surface term in the symplectic structure. As an example, consider a modification to the Einstein–Hilbert Lagrangian consisting in adding the Euler density

$$\mathcal{L}_\chi = R^2 - 4R_{ab}R^{ab} + R_{abcd}R^{abcd}\,. \tag{8}$$

The resulting extra term in the action is a topological invariant of $\mathcal{M}$, and thus only serves to shift the value of the action by a number which is locally constant in the space of histories. Nevertheless, it contributes to the gravitational surface term in the symplectic structure and therefore also shifts the value of the entropy by a number depending on the topology of $\mathcal{M}$. Though usually non-dynamical, for black hole mergers, this term will in general be dynamical, due to the fact that the topology of $\mathcal{M}$ changes during the merging process.[40,41]

2.3. *Global equilibrium: limitations*

The standard definition of a black hole event horizon is teleological in the sense that we need to know the structure of the entire space-time in order to construct the event horizon. This is a major drawback. The usual way to resolve this is to consider solutions to the field equations that are stationary, as we have done above. These are solutions that admit a time translation Killing field *everywhere*, not just in a small neighborhood of the black hole region. While this simple idealization is a natural starting point, it seems to be overly restrictive.

Physically, it should be sufficient to impose boundary conditions at the horizon which ensure that *only the black hole itself is in equilibrium.* This viewpoint is consistent with what is usually done in thermodynamics: for the laws of equilibrium thermodynamics to hold in other situations, one need only assume the system in question is in equilibrium, not the whole universe. An approach to quasi-local black hole horizons that achieves this is the *isolated horizon* (IH) framework. More precisely, an isolated horizon models a portion of an event horizon in which the intrinsic geometric structures are 'time independent', and in this sense are in 'equilibrium', while the geometry outside may be dynamical, even in an arbitrarily small neighborhood of the horizon. In terms of physical processes, an isolated horizon can be characterized as having *no flux of matter or gravitational energy* through it. In realistic situations of gravitational collapse, such an assumption will be approximately valid only for certain finite intervals of time, as represented in Figure 1. In the following sections, we define and review the isolated horizon framework in detail, and review a selection of its applications.

3. Isolated Horizons

3.1. *Null hypersurfaces in equilibrium: non-expanding horizons*

A null surface $\mathcal{N}$ has a normal ℓ_a which, when raised with the space-time metric, is *tangent* to $\mathcal{N}$. Given $\mathcal{N}$, the null normal is of course not uniquely determined,

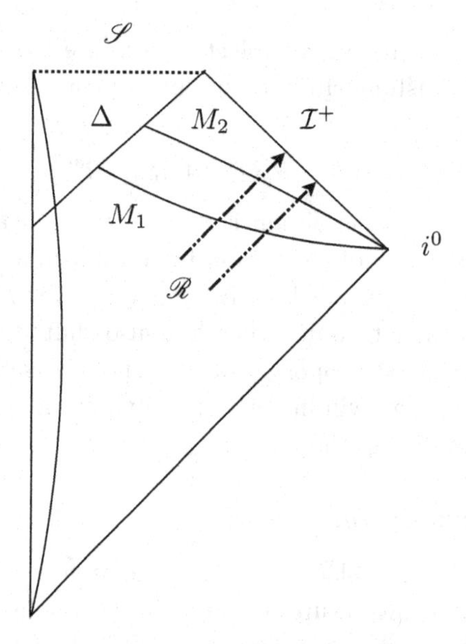

Figure 1. A Penrose–Carter diagram of gravitational collapse of an object that forms a singularity $\mathscr{S}$ and a horizon Δ, in the presence of external matter fields. The region of the space-time $\mathcal{M}$ being considered contains the quasi-local black hole boundary Δ intersecting the partial Cauchy surfaces M_1 and M_2 that extend to spatial infinity i^0. Here, Δ is in equilibrium with dynamical radiation fields $\mathscr{R}$ in the exterior region $\mathcal{M} \setminus \Delta$; these fields are not just in the exterior but can exist within an arbitrarily small neighbourhood of Δ.

but rather one has the freedom to rescale by a positive smooth function: $\ell_a \mapsto \ell'_a = f\ell_a$. The intrinsic metric q_{ab} on $\mathcal{N}$ is the pullback of the space-time metric; because $q_{ab}\ell^a \mathrel{\widehat{=}} g_{a\underleftarrow{b}}\ell^a \mathrel{\widehat{=}} \ell_{\underleftarrow{b}} \mathrel{\widehat{=}} 0$ (with "$\widehat{=}$" denoting equality restricted to $\mathcal{N}$ and "$\leftarrow$" denoting pullback to $\mathcal{N}$) it follows that q_{ab} is degenerate, i.e. l^a is the degenerate direction of q_{ab}. The signature of q_{ab} is then $(0,+,+)$ which means that the determinant of q_{ab} is zero. As a result, q_{ab} is non-invertable. However, an inverse metric q^{ab} can be defined such that $q^{ab}q_{ac}q_{bd} \mathrel{\widehat{=}} q_{cd}$; *any tensor* that satisfies this identity is said to be an inverse of q_{ab}.

If q^{ab} is an inverse, then so is $\tilde{q}^{ab} = q^{ab} + \ell^{(a}X^{b)}$ for any X^b tangent to $\mathcal{N}$. To see this, we observe that

$$\begin{aligned} \tilde{q}^{ab}q_{am}q_{bn} &\mathrel{\widehat{=}} (q^{ab} + \ell^{(a}X^{b)})q_{am}q_{bn} \\ &\mathrel{\widehat{=}} q^{ab}q_{am}q_{bn} + \frac{1}{2}\ell^a X^b q_{am}q_{bn} + \frac{1}{2}\ell^b X^a q_{am}q_{bn} \\ &\mathrel{\widehat{=}} q_{mn} \,. \end{aligned} \tag{9}$$

In going from the second to the third line, we used the property that $\ell^a q_{ab} = 0$.

Because ℓ_a is hypersurface orthogonal, it satisfies $\underleftarrow{\ell_{[a}\nabla_b\ell_{c]}} = 0$. From this alone, one can furthermore deduce that ℓ^a is geodesic, so that ℓ^a generates a *geodesic null congruence* on $\mathcal{N}$. The twist, expansion, and shear of this null congruence are given respectively by the anti-symmetric, trace, and trace-free symmetric parts of $\underleftarrow{\nabla_a \ell_b}$. Because ℓ_a is surface-forming, it is twist free. This leaves the expansion and shear,

$$\theta_{(\ell)} = q^{ab}\nabla_a l_b \quad \text{and} \quad \sigma_{ab} = \underleftarrow{\nabla_{(a}\ell_{b)}} - \frac{1}{2}q_{ab}q^{cd}\nabla_c l_d \,. \tag{10}$$

These both vanish iff $\underleftarrow{\nabla_a \ell_b}$ vanishes.

An important property of the expansion and shear is that both are independent of the choice of inverse metric q^{ab}. This follows from the following:

$$\begin{aligned}(\tilde{q}^{ab} - q^{ab})\nabla_a \ell_b &\,\hat{=}\, \ell^{(a}X^{b)}\nabla_a \ell_b \\ &\,\hat{=}\, \frac{1}{2}[\ell^a(\nabla_a \ell_b)X^b + (\ell^b \nabla_a \ell_b)X^a] \\ &\,\hat{=}\, \frac{1}{2}\kappa_{(\ell)}\ell_a X^a + \frac{1}{4}X^a \nabla_a(\ell_b \ell^b) = 0\,;\end{aligned} \tag{11}$$

the first term vanishes because $\ell_a X^a = 0$ and the second term vanishes because $\ell_a \ell^a = 0$.

A Killing horizon for stationary space-times is a null hypersurface, and so its null normal generates a twist-free geodesic null congruence on the horizon. This null congruence is furthermore *expansion free*. The condition that a general null hypersurface $\mathcal{N}$ be expansion free is in fact independent of the null normal used to define the expansion: If $\ell' = f\ell$ are two null normals related by some positive, smooth function f, we have

$$\theta_{(\ell')} \,\hat{=}\, q^{ab}\nabla_a f \ell_b \,\hat{=}\, q^{ab}\ell_b \nabla_a f + f\theta_{(\ell)} \,\hat{=}\, f\theta_{(\ell)}\,, \tag{12}$$

so that $\theta_{(\ell')} = 0$ iff $\theta_{(\ell)} = 0$. The vanishing of the expansion is therefore *an intrinsic property of a null surface* $\mathcal{N}$. This enables us to incorporate this *local* property of a Killing horizon into the following definition.

Definition 1. (Non-Expanding Horizon). *A three-dimensional null hypersurface $\Delta \subset \mathcal{M}$ of a space-time $(\mathcal{M}, g_{ab})$ is said to be a non-expanding horizon (NEH) if the following conditions hold: (i) Δ is topologically $R \times S$ with S a compact two-dimensional manifold; (ii) the expansion $\theta_{(\ell)}$ of any null normal ℓ to Δ vanishes; (iii) the field equations hold at Δ; and (iv) the stress-energy tensor T_{ab} of external matter fields is such that, at Δ, $-T^a_b\ell^b$ is a future-directed and causal vector for any future-directed null normal ℓ.*

For now we leave the horizon cross section S arbitrary. In Section 3.4 we will prove that, once the definition is strengthened a bit more, for vanishing cosmological constant, S has to be a two-sphere, thereby generalizing the Hawking Topology Theorem to non-stationary space-times.

The weakest notion of equilibrium for a NEH is the requirement that $\mathcal{L}_\ell q_{ab} \,\hat{=}\, 0$. This means that the intrinsic geometry of Δ is invariant under time translations. The condition is equivalent to

$$\mathcal{L}_\ell q_{ab} \,\hat{=}\, \underleftarrow{\mathcal{L}_\ell g_{ab}} = 2\underleftarrow{\nabla_{(a}\ell_{b)}} \,\hat{=}\, 0\,. \tag{13}$$

If $\pounds_\ell q_{ab} \mathrel{\hat{=}} 0$ for one null normal ℓ then it is true for any other null normal $\ell' = f\ell$:

$$\pounds_{\ell'} q_{ab} \mathrel{\hat{=}} 2\underleftarrow{\nabla_{(a}(\ell'_{b)})} \mathrel{\hat{=}} 2\underleftarrow{\nabla_{(a}(f\ell_{b)})} \mathrel{\hat{=}} 2\underleftarrow{(\nabla_{(a}f)\ell_{b)}} + 2f\underleftarrow{\nabla_{(a}\ell_{b)}} \mathrel{\hat{=}} 2f\underleftarrow{\nabla_{(a}\ell_{b)}} \mathrel{\hat{=}} f\pounds_\ell q_{ab}\,.$$

It follows that $\pounds_{\ell'} q_{ab} = 0$ if $\pounds_\ell q_{ab} = 0$.

3.2. *Intrinsic geometry of non-expanding horizons*

Let us now discuss the restrictions on the Riemann curvature tensor for space-times in the presence of a NEH Δ. The Riemann tensor is defined by the condition $2\nabla_{[a}\nabla_{b]}X^c = -R_{abd}{}^c X^d$; the tensor R_{abcd} decomposes into a trace part determined by the Ricci tensor $R_{ab} = R_{acb}{}^c$ and a trace-free part C_{abcd} such that:

$$R_{abcd} = C_{abcd} + \frac{2}{D-2}\left(g_{a[c}R_{d]b} - g_{b[c}R_{d]a}\right) - \frac{2}{(D-1)(D-2)}R g_{a[c}g_{d]b}\,. \quad (14)$$

The tensor C_{abcd} is called the Weyl tensor. The Ricci tensor is determined by the matter fields through the Einstein–Maxwell equations. The remaining trace-free part of R_{abcd} therefore corresponds to the gravitational degrees of freedom.

Next, let us introduce a null basis adapted to Δ. This can be done by partially gauge-fixing the tetrad so that $(e_0^a + e_1^a)/\sqrt{2}$ is a null normal to Δ; we then define

$$\ell^a := \frac{1}{\sqrt{2}}(e_0^a + e_1^a), \qquad n^a := \frac{1}{\sqrt{2}}(e_0^a - e_1^a),$$
$$m^a := \frac{1}{\sqrt{2}}(e_2^a + ie_3^a), \qquad \overline{m}^a := \frac{1}{\sqrt{2}}(e_2^a - ie_3^a)\,. \quad (15)$$

These are all null, and satisfy the usual normalizations $\ell^a n_a = 1, m^a\overline{m}_a = 1$ for a complex Newman–Penrose tetrad. In terms of this basis, the Riemann tensor can be decomposed into 15 scalar quantities:

$$\begin{aligned}
&\Psi_0 = C_{abcd}\ell^a m^b \ell^c m^d, && \Phi_{00} = \tfrac{1}{2}R_{ab}\ell^a\ell^b, && \Phi_{12} = \tfrac{1}{2}R_{ab}m^a n^b,\\
&\Psi_1 = C_{abcd}\ell^a m^b \ell^c n^d, && \Phi_{01} = \tfrac{1}{2}R_{ab}\ell^a m^b, && \Phi_{20} = \tfrac{1}{2}R_{ab}\overline{m}^a\overline{m}^b,\\
&\Psi_2 = C_{abcd}\ell^a m^b \overline{m}^c n^d, && \Phi_{02} = \tfrac{1}{2}R_{ab}m^a m^b, && \Phi_{21} = \tfrac{1}{2}R_{ab}\overline{m}^a n^b,\\
&\Psi_3 = C_{abcd}\ell^a n^b \overline{m}^c n^d, && \Phi_{10} = \tfrac{1}{2}R_{ab}\ell^a\overline{m}^b, && \Phi_{22} = \tfrac{1}{2}R_{ab}n^a n^b,\\
&\Psi_4 = C_{abcd}\overline{m}^a n^b \overline{m}^c n^d, && \Phi_{11} = \tfrac{1}{4}R_{ab}(\ell^a n^b + m^a\overline{m}^b), && \Lambda = \tfrac{R}{24}\,.
\end{aligned}$$

The four real scalars $(\Phi_{00}, \Phi_{11}, \Phi_{22}, \lambda)$ and three complex scalars $(\Phi_{10}, \Phi_{20}, \Phi_{21})$ correspond to the Ricci tensor, and the five complex scalars $(\Psi_0, \Psi_1, \Psi_2, \Psi_3, \Psi_4)$ correspond to the Weyl tensor. Of these, Φ_{00}, Φ_{01}, Φ_{10}, Φ_{02}, Φ_{20}, Ψ_0 and Ψ_1 are all identically zero at Δ by the boundary conditions and the Raychaudhuri equation; in particular, the vanishing of Ψ_0 and Ψ_1 imply that the Weyl tensor is algebraically special at the horizon. The remaining components of R_{abcd} are unconstrained at Δ. Furthermore, one can show that Ψ_2 is independent of the null tetrad provided that ℓ is a null normal to Δ.

The above properties of a NEH furthermore ensure that the space-time derivative operator ∇_a induces a natural *intrinsic* derivative operator on Δ. On a space-like or time-like hypersurface, an intrinsic derivative operator is *always* induced by ∇_a, but on a null hypersurface this is not always the case. Recall that a covariant derivative operator $\mathcal{D}$ intrinsic to a manifold Σ can be specified by giving a map from each pair of vector fields X, Y on Σ to a vector field $\mathcal{D}_X Y \equiv X^b \mathcal{D}_b Y^a$, also on Σ, satisfying the axioms

(1) $\mathcal{D}_Z(X+Y) = D_Z X + D_Z Y$,
(2) $\mathcal{D}_Z(fX) = f D_Z X + X \pounds_Z f$.

The action of $\mathcal{D}$ on tensors of other types is then determined by linearity $\mathcal{D}_X(T^{\mathcal{A}} + S^{\mathcal{A}}) = \mathcal{D}_X T^{\mathcal{A}} + \mathcal{D}_X S^{\mathcal{A}}$ and the Leibnitz rule $\mathcal{D}_X(T^{\mathcal{AC}} S_{\mathcal{BC}}) = (\mathcal{D}_X T^{\mathcal{AC}}) S_{\mathcal{BC}} + T^{\mathcal{AC}}(\mathcal{D}_X S_{\mathcal{BC}})$, where $\mathcal{A}, \mathcal{B}, \mathcal{C}$ each denote any combination of tensor indices.

Given a space-like or time-like hypersurface Σ, the derivative operator $\mathcal{D}$ induced by ∇_a is usually defined by

$$(\mathcal{D}_X Y)^a = h^a_b (\nabla_X Y)^b$$

where $h_a^b := \delta_a^b + s n^b n_a$ is the projector onto the tangent space of Σ, with n^a a unit normal to Σ satisfying $n^a n_a \equiv s = \pm 1$. On a *null* hypersurface such as Δ, however, no such canonical projector exists, and so the above standard prescription fails. In order for ∇_a to induce a natural derivative operator on Δ, one therefore needs the property that, if X and Y are vector fields tangent to Δ, then $\nabla_X Y$ is *already also* tangent to Δ, so that no projector is needed. For a NEH, remarkably, this is in fact the case: Given X and Y tangent to Δ, we have

$$\begin{aligned} Y^b \nabla_b (X^a \ell_a) &\hat{=} Y^b X^a \nabla_b \ell_a + Y^b \ell_a \nabla_b X^a \\ &\hat{=} Y^b X^a \underleftarrow{\nabla_b \ell_a} + \ell_a Y^b \nabla_b X^a . \end{aligned}$$

On Δ, however, $X^a \ell_a = 0$. In addition, $\underleftarrow{\nabla_{(a} \ell_{b)}} \hat{=} 0$ and ℓ is twist-free so that $\underleftarrow{\nabla_a \ell_b} \hat{=} 0$. Thus $\ell_a Y^b \nabla_b X^a \hat{=} 0$, so that $Y^b \nabla_b X^a$ is tangent to Δ, as required. It follows that ∇_a induces a well-defined derivative operator on Δ, which we denote by D_a. Because ∇_a is metric ($\nabla_a g_{bc} = 0$) and torsion-free, one deduces that D_a is also metric ($D_a q_{bc} = 0$) and torsion-free. However, because q_{ab} is degenerate, these two conditions do *not* uniquely determine D_a. Thus, in contrast to the situation on space-like or time-like hypersurfaces, the derivative operator D_a contains *more information* than is contained in q_{ab}.

On a NEH, one can show that the vanishing of the expansion, shear and twist of ℓ^a implies that certain components of D_a are reduced to a single intrinsic one-form ω_a, known as the induced normal connection:

$$\nabla_{\underleftarrow{a}} \ell_b \hat{=} D_a \ell_b \hat{=} \omega_a \ell_b . \tag{16}$$

This quantity is gauge-dependent under rescalings of the null normal on Δ. Under transformations $\ell \to \ell' = f\ell$ with f some smooth function, we find that

$$\begin{aligned} D_a \ell'_b &\mathrel{\hat{=}} f D_a \ell_b + (D_a f)\ell_b \\ &\mathrel{\hat{=}} \omega_a \ell'_b + \frac{(D_a f)}{f}\ell'_b \\ &\mathrel{\hat{=}} (\omega_a + D_a \ln f)\ell'_b \,. \end{aligned}$$

The induced normal connection ω'_a associated to ℓ'_a is thus

$$\omega'_a = \omega_a + D_a \ln f \,. \tag{17}$$

From (16), an expression for the surface gravity $\kappa_{(\ell)}$ can be isolated. Contracting both sides of equation (16) with ℓ^a,

$$\ell^a \nabla_a \ell^b = (\ell^a \omega_a)\ell^b \,. \tag{18}$$

This equation is the geodesic equation for ℓ with non-zero acceleration. Therefore the surface gravity of Δ is

$$\kappa_{(\ell)} = \ell^a \omega_a \,. \tag{19}$$

This quantity is likewise gauge-dependent under rescalings of ℓ. For $\ell' = f\ell$, $\kappa_{(\ell)}$ transforms as

$$\kappa_{(\ell')} = f\ell^a(\omega_a + D_a \ln f) = f\kappa_{(\ell)} + \pounds_\ell f. \tag{20}$$

However, from (17), one sees that the curvature $d\omega$ of ω is *gauge-invariant.* In fact one can find an explicit expression for it. We have the following.

Proposition 1. *The curvature of the induced normal connection ω on Δ is given by*

$$d\omega \mathrel{\hat{=}} 2\,(\mathrm{Im}[\Psi_2])\,\tilde{\epsilon}\,, \tag{21}$$

with Ψ_2 the second Weyl scalar, and $\tilde{\epsilon} = im \wedge \overline{m}$ the area element on Δ.

Proof. We follow the proof that is presented in.[6] Recall the definition of curvature, $2\nabla_{[a}\nabla_{b]}X^c = -R_{abd}{}^c X^d$. Applied to the case $X^a = \ell^a$, one has $2\nabla_{[a}\nabla_{b]}\ell^c = -R_{abd}{}^c \ell^d$. Pulling back the a, b indices and using $\nabla_{\underleftarrow{a}} \ell_b \mathrel{\hat{=}} \omega_a \ell_b$ from (16), one obtains

$$2\ell^c D_{[a}\omega_{b]} \mathrel{\hat{=}} -R_{\underleftarrow{ab}d}{}^c \ell^d. \tag{22}$$

Furthermore, from (14) one has $R_{\underleftarrow{ab}d}{}^c \ell^d \mathrel{\hat{=}} C_{\underleftarrow{ab}d}{}^c \ell^d$ because $R_{\underleftarrow{ab}}\ell^b \mathrel{\hat{=}} 0$. Combining this with (22) and contracting with n_c gives $2D_{[a}\omega_{b]} \mathrel{\hat{=}} C_{\underleftarrow{ab}cd}\ell^c n^d$. The Weyl tensor can be expanded in terms of the complex scalars $\{\Psi\}$ as

$$\begin{aligned} C_{abcd}\ell^c n^d \mathrel{\hat{=}}\; & 4\,(\mathrm{Re}[\Psi_2])\, n_{[a}\ell_{b]} + 2\Psi_3 \ell_{[a}m_{b]} + 2\bar{\Psi}_3 \ell_{[a}\bar{m}_{b]} \\ & -2\Psi_1 n_{[a}\bar{m}_{b]} - 2\bar{\Psi}_1 n_{[a}m_{b]} + 4\mathrm{Im}[\Psi_2] m_{[a}\bar{m}_{b]} \,. \end{aligned} \tag{23}$$

Pulling back the a, b indices and substituting in the curvature, the expression (21) follows. □

3.3. *Isolated horizons*

It is clear from the transformation law (20) for surface gravity that, on a given NEH, the zeroth law of black hole mechanics cannot hold for all possible null normals. We now ask the question: for which, if any, null normals does it hold? The Cartan identity reads

$$\pounds_\ell \omega_b \,\hat{=}\, 2\ell^a D_{[a}\omega_{b]} + D_b(\ell^a \omega_a)\,.$$

Combining this with (21), we have

$$0 \,\hat{=}\, 4\ell^a \mathrm{Im}[\Psi_2] m_{[a}\bar{m}_{b]} \,\hat{=}\, \pounds_\ell \omega_b - D_b(\ell^a \omega_a)\,, \tag{24}$$

so that $D_b \kappa_{(\ell)} \,\hat{=}\, \pounds_\ell \omega_b$. It follows that the surface gravity is constant over the entire NEH iff $\pounds_\ell \omega_b \,\hat{=}\, 0$, i.e., iff ω_a is in 'equilibrium' with respect to ℓ.

The condition $\pounds_\ell \omega_a \,\hat{=}\, 0$ can also be interpreted in terms of 'extrinsic curvature'. Strictly speaking, because Δ is null, it does not have an extrinsic curvature in the usual sense. Nevertheless, one can define an analogue of extrinsic curvature by using the same formula that is used for space-like surfaces involving the Levi–Civita derivative operator and normal to the surface:

$$K_a{}^b \equiv \nabla_{\underleftarrow{a}} \ell^b \,\hat{=}\, D_a \ell^b\,. \tag{25}$$

Note that, because Δ is a null surface, this analogue of extrinsic curvature has the curious property that it is fully determined by the *intrinsic* structures D_a and ℓ_a, so that the nomenclature "extrinsic geometry" is only appropriate by analogy, and should not be taken too literally. From (25), one sees that, on a NEH, fixing the extrinsic geometry of Δ to be time independent is equivalent to fixing the induced normal connection ω_a of Δ to be time independent, which in turn, as we saw above, is the necessary and sufficient condition for the (gravitational) zeroth law to hold.

Note furthermore that if this condition holds for a single null normal ℓ, then it will hold for all other null normals $\ell' = c\ell$ related by a *constant* rescaling. That is, if we wish the zeroth law to hold, it is sufficient to restrict to an *equivalence class* of null normals, where two normals are equivalent if they are related by a constant rescaling. Note the similarity to Killing horizons, where one has the freedom to rescale the Killing field ξ^a only by a *constant*.

In Einstein–Maxwell theory, one can establish a similar zeroth law for the electric potential $\Phi(\ell) := \ell^a \boldsymbol{A}_a$ as follows. First, the energy condition imposed in Definition 1 implies that the Maxwell field satisfies $\underleftarrow{\ell \lrcorner \boldsymbol{F}} \,\hat{=}\, 0$.[a] Using this, together with the Cartan identity and the Bianchi identity ($d\boldsymbol{F} = 0$), this implies $\pounds_\ell \underleftarrow{\boldsymbol{F}} = d(\underleftarrow{\ell \lrcorner \boldsymbol{F}}) + \underleftarrow{\ell \lrcorner d\boldsymbol{F}} \,\hat{=}\, 0$. It follows that the electromagnetic potential $\boldsymbol{A}$ can be partially gauge-fixed such that $\pounds_\ell \underleftarrow{\boldsymbol{A}} \,\hat{=}\, 0$. This is referred to as a *gauge adapted to the horizon*. When this is satisfied, it follows that $0 = \pounds_\ell \underleftarrow{\boldsymbol{A}} = d(\underleftarrow{\ell \lrcorner \boldsymbol{A}}) + \underleftarrow{\ell \lrcorner d\boldsymbol{A}} = \underleftarrow{d\Phi_{(\ell)}}$,

[a]Here and throughout this chapter $\lrcorner$ denotes contraction of a vector with the first index of a form.

where we used the condition $\underleftarrow{\ell \lrcorner \boldsymbol{F}} \,\hat{=}\, 0$. It follows that the electric potential is constant over the entire NEH, so that the zeroth law also holds for the electric potential $\Phi(\ell)$. The above observations lead to the following definition.

Definition 2. (Weakly Isolated Horizon). *A NEH Δ equipped with an equivalence class $[\ell]$ of future-directed null normals, with $\ell' \sim \ell$ if $\ell' = c\ell$ ($c > 0$ a constant), such that the $\pounds_\ell \omega_a \,\hat{=}\, 0$ and $\pounds_\ell \underleftarrow{\boldsymbol{A}} = 0$ for all $\ell \in [\ell]$, is said to be a weakly isolated horizon (WIH).*

Note that because c is now a constant, ω_a is uniquely determined by the equivalence class $[\ell]$. Under the rescaling $\ell' = c\ell$, the surface gravity $\kappa_{(\ell)}$ transforms as $\kappa_{\ell'} = c\kappa_{(\ell)}$. However, the condition $\kappa_{(\ell)} = 0$ is independent of the choice of $\ell \in [\ell]$. If $\kappa_{(\ell)} = 0$, we say the WIH is *extremal*, in analogy with the nomenclature for Killing horizons. When discussing a WIH, it is convenient at this point to strengthen the partial gauge-fixing of the tetrad so that $\ell^a \in [\ell]$, and additionally $dn = 0$. Via the Cartan identity, the latter condition implies $\pounds_\ell n_a = 0$.

Let us now ask: Given a NEH, under what conditions does there exist an equivalence class $[\ell]$ of null normals such that it becomes a WIH? The answer to this question turns out to be *always*.[8] Thus, as far as geometry is concerned, the WIH definition is not more restrictive than the NEH definition. Rather, the importance of the WIH definition, as compared to the NEH definition, lies in the selection of an equivalence class $[\ell]$.

There is also a stronger notion of isolated horizon that one can introduce. On a WIH, while the normal connection ω_a on a WIH is time independent, the other components of the connection D_a can in general vary. Thus, a stronger condition would be to impose that the *entire* connection D_a on Δ possess ℓ^a as a symmetry. This condition is equivalent to $[\pounds_\ell, D_a] = 0$. By contrast, the condition $\pounds_\ell \omega_a = 0$ in Definition 2 is equivalent to the weaker condition $[\pounds_\ell, D_a]\ell^b = 0$. We have the following.

Definition 3. (Strongly Isolated Horizon). *A NEH Δ equipped with an equivalence class $[\ell]$ of future-directed null normals such that $[\pounds_\ell, D_a]X^b \,\hat{=}\, 0$ for every X^b tangent to Δ is said to be a strongly isolated horizon (SIH).*

In particular, every Killing horizon is also a SIH because all of the geometry at Δ (including the connection) possesses ℓ^a as a symmetry.

Given a NEH, the above stronger condition cannot always be met by simply choosing the equivalence class $[\ell]$ appropriately — to be a SIH involves a genuinely stronger restriction on the geometry of the horizon. When an equivalence class $[\ell]$ does exist making a NEH into a SIH, it is furthermore *unique*.[8] The sole remaining ambiguity in the choice of ℓ^a is then to rescale it by a constant, similar to the ambiguity to rescale by a constant the Killing vector field in a stationary space-time. In the case of a stationary space-time, this rescaling freedom can be eliminated by requiring the Killing vector field to be unit at spatial infinity. As we shall see in the next section, for a SIH in the context of Einstein-Maxwell theory, one can similarly remove the remaining constant rescaling ambiguity in ℓ^a, but this time in a way that is purely *quasi-local.* For the moment, we leave the freedom intact.

For most of the rest of this chapter, the results we consider will require only weakly isolated horizons. For this reason, henceforth, the term 'isolated horizon'(IH), when not otherwise qualified, shall refer specifically to a *weakly* isolated horizon.

3.4. *The topology of strongly isolated horizons*

Let us make a couple of further definitions. If the expansion of n^a is negative on Δ, $\theta_n := q^{ab}\nabla_a n_b < 0$, let us call Δ a *future isolated horizon.* In this case Δ describes a black hole horizon and not a white hole horizon. Furthermore, recall that, although $\kappa_{(\ell)}$ in general depends on the choice of ℓ in the equivalence class $[\ell]$, the *sign* of $\kappa_{(\ell)}$ does not. We have already defined Δ to be *extremal* if $\kappa_{(\ell)} = 0$. Furthermore, for future isolated horizons, if $\kappa_{(\ell)} > 0$, we call Δ *sub-extremal,* and if $\kappa_{(\ell)} < 0$ we call Δ *super-extremal.* We then have the following result:

Proposition 2. *Suppose Δ is a future SIH in a space-time with zero cosmological constant. If Δ is sub-extremal, then its cross-sections have 2-sphere topology. If Δ is extremal, then its cross-sections can have either 2-sphere or 2-torus topology, with the 2-torus topology occurring iff both $\tilde{\omega} = 0$ and $T_{ab}\ell^a n^b \,\hat{=}\, 0$.*

Proof.

Consider the evolution equation for the expansion of the auxiliary null normal n^a (see,[42] or equivalently (3.7) in[8]):

$$\pounds_\ell \theta_{(n)} + \kappa\theta_{(n)} + \frac{1}{2}\mathcal{R} = \mathcal{D}_a\tilde{\omega}^a + \|\tilde{\omega}\|^2 + (\Lambda - 8\pi G T_{ab}\ell^a n^b)\,. \tag{26}$$

Here, $\mathcal{R}$ is the scalar curvature of $\mathcal{S}$, $\mathcal{D}_a$ is the covariant derivative operator that is compatible with the metric $\tilde{q}_{ab} = g_{ab} + \ell_a n_b + \ell_b n_a$ on $\mathcal{S}$, and $\|\tilde{\omega}\|^2 = \tilde{\omega}_a\tilde{\omega}^a$ with $\tilde{\omega}_a = \tilde{q}_a{}^b\omega_b = \omega_a + \kappa_{(\ell)}n_a$ the projection of ω onto $\mathcal{S}$. From strong isolation, and $\pounds_\ell n = 0$, one has $\pounds_\ell\theta_{(n)} = \pounds_\ell q^{ab}D_a n_b = q^{ab}D_a\pounds_\ell n_b = 0$. Using this fact with $\Lambda = 0$ and $\theta_{(n)} < 0$, (26) becomes

$$-\kappa|\theta_{(n)}| = -\frac{1}{2}\mathcal{R} + \mathcal{D}_a\tilde{\omega}^a + \|\tilde{\omega}\|^2 + T_{ab}\ell^a n^b\,. \tag{27}$$

Integrating both sides over the surface $\mathcal{S}$, and finally using $\oint_{\mathcal{S}} \tilde{\epsilon}\mathcal{D}_a\tilde{\omega}^a = 0$, one has

$$\oint_{\mathcal{S}} \tilde{\epsilon}\mathcal{R} \geq 2\oint_{\mathcal{S}} \tilde{\epsilon}(T_{ab}\ell^a n^b + \kappa|\theta_{(n)}| + \|\tilde{\omega}\|^2)\,. \tag{28}$$

The dominant energy condition requires that $T_{ab}\ell^a n^b \geq 0$. In addition, because we have excluded the super-extremal case, the second term is non-negative. Lastly, $\|\tilde{\omega}\|^2$ is manifestly non-negative. It follows that $\oint_{\mathcal{S}} \tilde{\epsilon}\mathcal{R} \geq 0$ with equality iff $\kappa_{(\ell)} = 0$, $T_{ab}\ell^a n^b = 0$ and $\tilde{\omega} = 0$. On the other hand, the Gauss–Bonnet theorem gives

$$\oint_{\mathcal{S}} \tilde{\epsilon}\mathcal{R} = 8\pi(1-g)$$

where g is the genus of $\mathcal{S}$, so that $\oint_{\mathcal{S}} \tilde{\epsilon}\mathcal{R} > 0$ iff $\mathcal{S}$ is a 2-sphere, and $\oint_{\mathcal{S}} \tilde{\epsilon}\mathcal{R} = 0$ iff $\mathcal{S}$ is a 2-torus. Thus, if Δ is sub-extremal ($\kappa_{(\ell)} > 0$), one has $\oint_{\mathcal{S}} \tilde{\epsilon}\mathcal{R} > 0$, so that $\mathcal{S}$ is a 2-sphere, whereas if Δ is extremal, both 2-sphere and 2-torus topologies are possible, with the latter occurring iff $T_{ab}\ell^a n^b = 0$ and $\tilde{\omega} = 0$. □

3.5. *Existence of Killing spinors*

The extremal Kerr–Newman black hole is a solution to the $N = 2$ supergravity field equations with the fermion fields set to zero. The condition for this solution to have positive energy is that[43]

$$\mathfrak{M} = |\mathfrak{Q}| , \tag{29}$$

relating the mass $\mathfrak{M}$ and total charge $\mathfrak{Q} \equiv \sqrt{q_e^2 + q_m^2}$ (with q_e and q_m the electric and magnetic charges); this is the extremality condition for the Kerr-Newman black hole. This is also the saturated Bogomol'ny–Prasad–Sommerfeld (BPS) inequality.[43] This leads to an interesting question: If we try to define a supersymmetric *isolated horizon*, do we have a similar restriction to extremality?

Generally, Einstein–Maxwell theory (in four dimensions) can be viewed as the bosonic sector of $N = 2$ supergravity — that is, the sector in which the spin-3/2 gravitino field and its complex conjugate vanish in vacuum. In this sector, the defining property of supersymmetric configurations is the existence of a *Killing spinor*, whence we look for a restricted space of solutions having this property. In the presence of zero cosmological constant, a Killing spinor can be defined as a (Dirac) spinor ζ satisfying

$$\left[\nabla_a + \frac{i}{4}\boldsymbol{F}_{bc}\gamma^{bc}\gamma_a\right]\zeta = 0 . \tag{30}$$

Here, γ^a are a set of gamma matrices that satisfy the usual anticommutation rule

$$\gamma^a\gamma^b + \gamma^b\gamma^a = 2g^{ab} \tag{31}$$

and the antisymmetry product

$$\gamma_{abcd} = \epsilon_{abcd} . \tag{32}$$

$\gamma_{a_1 \ldots a_D}$ denotes the antisymmetrized product of D gamma matrices. The complex conjugate $\bar{\zeta}$ of ζ is defined as

$$\bar{\zeta} = i(\zeta)^\dagger\gamma_0 ; \tag{33}$$

with † denoting Hermitian conjugation.

From ζ and $\bar{\zeta}$ one can construct five (real) bosonic bilinear covariants

$$f = \bar{\zeta}\zeta , \quad g = i\bar{\zeta}\gamma^5\zeta , \quad V^a = \bar{\zeta}\gamma^a\zeta , \quad W^a = i\bar{\zeta}\gamma^5\gamma^a\zeta , \quad \Psi^{ab} = \bar{\zeta}\gamma^{ab}\zeta . \tag{34}$$

These bilinear covariants are all related to each other via several algebraic conditions (from the Fierz identity) and differential equations (from the Killing spinor equation

(30)). In particular, the vector V satisfies the equations

$$V_a V^a = -(f^2 + g^2)\,, \tag{35}$$

$$\nabla_a V_b = -f\boldsymbol{F}_{ab} + \frac{g}{2}\epsilon_{abcd}\boldsymbol{F}^{cd}\,. \tag{36}$$

The above reviews the standard notion of a Killing spinor. What we now wish to consider is a Killing spinor which exists *on the horizon* Δ *only* (which we assume to be strongly isolated). Then, in place of (30), we can at most impose a version with the derivative pulled back to Δ:

$$\left[\nabla_{\underleftarrow{a}} + \frac{i}{4}\boldsymbol{F}_{bc}\gamma^{bc}\gamma_{\underleftarrow{a}}\right]\zeta = 0\,. \tag{37}$$

which, in place of (36), leads to

$$\nabla_{\underleftarrow{a}} V_b = -f\boldsymbol{F}_{\underleftarrow{a}b} + \frac{g}{2}\epsilon_{\underleftarrow{a}bcd}\boldsymbol{F}^{cd}.$$

If one additionally pulls back b to Δ and symmetrizes, one sees that

$$\pounds_V q_{ab} = 2\underleftarrow{\nabla_{(a}V_{b)}} = 0,$$

which is the same as the condition (13) satisfied by ℓ on Δ as a NEH. If we furthermore stipulate that V be *equal* to a null normal in the equivalence class $[\ell]$ making Δ a SIH, we have the notion of a *supersymmetric isolated horizon.*

Definition 4. (Supersymmetric Isolated Horizon). *A SIH Δ equipped with a Killing spinor ζ and complex conjugate $\bar{\zeta} = i(\zeta)^\dagger\gamma_0$, such that $V := \bar{\zeta}\gamma^a\zeta \in [\ell]$, is said to be a supersymmetric isolated horizon (SSIH).*

For SSIHs, the following can be proved.

Proposition 3. *An SSIH of Einstein–Maxwell theory with zero cosmological constant is necessarily extremal and 'non-rotating' — that is, has zero gravitational angular momentum as defined in equation (55) below.*

Proof. For a SSIH, V^a and ℓ^a are identified. Hence, at Δ, V^a is null so that from (35) we have $f = g = 0$. Equation (16) together with (36) then gives

$$\nabla_{\underleftarrow{a}}\ell_b = \omega_a\ell_b = 0\,, \tag{38}$$

so that $\omega \,\hat{=}\, 0$. This condition implies that the gravitational angular momentum, defined in equation (55) below, is identically zero. However, in general $\boldsymbol{A}$ is non-zero, which means that there may be non-zero angular momentum stored in the electromagnetic fields. The condition also implies that $\kappa_{(\ell)} = \ell\lrcorner\omega = 0$. Therefore, *SSIHs are extremal and non-rotating.* □

4. Hamiltonian Mechanics

Up until now we have been considering geometric properties of isolated horizons. At this point we show how the isolated horizon boundary conditions lead to a *consistent variational and canonical framework.* Specifically, we start with an action principle, derive from this the covariant phase space, and then finally discuss the definition of angular momentum and mass of isolated horizons as generators of rotations and time translations. This will lead directly to a derivation of the first law of black hole mechanics involving only quantities which are *quasi-locally defined* in terms of fields intrinsic to the horizon.

For conceptual clarity, in this presentation we focus on Einstein–Maxwell theory. Furthermore, we use the Palatini formulation of gravity in terms of a co-tetrad e_a^I and associated internal Lorentz connection A_a^{IJ}, where $I, J = 0, 1, 2, 3$ are internal indices. This will greatly simplify the necessary formulae, and allow an easier transition to the brief discussion on quantum theory at the end of the chapter. In this formulation, the space-time metric is constructed from the co-tetrad as $g_{ab} = \eta_{IJ} e_a^I e_b^J$ where $\eta_{IJ} = \mathrm{diag}(-1,1,1,1)$ is the internal Lorentz metric, and the curvature F_{ab}^{IJ} of A_a^{IJ} determines the Riemann tensor via $F_{ab}^{IJ} = \left(dA^{IJ} + A^I{}_K \wedge A^K{}_J\right)_{ab} = R_{abc}{}^d e^{cI} e_d^J$. Internal indices are raised and lowered with η^{IJ}, η_{IJ}. We denote the Maxwell vector potential by $\boldsymbol{A}$, so that the field strength is $\boldsymbol{F} = d\boldsymbol{A}$.

4.1. *Action principle*

Let us consider the action for Einstein–Maxwell theory in the Palatini formulation on a four-dimensional manifold $\mathcal{M}$ with boundary $\partial\mathcal{M} \cong M_1 \cup M_2 \cup \Delta \cup \mathcal{I}$. Here $\mathcal{I}$ represents the two-sphere at spatial infinity crossed with time, where appropriate asymptotically flat boundary conditions are imposed. Δ is a three-dimensional manifold with topology $S^2 \times [0,1]$ constrained to be an isolated horizon equipped with a fixed equivalence class of null normals $[\ell_a]$, with all equations of motion holding at Δ. M_1, M_2 play the role of partial Cauchy surfaces. As before, the isolated horizon boundary conditions are understood to include the requirement that the Maxwell potential be in a gauge adapted to the horizon, $\pounds_\ell \underleftarrow{\boldsymbol{A}} = 0$. The space-time region $\mathcal{M}$ thus described is shown in Figure 2.

The action is then given by

$$S = \frac{1}{16\pi G}\int_{\mathcal{M}} \Sigma_{IJ} \wedge F^{IJ} - \frac{1}{16\pi G}\int_{\mathcal{I}} \Sigma_{IJ} \wedge A^{IJ} - \frac{1}{8\pi}\int_{\mathcal{M}} \boldsymbol{F} \wedge \star \boldsymbol{F} \,. \tag{39}$$

where $\boldsymbol{F} = d\boldsymbol{A}$ is the field strength of the Maxwell field, $\star$ denotes Hodge dual, and $\Sigma^{IJ} := \frac{1}{2}\epsilon^{IJ}{}_{KL} e^K \wedge e^L$ with ϵ_{IJKL} the internal alternating tensor. The boundary term at $\mathcal{I}$ is necessary to make the action differentiable. By contrast, as we will show below, *no* term at Δ is necessary for the action to be differentiable.[44–46]

Let us denote the dynamical variables $(e, A, \boldsymbol{A})$ collectively by Ψ. Application of an arbitrary variation δ to the action then yields the form

$$\delta S = \int_{\mathcal{M}} E[\Psi] \cdot \delta\Psi - \int_{\partial\mathcal{M}} \theta(\delta)[\Psi] \,. \tag{40}$$

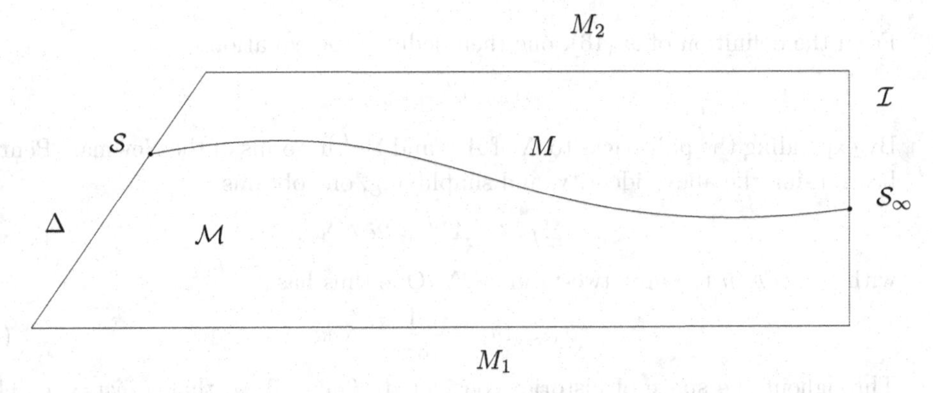

Figure 2. The region of space-time $\mathcal{M}$ considered has an internal boundary Δ which will be the isolated horizon, and is bounded to the past and future by two three-dimensional Cauchy surfaces M_1 and M_2. M is a partial Cauchy surface that intersects Δ and $\mathcal{I}$ each in a two-sphere.

Here $E[\Psi] = 0$ denotes the equations of motion. Specifically, these are:

$$\epsilon_{IJKL} e^J \wedge F^{KL} = T_I \,, \tag{41}$$

$$\mathrm{d}\Sigma_{IJ} - A_I{}^K \wedge \Sigma_{KJ} - A_J{}^K \wedge \Sigma_{IK} = 0 \,, \tag{42}$$

$$\mathrm{d} \star \boldsymbol{F} = 0 \,. \tag{43}$$

where in the first of these equations, T_I denotes the electromagnetic stress-energy three-form. The second equation imposes that A_a^{IJ} be the unique torsion-free connection compatible with e_a^I. Together the above equations are equivalent to the Einstein–Maxwell equations in metric variables, with the components of T_I identified with appropriate components the electromagnetic stress-energy tensor.

The integrand $\theta(\delta)$ of the surface term in (40) is given by

$$\theta(\delta) = \theta_{\mathrm{Grav}}(\delta) + \theta_{\mathrm{EM}}(\theta) \,, \tag{44}$$

where we have defined $\theta_{\mathrm{Grav}}(\delta) := (1/16\pi G)\Sigma_{IJ} \wedge \delta A^{IJ}$ and $\theta_{\mathrm{EM}}(\delta) := (1/16\pi G) \star \boldsymbol{F} \wedge \delta \boldsymbol{A}$, and where we suppress explicit representation of the dependence on $\Psi = (e, A, \boldsymbol{A})$.

The action S is said to be differentiable if, when the configuration fields $e, A, \boldsymbol{A}$ are fixed on M_1 and M_2, the boundary term in (40) vanishes. Let us show that this is the case. Because A and $\boldsymbol{A}$ are held fixed at M_1, M_2, $\theta(\delta)$ vanishes there. In addition, the boundary term in the action (39) is constructed precisely such that the boundary terms at $\mathcal{I}$ coming from the variation of the action cancel. Therefore it remains only to show that the integral of $\theta(\delta)$ over Δ vanishes.

To see that this is true, let us define an *internal* Newman–Penrose basis $\ell^I = (1,1,0,0)/\sqrt{2}$, $n^I = (1,-1,0,0)/\sqrt{2}$, $m^I = (0,0,1,i)/\sqrt{2}$, $\overline{m} = (0,0,1,-i)/\sqrt{2}$. The relations (15) then become

$$\ell^a = e^a_I \ell^I, \; n^a = e^a_I n^I, \;\; m^a = e^a_I m^I, \;\;\; \overline{m}^a = e^a_I \overline{m}^I.$$

From the definition of ω (16), one then deduces the equation

$$A_{\underleftarrow{a}\,IJ}\ell^J \mathrel{\hat{=}} \omega_a \ell_I\,.$$

By expanding the pull-backs to Δ of A^{IJ} and Σ^{IJ} in terms of the Newman–Penrose basis, using the above identity, and simplifying, one obtains

$$\underleftarrow{\Sigma}_{IJ} \wedge \delta \underleftarrow{A}^{IJ} \mathrel{\hat{=}} 2\tilde{\epsilon} \wedge \delta\omega\,,$$

with $\tilde{\epsilon} = m \wedge \overline{m}$ the area two-form on Δ. One thus has

$$\underleftarrow{\theta}_{\mathrm{Grav}}(\delta) \mathrel{\hat{=}} \frac{1}{16\pi G}\tilde{\epsilon} \wedge \delta\omega\,. \tag{45}$$

Throughout the space of histories considered, $\pounds_\ell \omega = 0$, so that $\pounds_\ell \delta\omega = 0$. This, combined with the fact that A^{IJ}_a, e^I_a, and hence ω, is fixed on $\mathcal{S}_1 := M_1 \cap \Delta$ and $\mathcal{S}_2 := M_2 \cap \Delta$, implies $\delta\omega = 0$ on all of Δ, so that $\underleftarrow{\theta}_{\mathrm{Grav}}(\delta) = 0$. The argument for the electromagnetic part $\underleftarrow{\theta}_{\mathrm{EM}}(\delta)$ is similar: Because $\boldsymbol{A}$ is in a gauge adapted to the horizon, so that $\pounds_\ell \underleftarrow{\boldsymbol{A}} = 0$ throughout the space of histories, we have $\pounds_\ell \delta \underleftarrow{\boldsymbol{A}} = 0$. This combined with the fact that $\boldsymbol{A}$ is fixed on $\mathcal{S}_1$ and $\mathcal{S}_2$ implies $\delta \underleftarrow{\boldsymbol{A}} = 0$, implying $\underleftarrow{\theta}_{\mathrm{EM}}(\delta) = 0$. This shows that, when the configuration variables $(e, A, \boldsymbol{A})$ are fixed on M_1 and M_2, the boundary term in the variation of the action (40) vanishes, so that $\delta S = 0$ implies the equations of motion, as required.

4.2. *Covariant phase space*

The covariant phase space of a theory is the space of solutions of the equations of motion. Because it is not formulated in terms of initial data, such a formulation of phase space enables a *space-time covariant* version of the canonical framework. Furthermore, in this approach, one can derive the symplectic structure directly through (anti-symmetrized) second variations of the action, as we shall review. The notion of the covariant phase space as the space of solutions, and the corresponding method of determining the symplectic structure can be traced back to the work of Lagrange himself (see[47,48]).

The covariant phase space $\boldsymbol{\Gamma}_{\mathrm{Cov}}$ and its symplectic structure Ω_{Cov} are directly related to the more standard canonical phase space $\boldsymbol{\Gamma}$ and its symplectic structure Ω, through the fact that the space of initial data is in one-to-one correspondence with the space of solutions. This isomorphism furthermore maps the covariant symplectic structure into the canonical symplectic structure, so that the two frameworks are completely equivalent.

In the present context, the situation is more subtle, for two reasons: First, general relativity is a *constrained* theory, and usually one works with the phase space of unconstrained initial data. By contrast, in the covariant phase space, all equations of motion are by definition satisfied. Second, the space-times under consideration are not globally hyperbolic, but rather only admit partial Cauchy surfaces, with inner boundary at the isolated horizon Δ. Consequently, solutions in the covariant phase space have more information than is present in the initial data on any given spatial hypersurface — solutions in the covariant phase space "know" what fell into the black hole in the past, whereas the initial data does not. Nevertheless, the two frameworks can be related. If we let $\overline{\mathbf{\Gamma}}$ denote the space of constrained initial data, one has a projector $\pi_M : \mathbf{\Gamma}_{\mathrm{Cov}} \to \overline{\mathbf{\Gamma}}$ mapping a given solution to the initial data which it induces on a given fixed hypersurface M. Using this projector, and the inclusion map $\iota : \overline{\mathbf{\Gamma}} \hookrightarrow \mathbf{\Gamma}$, one can show that

$$(\iota \circ \pi_M)^* \boldsymbol{\Omega} = \boldsymbol{\Omega}_{\mathrm{Cov}}$$

so that the symplectic structure on the usual unconstrained phase space of initial data is exactly mapped into that on the covariant phase space computed using the second variations of the action.

In the following we will use the covariant phase space framework. This will not only make the space-time geometry more transparent, but will also simplify many calculations. More specifically, let $\mathbf{\Gamma}_{\mathrm{IH}}$ denote *the covariant phase space of possible fields* $(e, A, \boldsymbol{A})$ *on* $\mathcal{M}$ *(1.) satisfying the Einstein–Maxwell equations, (2.) possessing appropriate asymptotically flat fall off conditions at infinity, and (3.) such that* Δ *is an isolated horizon with fixed equivalence class of null normals* $[\ell]$.

4.3. *Symplectic structure on* $\mathbf{\Gamma}_{\mathrm{IH}}$

The boundary term $\theta(\delta)$ in the variation of the action (40) also determines the symplectic structure. Though $\theta(\delta)$ is a 3-form on space-time, as it is a linear function of a single variation δ, it is also a *1-form on phase space.* One can therefore take its exterior derivative to obtain a *2-form* on phase space, called the *symplectic current*, $\omega := \mathbb{d}\boldsymbol{\theta}$, where $\mathbb{d}$ denotes the exterior derivative on (the infinite dimensional space) $\mathbf{\Gamma}_{\mathrm{IH}}$. Explicitly, one obtains

$$\boldsymbol{\omega}(\delta_1, \delta_2) = \frac{1}{16\pi G}\left[\delta_1 A^{IJ} \wedge \delta_2 \Sigma_{IJ} - \delta_2 A^{IJ} \wedge \delta_1 \Sigma_{IJ}\right] + \frac{1}{4\pi}\left[\delta_1 \boldsymbol{A} \wedge \delta_2(\star \boldsymbol{F}) - \delta_2 \boldsymbol{A} \wedge \delta_1(\star \boldsymbol{F})\right] . \tag{46}$$

The symplectic current has the property that for any two variations δ_1, δ_2 tangent to the space of solutions, $\boldsymbol{\omega}(\delta_1, \delta_2)$ is closed: $d\boldsymbol{\omega}(\delta_1, \delta_2) = 0$. Normally one would then define the symplectic structure for a given Cauchy hypersurface M to be

$$\boldsymbol{\Omega}^B_M(\delta_1, \delta_2) := \int_M \boldsymbol{\omega}(\delta_1, \delta_2) .$$

However, in the present case, because of the presence of the inner boundary Δ, this symplectic structure is not conserved in time. That is: for two different partial Cauchy surfaces M_1, M_2, $\boldsymbol{\Omega}^B_{M_1} \neq \boldsymbol{\Omega}^B_{M_2}$. The reason for this can be seen in figure 3: symplectic current is escaping across the horizon Δ. More precisely, because the symplectic current is closed ($d\boldsymbol{\omega}(\delta_1, \delta_2) = 0$), Stokes theorem implies $\oint_{\partial \mathcal{M}} \boldsymbol{\omega}(\delta_1, \delta_2) = 0$, so that

$$\int_{M_2} \boldsymbol{\omega}(\delta_1, \delta_2) = \int_{M_1} \boldsymbol{\omega}(\delta_1, \delta_2) - \int_{\Delta} \boldsymbol{\omega}(\delta_1, \delta_2) .$$

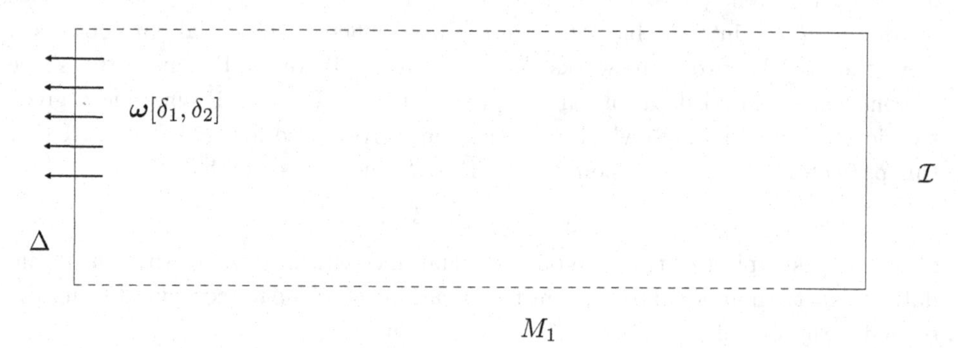

Figure 3. Symplectic current escaping across the horizon.

(The part of the boundary integral at spatial infinity $\mathcal{I}$ vanishes due to the imposition of fall-off conditions.) The solution to this problem is to use the IH boundary conditions to rewrite the integral over Δ as an integral over the two-sphere intersections $\mathcal{S}_1$ and $\mathcal{S}_2$ of Δ with M_1 and M_2:

$$\int_\Delta \boldsymbol{\omega}(\delta_1, \delta_2) = \left(\oint_{\mathcal{S}_2} - \oint_{\mathcal{S}_1}\right) \boldsymbol{\lambda}(\delta_1, \delta_2) \tag{47}$$

for some two-form $\boldsymbol{\lambda}(\delta_1, \delta_2)$ on Δ. If one then defines the full symplectic structure on a spatial slice M to be

$$\boldsymbol{\Omega}_M(\delta_2, \delta_1) = \int_M \boldsymbol{\omega}(\delta_1, \delta_2) + \oint_{\mathcal{S}} \boldsymbol{\lambda}(\delta_1, \delta_2)\,,$$

with $\mathcal{S} = M \cap \Delta$, then conservation of the symplectic structure is restored, giving $\boldsymbol{\Omega}_{M_1}(\delta_2, \delta_2) = \boldsymbol{\Omega}_{M_2}(\delta_1, \delta_2)$.

To carry this out for the present case, one defines potentials ψ and χ for the surface gravity $\kappa_{(\ell)}$ and electric potential $\Phi_{(\ell)}$ such that

$$\pounds_\ell \psi \mathrel{\widehat{=}} \ell \lrcorner \omega = \kappa_{(\ell)} \quad \text{and} \quad \pounds_\ell \chi \mathrel{\widehat{=}} \ell \lrcorner \boldsymbol{A} = -\Phi_{(\ell)}\,. \tag{48}$$

The pullback to Δ of the symplectic current will then be exact,[6] so that, using Stokes theorem, the integral over Δ decomposes as in (47). The final symplectic structure turns out to be

$$\begin{aligned}\boldsymbol{\Omega}_{\mathrm{Cov}}(\delta_1, \delta_2) = {} & \frac{1}{16\pi G}\int_M [\delta_1 A^{IJ} \wedge \delta_2 \Sigma_{IJ} - \delta_2 A^{IJ} \wedge \delta_1 \Sigma_{IJ}] \\ & + \frac{1}{4\pi}\int_M [\delta_1 \boldsymbol{A} \wedge \delta_2(\star \boldsymbol{F}) - \delta_2 \boldsymbol{A} \wedge \delta_1(\star \boldsymbol{F})] \\ & - \frac{1}{8\pi G}\oint_{\mathcal{S}} [\delta_1 \psi \delta_2 \tilde{\epsilon} - \delta_2 \psi \delta_1 \tilde{\epsilon}] - \frac{1}{4\pi}\oint_{\mathcal{S}} [\delta_1 \chi \delta_2(\star \boldsymbol{F}) - \delta_2 \chi \delta_1(\star \boldsymbol{F})]\,.\end{aligned} \tag{49}$$

4.3.1. *Symmetry classes and the phase space of rigidly rotating horizons*

For the purposes of black hole mechanics and the quantum theory of black holes, it is useful to categorize different black hole geometries according to their symmetry groups. A *symmetry* of an isolated horizon is an infinitesimal diffeomorphism on Δ which preserves q_{ab} and D, and at most rescales ℓ by a positive constant. (We restrict consideration to infinitesimal symmetries because the symmetry vector fields we consider will not always be complete.) By definition of an isolated horizon, diffeomorphisms generated by ℓ are symmetries, so that the symmetry group G_Δ of Δ is at least 1-dimensional. But beyond this, the group of symmetries of an isolated horizon, unlike the symmetries of the asymptotic-flat metric at spatial infinity, is not universal. The symmetry group can be classified according to its dimension into three categories:

(1) Type I: (q, D) is spherically symmetric, and G_Δ is four-dimensional.
(2) Type II: (q, D) is axisymmetric, and G_Δ is two-dimensional.
(3) Type III: (q, D) has no symmetry other than ℓ, and G_Δ is one-dimensional.

Note that these symmetries refer *only* to geometric structures intrinsic to the horizon. No assumption is made about symmetries of fields outside the horizon, even in an arbitrarily small neighborhood of Δ.

Type II horizons are the most interesting ones, because they allow rotation and distortion, but still have enough structure to admit a clear notion of (quasi-local) angular momentum, as we shall see. They include the Kerr–Newman family of horizons as well as generalizations possessing distortion due to external matter, other black holes, or even 'hair' due to matter such as Yang–Mills for which the black hole uniqueness theorem does not apply. In the rest of this section, we will focus on the type II case.

Let us fix an axial vector field ϕ on Δ such that it commutes with any one and hence all members of $[\ell]$. Define $\mathbf{\Gamma}^\phi$ to be the set of all data $(e, A, \boldsymbol{A})$ in $\mathbf{\Gamma}_{\text{IH}}$ such that (1.) the induced q_{ab} and D on Δ possess ϕ as a symmetry: $\pounds_\phi q_{ab} = 0$, and $[\pounds_\phi, D_a] = 0$, and (2.) $\pounds_\phi \star \underleftarrow{\boldsymbol{F}} = \pounds_\phi \underleftarrow{\boldsymbol{F}} = 0$. Endow $\mathbf{\Gamma}^\phi$ with the pull-back, via inclusion, of the symplectic structure on $\mathbf{\Gamma}_{\text{IH}}$ derived above — we denote the resulting symplectic structure by $\mathbf{\Omega}^\phi$. The resulting phase space $(\mathbf{\Gamma}^\phi, \mathbf{\Omega}^\phi)$ is called the *phase space of rigidly rotating horizons*. This will form the basis of the discussion for the rest of this section. Furthermore, whenever a partial Cauchy surface M is used in the following, we assume it is chosen such that its intersection S with Δ is everywhere tangent to ϕ.

4.4. *Conserved charges and the first law*

4.4.1. *Symmetries and Hamiltonian flow*

In $\mathbf{\Gamma}^{\phi}$, all space-times possess as isolated horizon symmetries the fixed vector fields ϕ and ℓ. A general symmetry at Δ thus takes the form

$$W^a \,\hat{=}\, B_{(W,\ell)}\ell^a + A_{(W)}\phi^a \,, \tag{50}$$

where $B_{(W,\ell)}$ depends on which $\ell \in [\ell]$ is used. Both $B_{(W,\ell)}$ and $A_{(W)}$ are constants on Δ, but may be 'q-numbers' — i.e., may vary on *phase space.* Suppose we extend W^a arbitrarily to the rest of the space-time, such that at spatial infinity it approaches a symmetry of the asymptotic flat metric there. One can then ask: Is the flow generated by W^a Hamiltonian? Because the flow generated by W^a preserves the boundary conditions at Δ and at infinity involved in the definition of $\mathbf{\Gamma}^{\phi}$, there exists a corresponding well-defined variation δ_W on $\mathbf{\Gamma}^{\phi}$. The flow will be Hamiltonian if and only if there exists a Hamiltonian H on $\mathbf{\Gamma}^{\phi}$ such that $\delta H = \mathbf{\Omega}^{\phi}(\delta, \delta_W)$ for all variations δ. Using (49), explicitly this becomes

$$\begin{aligned}\delta H = \mathbf{\Omega}^{\phi}(\delta, \delta_W) = &\frac{-1}{8\pi G}\oint_{S} A_{(W)}\delta[(\phi\lrcorner\omega)\tilde{\epsilon}] + \kappa_{(W)}\delta(\tilde{\epsilon}) \\ &-\frac{1}{4\pi}\oint_{S} A_{(W)}\delta[(\phi\lrcorner\boldsymbol{A})\star\boldsymbol{F}] - \Phi_{(W)}\delta(\star\boldsymbol{F}) \\ &+\frac{1}{16\pi G}\oint_{S_\infty} \mathrm{tr}[\delta A\wedge(W\lrcorner\Sigma) + (W\lrcorner A)\delta\Sigma] \\ &-\frac{1}{4\pi}\oint_{S_\infty} \delta\boldsymbol{A}\wedge(W\lrcorner\star\boldsymbol{F}) + (W\lrcorner\boldsymbol{A})\delta(\star\boldsymbol{F})\,,\end{aligned} \tag{51}$$

where $\kappa_{(W)} = W^a\omega_a$ and $\Phi_{(W)} := -W^a\boldsymbol{A}_a$ respectively denote the surface gravity and electric potential 'in the frame defined by W'.

In the following, we will find such an H first for W a *pure rotation*, and then an appropriate *time translation.* This will lead to the definition of the horizon angular momentum, horizon mass, and a proof of a version of the first law of black hole mechanics involving *only quasi-locally defined quantities.* In passing, the definition of angular momentum multipoles and the associated mass multipoles will also be introduced.

4.4.2. *Horizon angular momentum as generator of rotations*

Angular momentum is the generator of rotations. In terms of equation (50), the case of spatial rotations corresponds to $B_{(W,\ell)} = 0$ and $A_{(W)} = 1$, in which case equation (51) reduces to

$$\mathbf{\Omega}^{\phi}(\delta, \delta_\phi) = -\frac{1}{8\pi G}\oint_{S}\delta[(\phi\lrcorner\omega)\tilde{\epsilon}] - \frac{1}{4\pi}\oint_{S}\delta[(\phi\lrcorner\boldsymbol{A})\star\boldsymbol{F}] - \delta J_{\mathrm{ADM}}\,,$$

so that the boundary integral at infinity is exact, equal to the variation of the ADM angular momentum. One can see that the terms associated with S also form an exact variation. If we set

$$J_\Delta := -\frac{1}{8\pi G}\oint_{\mathcal{S}}(\phi\lrcorner\omega)\tilde{\epsilon} - \frac{1}{4\pi}\oint_{\mathcal{S}}(\phi\lrcorner\boldsymbol{A})\star\boldsymbol{F}\,, \tag{52}$$

then

$$\boldsymbol{\Omega}^\phi(\delta,\delta_W) = \delta(J_\Delta - J_{\mathrm{ADM}})\,; \tag{53}$$

this means that $J_\Delta - J_{\mathrm{ADM}}$ is the Hamiltonian generating rotations in $\boldsymbol{\Gamma}^\phi$. Whereas J_{ADM} represents the angular momentum of the entire space-time, J_Δ can be interpreted as the angular momentum of the black hole itself, so that $J_{\mathrm{ADM}} - J_\Delta$ is then the angular momentum outside of the horizon. Note how the extension of ϕ^a in the bulk between Δ and infinity did not matter precisely because the expression for the symplectic structure consists only in boundary terms.

One can show that the gravitational contribution to the horizon angular momentum (that is, the first term in (52)) is equivalent to the Komar integral. To show that this statement is true, note that the normalization $\ell^a n_a = -1$ implies that the rotation one-form is given by $\omega_a = -n^b\nabla_a\ell_b = \ell^b\nabla_b n_a$. Substituting this into the first term in (52), integrating by parts and using the Killing property of ϕ, we have:

$$\begin{aligned} J_{\mathrm{Grav}} &= \frac{1}{8\pi G}\oint_{\mathcal{S}}(\phi\lrcorner\nabla_\ell n)\tilde{\epsilon} = -\frac{1}{8\pi G}\oint_{\mathcal{S}}(\nabla_\ell\phi\lrcorner n)\tilde{\epsilon} \\ &= -\frac{1}{16\pi G}\oint_{\mathcal{S}}(\ell\lrcorner d\phi)\lrcorner n\tilde{\epsilon} = -\frac{1}{8\pi G}\oint_{\mathcal{S}}\star d\phi\,. \end{aligned} \tag{54}$$

This is the Komar integral for the gravitational contribution to the angular momentum of Δ, evaluated at $\mathcal{S}$. Note that J_{Grav} is equivalent to the Komar integral even in the presence of Maxwell fields.

4.4.3. *Multipoles*

The gravitational angular momentum can also be written as a particular moment of the *imaginary part of* Ψ_2. This then leads to a way to define higher order angular momentum *multipoles*.[11] Let us see how this comes about. ϕ is a symmetry of the intrinsic geometry of Δ and therefore also a symmetry of $\tilde{\epsilon}$. This implies $\mathcal{L}_\phi\tilde{\epsilon} \,\hat{=}\, \mathrm{d}(\phi\lrcorner\tilde{\epsilon}) \,\hat{=}\, 0$ so that $\phi\lrcorner\tilde{\epsilon} = \mathrm{d}g$ for some smooth function g. Fix the freedom to add a constant to g by imposing $\oint_{\mathcal{S}} g\tilde{\epsilon} = 0$, so that g is unique. Some manipulation then gives

$$J_{\mathrm{Grav}} = -\frac{1}{8\pi G}\oint_{\mathcal{S}}(\phi\lrcorner\omega)\tilde{\epsilon} = -\frac{1}{8\pi G}\oint_{\mathcal{S}}(\mathrm{d}\omega)g = -\frac{1}{4\pi G}\oint_{\mathcal{S}} g\,\mathrm{Im}\Psi_2\tilde{\epsilon}\,, \tag{55}$$

where (21) has been used in the last line. One can show that g always has the range $(-R_\Delta^2, R_\Delta^2)$ where $4\pi R_\Delta^2 := a_\Delta$ is the areal radius of the horizon. In fact, there exists a canonical round metric q^o_{ab} determined by the axisymmetric metric q_{ab}, sharing the same area element and Killing field ϕ. In terms of the standard spherical coordinates (θ, ϕ) adapted to q^o_{ab} (with $\phi^a = \left(\frac{\partial}{\partial\phi}\right)^a$), one has $g = R^2_\Delta \cos\theta$, and the angular momentum can be written

$$J_{\text{Grav}} = -\frac{R^4_\Delta}{4\pi G}\oint_{\mathcal{S}} (\cos\theta)\Psi_2 \mathrm{d}\Omega\ , \tag{56}$$

where $\mathrm{d}\Omega = \sin\theta \mathrm{d}\theta \wedge \mathrm{d}\phi$. This leads to a more general definition of angular momentum *multipoles*

$$J_n := -\frac{R^{n+3}_\Delta}{4\pi G}\oint_{\mathcal{S}} P_n(\cos\theta)\mathrm{Im}\Psi_2 \mathrm{d}\Omega\ , \tag{57}$$

where P_n are the Legendre polynomials, and where the power of R_Δ is chosen to give the correct dimensions. One can similarly take moments of the scalar curvature $\mathcal{R}$ of the 2-metric on any slice of the horizon to give the *mass* multipoles

$$M_n := \frac{M_\Delta R^{n+2}_\Delta}{8\pi G}\int_{\mathcal{S}} P_n(\cos\theta)\mathcal{R}\mathrm{d}\Omega\ , \tag{58}$$

where M_Δ is the horizon mass, to be derived in the next section. These two sets of multipoles satisfy $J_0 = 0$ (no angular momentum monopole), $J_1 = J_\Delta$, $M_0 = M_\Delta$, and $M_1 = 0$ (one is 'automatically in the center of mass frame'). More importantly, the multipoles are *diffeomorphism invariant* and together *uniquely determine both* q_{ab} *and the horizon derivative operator* D_a *up to diffeomorphism.*[11] They have been used both in numerical relativity,[49–51] as well as to extend the black hole entropy calculation in loop quantum gravity (to be reviewed later in this chapter) to include rotation and distortion compatible with axisymmetry.[52]

4.4.4. *Horizon energy and the first law*

Energy is the generator of time-translations. To derive a notion of energy, we therefore seek a linear combination of the symmetry vector fields ℓ and ϕ

$$t \mathrel{\widehat{=}} B_{(t,\ell)}\ell - \Omega_{(t)}\phi \tag{59}$$

to play the roll of time translation, such that the corresponding flow is Hamiltonian. Here, as in the case of the Kerr–Newman family of solutions, $\Omega_{(t)}$ is interpreted as the *angular velocity of the horizon* relative to t. The generator of this flow will then provide us with the horizon energy. In turns out that, in order to accomplish this, unlike for the case of rotational symmetry considered above, the coefficients $B_{(t,\ell)}$, $\Omega_{(t)}$, and hence the vector field t^a, *must be allowed to vary from point to point in phase space.* One can see a hint that this would be necessary already in the Kerr–Newman family of solutions: the stationary Killing vector field t, determined by the condition that it approach a fixed unit time-translation at infinity, as a linear combination of ℓ and ϕ, is not constant over the family. For example, in the Reissner–Nordström sub-family of solutions, $\Omega_{(t)}$ is zero, and otherwise it is not.

Fix a unit time translation field of the asymptotic flat metric. At each point of $\mathbf{\Gamma}^\phi$, we introduce a vector field t on the entire space-time such that (1.)

$t \hat{=} B_{(t,\ell)}\ell - \Omega_{(t)}\phi$ at Δ and (2.) t approaches the fixed unit time translation at infinity. Evaluating the symplectic structure (51) at (δ, δ_t), one obtains

$$\mathbf{\Omega}^{\phi}(\delta, \delta_t) = \mathbf{\Omega}_{\Delta}(\delta, \delta_t) + \delta E_{\mathrm{ADM}}\,, \tag{60}$$

where E_{ADM} is the usual ADM energy, given by an integral at spatial infinity $\mathcal{S}_{\infty} = M \cap \mathcal{I}$. The integral at Δ is given by

$$\mathbf{\Omega}_{\Delta}(\delta, \delta_t) = \frac{\kappa_{(t)}}{8\pi G}\delta \oint_{\mathcal{S}} \tilde{\epsilon} + \frac{\Phi_{(t)}}{8\pi G}\delta \oint_{\mathcal{S}} \star \boldsymbol{F} + \frac{\Omega_{(t)}}{8\pi G}\delta \oint_{\mathcal{S}} [(\phi \lrcorner \omega)\tilde{\epsilon} + (\phi \lrcorner \mathbf{A}) \star \boldsymbol{F}]\,, \tag{61}$$

where we used $\kappa_{(t)} = \pounds_t \psi = t \lrcorner \omega$ and $\Phi_{(t)} = \pounds_t \chi = t \lrcorner \boldsymbol{A}$. The flow determined by t will be Hamiltonian iff $\mathbf{\Omega}^{\phi}(\delta, \delta_t)$ is an exact variation. From (60) and (61), this will be the case iff there exists a phase space function E_{Δ} such that for all variations δ,

$$\delta E_{\Delta}^{(t)} = -\frac{\kappa_{(t)}}{8\pi G}\delta \oint_{\mathcal{S}} \tilde{\epsilon} - \frac{\Phi_{(t)}}{8\pi G}\delta \oint_{\mathcal{S}} \star \boldsymbol{F} - \frac{\Omega_{(t)}}{8\pi G}\delta \oint_{\mathcal{S}} [(\phi \lrcorner \omega)\tilde{\epsilon} + (\phi \lrcorner \mathbf{A}) \star \boldsymbol{F}]\,. \tag{62}$$

If this is true, one has

$$\mathbf{\Omega}^{\phi}(\delta, \delta_t) = \delta(E_{\mathrm{ADM}} - E_{\Delta})\,. \tag{63}$$

The above equation tells us that, if E_{Δ} exists, $E_{\mathrm{ADM}} - E_{\Delta}$ will be the hamiltonian generating the flow determined by t. E_{ADM} is interpreted as the energy of the entire space-time, whereas E_{Δ} will have the interpretation of the energy of the black hole proper. $E_{\mathrm{ADM}} - E_{\Delta}$ is therefore the energy of the gravitational radiation and matter present in the entire intervening region between the horizon and infinity.

We now ask: What are the conditions which t must satisfy in order for E_{Δ} to exist? How many such vector fields t are there? This will be clarified in the following proposition. We shall see that there are an infinite number of evolution vectors leading to a Hamiltonian flow, and hence for which E_{Δ} exists.

Proposition 4. *There exists a function E_{Δ} such that (62) holds if and only if $\kappa_{(t)}$, $\Phi_{(t)}$ and $\Omega_{(t)}$ can be expressed as functions of the 'charges' a_{Δ}, Q_{Δ} and J_{Δ} defined by*

$$a_{\Delta} = \oint_{\mathcal{S}} \tilde{\epsilon} \tag{64}$$

$$Q_{\Delta} = \frac{1}{8\pi G}\oint_{\mathcal{S}} \star \boldsymbol{F} \tag{65}$$

$$J_{\Delta} = \frac{1}{8\pi G}\oint_{\mathcal{S}} [(\phi \lrcorner \omega)\tilde{\epsilon} + (\phi \lrcorner \mathbf{A}) \star \boldsymbol{F}]\,, \tag{66}$$

and satisfy the integrability conditions

$$\frac{1}{8\pi G}\frac{\partial \kappa_{(t)}}{\partial J_{\Delta}} = \frac{\partial \Omega_{(t)}}{\partial a_{\Delta}}, \quad \frac{1}{8\pi G}\frac{\partial \kappa_{(t)}}{\partial Q_{\Delta}} = \frac{\partial \Phi_{(t)}}{\partial a_{\Delta}}, \quad \frac{\partial \Omega_{(t)}}{\partial Q_{\Delta}} = \frac{\partial \Phi_{(t)}}{\partial J_{\Delta}}\,. \tag{67}$$

Proof. Let $\mathbb{d}$ and $\mathbb{\wedge}$ denote exterior derivative and exterior product on the infinite dimensional space $\mathbf{\Gamma}^{\phi}$. Suppose there exists a phase space function E_{Δ} such that (62) holds. Equation (62) is equivalent to

$$\mathbb{d}E_{\Delta} = \frac{\kappa_{(t)}}{8\pi G}\mathbb{d}a_{\Delta} + \Phi_{(t)}\mathbb{d}Q_{\Delta} + \Omega_{(t)}\mathbb{d}J_{\Delta}\,. \tag{68}$$

Because the gradient of E_Δ is a linear combination of the gradients of a_Δ, Q_Δ and J_Δ, E_Δ is a function only of these parameters, $E_\Delta = E_\Delta(a_\Delta, Q_\Delta, J_\Delta)$. The chain rule then implies

$$\mathbb{d}E_\Delta = \frac{\partial E_\Delta}{\partial a_\Delta}\mathbb{d}a_\Delta + \frac{\partial E_\Delta}{\partial Q_\Delta}\mathbb{d}Q_\Delta + \frac{\partial E_\Delta}{\partial J_\Delta}\mathbb{d}J_\Delta\,. \tag{69}$$

From, for example, the Kerr–Newman family of solutions, we know a_Δ, Q_Δ and J_Δ are independent quantities and hence $\mathbb{d}a_\Delta, \mathbb{d}Q_\Delta, \mathbb{d}J_\Delta$ are linearly independent. From (68) and (69) one then has

$$\frac{\kappa_{(t)}}{8\pi G} = \frac{\partial E_\Delta}{\partial a_\Delta}, \qquad \Phi_{(t)} = \frac{\partial E_\Delta}{\partial Q_\Delta}, \qquad \Omega_{(t)} = \frac{\partial E_\Delta}{\partial J_\Delta}\,, \tag{70}$$

which imply in turn that $\kappa_{(t)}$, $\Phi_{(t)}$, and $\Omega_{(t)}$ similarly depend only on a_Δ, Q_Δ,and J_Δ. Commutativity of partials together with (70) then imply the integrability conditions (67).

Conversely, if $\kappa_{(t)}, \Phi_{(t)}, \Omega_{(t)}$ depend on a_Δ, Q_Δ, and J_Δ alone and satisfy the integrability conditions (67), there exists E_{ADM} such that (70), and hence such that (68) and (62) hold. □

If we replace the surface gravity $\kappa_{(t)}$ and the area a_Δ with the Hawking temperature $T_{(t)} = \kappa_{(t)}/2\pi$ and the Bekenstein–Hawking entropy $S_\Delta = a_\Delta/4G$, (62) becomes

$$\delta E_\Delta = T_{(t)}\delta S_\Delta + \Phi_{(t)}\delta Q_\Delta + \Omega\delta J_\Delta\,. \tag{71}$$

which is none other than the first law of black hole mechanics. One thus sees that *the necessary and sufficient condition for t to be Hamiltonian is that there exist a phase space function E_Δ such that the first law holds.*

One can shows that there are an infinite number of possible functions $\kappa_{(t)}$, $\Phi_{(t)}$, $\Omega_{(t)}$, of a_Δ, Q_Δ, and J_Δ satisfying the integrability conditions (67). For each of these, there is a corresponding t^a which by the foregoing proposition, is hamiltonian, and hence gives rise to a notion of horizon energy E_Δ. That there are an infinite number of hamiltonian time evolution vector fields t^a and corresponding notions of horizon energy is not so surprising: In the context of stationary space-times, the only thing which allows one to isolate a single time-translation vector field t at the horizon is the global stationarity of the space-time, which rigidly connects the choice of t at the horizon to the choice of t at infinity, where it can be fixed by requiring it to be a time translation of the asymptotic Minkowski metric. In the present context of isolated horizons, by contrast, one does not have global stationarity. Thus, physically, one expects an ambiguity in t and hence in the definition of horizon energy. What is remarkable is that *all* of these horizon energies satisfy the first law.

Furthermore, the foregoing proposition gives one tight control over this infinity of possible evolutions and corresponding energies. In the following subsection, we will see how this control can be exploited to select a *unique, canonical* t on the entire phase space, and hence a *unique, canonical* notion of horizon energy, which one calls the horizon mass.

4.4.5. *Mass*

For practical applications, such as in numerical relativity, it is useful to isolate a canonical notion of horizon mass. In the context of Einstein–Maxwell theory, the uniqueness theorem provides a way to do this. Specifically, one takes advantage of the fact that, although the phase space $\mathbf{\Gamma}^\phi$ is far larger than the Kerr–Newman family of solutions, the Kerr–Newman family nevertheless forms a subset of $\mathbf{\Gamma}^\phi$, and on this subset we can stipulate that t be equal to the standard stationary Killing field t for the Kerr–Newman space-time in question, determined in the usual way by requiring t to approach a unit time translation at infinity. Combined with the results of Proposition 4, *this suffices to uniquely determine t on the entire phase space $\mathbf{\Gamma}^\phi$.*

Let us see how this comes about. From Proposition 4, $\kappa_{(t)}$ must be a function of $a_\Delta, Q_\Delta, J_\Delta$ alone. The fact that we have stipulated that t and hence $\kappa_{(t)}$ take their standard canonical values on the Kerr–Newman family uniquely fixes this function to be

$$\kappa_{(t)} = \kappa_0(a_\Delta, Q_\Delta, J_\Delta) := \frac{R_\Delta^4 - G^2(Q_\Delta^2 + 4J_\Delta^2)}{2R_\Delta^3\sqrt{(R_\Delta^2 + GQ_\Delta^2)^2 + 4G^2J_\Delta^2}}, \tag{72}$$

where $R_\Delta = \sqrt{a_\Delta/(4\pi)}$ is areal radius of $\mathcal{S}$. With $\kappa_{(t)}$ uniquely determined, the integrability conditions (67) can be used to uniquely determine the associated $\Phi_{(t)}$ and $\Omega_{(t)}$ as well. One finds that these are given by

$$\Phi_{(t)} = \frac{Q_\Delta(R_\Delta^2 + GQ_\Delta^2)}{R_\Delta\sqrt{(R_\Delta^2 + GQ_\Delta^2)^2 + 4G^2J_\Delta^2}}, \tag{73}$$

$$\Omega_{(t)} = \frac{2GJ_\Delta}{R_\Delta\sqrt{(R_\Delta^2 + GQ_\Delta^2)^2 + 4G^2J_\Delta^2}}. \tag{74}$$

Finally, equation (70) can be integrated to uniquely determine the horizon energy up to an additive constant. This additive constant can be fixed by requiring the horizon energy to be equal to the usual value of the energy when evaluated on members of the Kerr–Newman family. The resulting horizon energy we denote M_Δ and is called the *horizon mass.* It is given by

$$M_\Delta = \frac{\sqrt{(R_\Delta^2 + GQ_\Delta^2)^2 + 4G^2J_\Delta^2}}{2GR_\Delta}. \tag{75}$$

Note that this definition of the horizon mass involves only quantities *intrinsic* to the horizon. This is in contrast to the ADM or Komar definitions of mass, in which reference to infinity is required.

Finally, note that this strategy for selecting a canonical notion of mass works in Einstein–Maxwell theory only because the uniqueness theorem holds. If we consider theories in which the uniqueness theorem fails to hold, such as Einstein–Yang–Mills theory, it is no longer possible to consistently stipulate that t^a (and hence $\kappa_{(t)}$, $\Phi_{(t)}$, and $\Omega_{(t)}$) be equal to its canonical value when evaluated on stationary solutions:

Because each triple $(a_\Delta, J_\Delta, Q_\Delta)$ corresponds to multiple stationary solutions, such a prescription now would contradict the requirement of Proposition 4 that $\kappa_{(t)}$, $\Phi_{(t)}$, and $\Omega_{(t)}$ must be functions of $a_\Delta, J_\Delta, Q_\Delta$ alone. In such theories, one still has an infinite family of first laws. It is just that one cannot select a single one as canonical.

5. Quantum Isolated Horizons and Black Hole Entropy

As first observed by Bekenstein,[32] if we identify the surface gravity κ of a black hole with its temperature and its area a with its entropy, then the first law of black hole mechanics takes the form of the standard first law of thermodynamics. With this identification, the zeroth and second laws of black hole mechanics, reviewed in the introduction, take the form of the standard zeroth and second laws of thermodynamics and furthermore provide a way to preserve the second law of thermodynamics in the presence of a black hole. These observations led Bekenstein to hypothesize that a black hole is a thermodynamic object, its area *is* its entropy, and its surface gravity *is* its temperature. This hypothesis was impressively confirmed by Hawking's calculation showing that black holes radiate thermally at a temperature equal to $T = \hbar\kappa/2\pi$, a calculation which also fixed the coefficient relating T and κ, which in turn allowed one, via the first law, to fix the coefficient relating the entropy S to the area a:

$$S = \frac{a}{4\ell_{\rm Pl}^2} \, .$$

This is the celebrated Bekenstein–Hawking entropy. It provides tantalizing evidence that black holes are thermodynamic objects with an entropy whose statistical mechanical origin lie in quantum theory. To account for the statistical mechanical origin of black hole entropy has become one of the challenges for any quantum theory of gravity.

In this section we will review such an account provided by *quantum isolated horizons* using the approach to quantum gravity known as *loop quantum gravity.* The key principle which guides loop quantum gravity is that of *background independence*, which is equivalent to the symmetry principle of *diffeomorphism covariance* at the foundation of classical general relativity.

For simplicity, we here review the more recent $SU(2)$ covariant account of this derivation of the entropy,[53,14] which built to a great extent on the original work[54,3,55] which used a $U(1)$ partial gauge-fixing to aid calculations. Furthermore, we will only present the *type I* case — i.e., the case in which the intrinsic geometry of the horizon is spherically symmetric. The type II and type III horizons are handled in;[52,56] see also.[57] Lastly, again in order to focus on the key ideas, we will handle only pure gravity. Inclusion of Maxwell and Yang-Mills fields, and even non-minimally coupled scalar fields can also be handled, and can be found in the references.[55,10,58]

5.1. *Phase space and symplectic structure*

In loop quantum gravity, one uses the *Ashtekar–Barbero* variables,[59] which we use in the form $({}^\gamma\! A, \Sigma)$ consisting in the Ashtekar–Barbero connection ${}^\gamma\! A^i_a$, and the 'flux-2-form' Σ^i_{ab}, where $i = 1, 2, 3$. These are related to the space-time variables used in the prior sections via

$$ {}^\gamma\! A^i = -\frac{1}{2}\,\epsilon^i_{jk}\,\underleftarrow{A}^{jk} + \beta\,\underleftarrow{A}^{i0} $$
$$ \Sigma^i = \epsilon^i{}_{jk}\,\underleftarrow{e}^j \wedge \underleftarrow{e}^k\,, $$

where the arrows $\leftarrow$ denote pull-back to the partial Cauchy surface M, ε^i_{jk} denotes the three dimensional alternating tensor, and $\beta \in \mathbb{R}^+$ is the *Barbero–Immirzi parameter*, a quantization ambiguity consisting in a single real number. Furthermore, in the above equation, it is understood that the space-time variables A^{IJ}_a, e^I_a are in the *time-gauge* in which e^a_0 is normal to the partial Cauchy surface M, reducing the internal local gauge group from $SO(1,3)$ to $SO(3)$. To give a sense of the meaning of these variables, in terms of generalized ADM variables, one has $\Sigma^i_{ab} = \underset{\sim}{\epsilon}_{abc}(\det e^d_j)^{-1} e^{ci}$ where e^a_i is the triad, and the two terms in the expression for ${}^\gamma\! A$ are given by $-\frac{1}{2}\,\epsilon^i_{jk}\,\underleftarrow{A}^{jk}_a = \Gamma^i_a$, the spin connection determined by the triad, and $\underleftarrow{A}^{i0} = K^i_a = K_{ab} e^{bi}$ where K_{ab} is the extrinsic curvature of M.

Let $\mathbf{\Gamma}^{(I)}$ denote the space of possible fields $({}^\gamma\! A, \Sigma)$ on M satisfying Einstein's equations, such that the inner boundary $\mathcal{S} := \Delta \cap M$ corresponds to a type I isolated horizon of a *fixed* area a_0, and such that appropriate asymptotically flat boundary conditions are satisfied at infinity. Thus, the intrinsic geometry of $\mathcal{S}$ is restricted to be round in $\mathbf{\Gamma}^{(I)}$. If one starts from the symplectic structure (49), imposes the above conditions, and rewrites the result using the Ashtekar–Barbero variables $({}^\gamma\! A^i, \Sigma^i)$, one obtains the following symplectic structure:

$$ \mathbf{\Omega}^{(I)}(\delta_1, \delta_2) = \frac{1}{8\pi G\beta}\int_M [\delta_1\Sigma^i \wedge \delta_2{}^\gamma\! A_i - \delta_2\Sigma^i \wedge \delta_1{}^\gamma\! A_i] \\ -\frac{1}{8\pi G\beta}\frac{a_o}{\pi(1-\beta^2)}\int_{\mathcal{S}} \delta_1{}^\gamma\! A_i \wedge \delta_2{}^\gamma\! A^i\,. \tag{76} $$

In deriving this, one makes key use of the fact that, in terms of the Ashtekar–Barbero variables, the isolated horizon boundary conditions take the form

$$ \Sigma^i = -\frac{a_o}{\pi(1-\beta^2)} F^i({}^\gamma\! A) \tag{77} $$

where $F^i({}^\gamma\! A) = d{}^\gamma\! A^i + \frac{1}{2}\,\epsilon^i_{jk}{}^\gamma\! A^j \wedge {}^\gamma\! A^k$ is the curvature of ${}^\gamma\! A^i$. This is referred to as the *quantum horizon boundary condition.* Also note that the symplectic structure (76) consists in *two terms* — a bulk term equal to the standard symplectic structure used in loop quantum gravity, and a surface term at $\mathcal{S}$, equal to the symplectic structure of an $SU(2)$ *Chern–Simons theory.* This observation, together with the horizon boundary condition, are the keys to the quantization of this system.

5.2. *Quantization*

As just remarked, the symplectic structure (76) consists in two terms, a bulk term identical to that used in LQG, and a horizon, $SU(2)$ Chern–Simons theory term. This motivates one to *quantize the bulk and horizon degrees of freedom separately.* After they have been separately quantized, the horizon boundary condition (77) will be used to couple them again. Let us start with the bulk. The bulk Hilbert space of states $\mathcal{H}^B$ is the standard one in loop quantum gravity, spanned by *spin-network states.* A spin-network state $|\gamma, \{j_e\}, \{i_v\}\rangle$ is labelled by a collection of edges γ called a graph with each edge labelled by a half-integer spin j, and each point at the end of an edge, called a vertex, labelled by an *intertwiner* among the irreducible representations on the edges meeting at the vertex. Spin-network states are eigenstates of *area*: For any 2-surface T, one classically has a corresponding area a_T, and hence in quantum theory a corresponding area operator $\hat{a}_T$, and one has

$$\hat{a}_T|\gamma, \{j_e\}, \{i_v\}\rangle = \left(8\pi G\beta \sum_{p\in T\cap\gamma} \sqrt{j_p(j_p+1)}\right)|\gamma, \{j_e\}, \{i_v\}\rangle\,. \tag{78}$$

In the present case, because the spatial manifold M has boundary $\mathcal{S}$, γ may have edges which *intersect* this boundary. We call such intersections punctures. Suppose we are given a finite set of points $\mathcal{P}$, and assignment of a spin j_p to each point. Let $\mathcal{H}^B_{\mathcal{P},\{j_p\}}$ denote the span of all spin-network states whose punctures are at the points $\mathcal{P}$ and the spins on the corresponding edges are $\{j_p\}$. In terms of these spaces, the full bulk Hilbert space can be expressed as a direct sum over spins and a direct limit over all finite subsets of $\mathcal{S}$:

$$\mathcal{H}^B = \varinjlim_{\mathcal{P}\subset\mathcal{S}} \bigoplus_{\{j_p\}_{p\in\mathcal{P}}} \mathcal{H}^B_{\mathcal{P},\{j_p\}}\,. \tag{79}$$

Consider one of these spaces $\mathcal{H}^B_{\mathcal{P},\{j_p\}}$. On this space, the action of the operator corresponding to the two form Σ^i_{ab} pulled back to $\mathcal{S}$ reduces to[20]

$$\epsilon^{ab}\hat{\Sigma}^i_{ab}(x) = 8\pi G\beta \sum_{p\in\mathcal{P}} \delta(x, x_p)\hat{J}^i(p)\,, \tag{80}$$

where at each puncture p the operators $J^i(p)$ satisfy the usual angular momentum algebra $[J^i(p), J^j(p)] = \epsilon^{ij}{}_k J^k(p)$. Substituting this into (77), we get

$$-\frac{a_o}{\pi(1-\beta^2)}\epsilon^{ab}\hat{F}^i_{ab} = 8\pi G\beta \sum_{p\in\mathcal{P}} \delta(x, x_p)\hat{J}^i(p)\,. \tag{81}$$

Here $\hat{F}^i_{ab}$ is the quantization of the curvature of ${}^\gamma A^i_a$ pulled back to $\mathcal{S}$, and hence the curvature of the $SU(2)$ *Chern–Simons connection.* This shows us that the generators $\hat{J}^i(p)$ act as point sources for the $SU(2)$ Chern–Simons theory. The quantization of $SU(2)$ Chern–Simons theory with such point sources is well-understood

— see for example Witten.[60] In the end, for each fixed set of spins j_p associated to the point sources, one obtains a Hilbert space of states $\mathcal{H}_k^{CS}(\{j_p\})$, where $k := a_o/(2\pi\beta(1-\beta^2)\ell_P^2)$ is the 'level' of the Chern–Simons theory, appearing in the coefficient of the surface symplectic structure in (76). This Hilbert space can be viewed as a subset of the tensor product of carrying spaces associated to the $SU(2)$ representations labeled by the spins j_p

$$\mathcal{H}_k^{CS}(\{j_p\}) \subset \otimes_{p\in\mathcal{P}} V_{j_p},$$

where $\hat{J}^i(p)$ acts on each V_{j_p} irreducibly. $\mathcal{H}^B_{\mathcal{P},\{j_p\}}$ can similarly be decomposed in a way that makes the action of the generators apparent

$$\mathcal{H}^B_{\mathcal{P},\{j_p\}} = \mathcal{H}_{\{j_p\}} \bigotimes \left(\otimes_{p\in\mathcal{P}} V_{j_p}\right),$$

where again V_{j_p} denotes the spin j_p carrying space on which $\hat{J}^i(p)$ acts irreducibly. For a given set of punctures $\mathcal{P}$ and spins $\{j_p\}$, one is therefore led to the following space of states satisfying the quantum version of the horizon boundary condition (77),

$$\mathcal{H}^{\text{Kin}}_{\mathcal{P},\{j_p\}} = \mathcal{H}_{\{j_p\}} \otimes \mathcal{H}_k^{CS}(\{j_p\}). \tag{82}$$

Upon reassembling these spaces using a direct sum over spins and direct limit over sets of punctures, one obtains the full space of states solving the horizon boundary condition (81):

$$\mathcal{H}^{\text{Kin}} = \varinjlim_{\mathcal{P}\subset\mathcal{S}} \bigoplus_{\{j_p\}_{p\in\mathcal{P}}} \mathcal{H}_{\{j_p\}} \otimes \mathcal{H}_k^{CS}(\{j_p\}). \tag{83}$$

Here 'Kin' indicates that the diffeomorphism and Hamiltonian constraints have not yet been imposed. Imposition of the diffeomorphism constraint roughly speaking leads to replacement of quantum states with their *diffeomorphism equivalence class.* See[55] for further details. The solution to the diffeomorphism constraint then takes the form

$$\mathcal{H}^{\text{Diff}} = \bigoplus_n \bigoplus_{(j_p)_1^n} \mathcal{H}_{(j_p)_1^n} \otimes \mathcal{H}_k^{CS}(\{j_p\}), \tag{84}$$

where now one is no longer summing over possible positions of punctures, but only over the total *number* of punctures and corresponding spins. Because lapse is restricted to vanish on the horizon,[3,14] the Hamiltonian constraint is imposed only in bulk, resulting in replacement of $\mathcal{H}_{(j_p)_1^n}$ with an appropriate subspace $\tilde{\mathcal{H}}_{(j_p)_1^n}$, yielding the final space of physical states

$$\mathcal{H}^{\text{Phys}} = \bigoplus_n \bigoplus_{(j_p)_1^n} \tilde{\mathcal{H}}_{(j_p)_1^n} \otimes \mathcal{H}_k^{CS}(\{j_p\}). \tag{85}$$

For further details, we refer to.[55,14]

5.3. *Ensemble and entropy*

We are interested in the ensemble of physical states in $\mathcal{H}^{\rm phys}$ consisting in eigenstates of the horizon area $\hat{a}_S$ with eigenvalue in the range $(a_o - \delta, a_o + \delta)$ for some tolerance δ. Let $\mathcal{H}^{\rm bh}$ denote the span of such states. One would like to define the entropy via the standard Boltzmann formula as the logarithm of the dimension of this space. However, the dimension of this space is infinite, for the simple reason that all of bulk degrees of freedom are represented in it — degrees of freedom which are irrelevant for the question which interests us.

To eliminate this infinity, one needs to talk more precisely about the horizon degrees of freedom. To this end, we define

$$\mathcal{H}^S := \bigoplus_n \bigoplus_{(j_p)_1^n} \mathcal{H}_k^{CS}(\{j_p\}) \, . \tag{86}$$

We say that a given horizon state $\psi^S \in \mathcal{H}_k^{CS}(\{j_p\}) \subset \mathcal{H}^S$ is 'compatible' with $\mathcal{H}^{\rm bh}$ if there exists $\psi^B \in \widetilde{\mathcal{H}}_{(j_p)_1^n}$ such that $\psi^B \otimes \psi^S \in \mathcal{H}^{\rm bh}$. Let $\mathcal{H}_S^{\rm bh}$ denote the span of all such compatible horizon states. The entropy is then given by the usual Boltzmann formula

$$S := \log \dim \mathcal{H}_S^{\rm bh} \, .$$

One finds[61,62] the final entropy to be

$$S = \frac{\beta_0 a_o}{4\beta \ell_P^2}, \qquad \beta_0 = 0.274067\ldots \tag{87}$$

so that if we choose $\beta = \beta_0 = 0.274067\ldots$, the Bekenstein–Hawking entropy formula results. Note the role played by β: it is a single quantization ambiguity, which in principle can be fixed by a single experiment. By requiring the Bekenstein–Hawking entropy law to hold for any black hole with spherical intrinsic geometry (type I) and a single value of the area, β becomes fixed. The fact that the entropy law then continues to be satisfied for black holes of other areas is already a non-trivial test. But, in fact, the present framework has been shown to pass much stronger tests as well: if one extends the framework to include arbitrary Maxwell and Yang–Mills fields,[55] or type II (axisymmetric) and type III (generic) isolated horizons,[52,56] one again obtains an entropy equal to one quarter times the area, for the *same* value of the Barbero–Immirzi parameter.[b] One can even include a scalar field which is non-minimally coupled to gravity.[10,58] As mentioned in the introduction, this leads to a modified entropy law. The present framework, when extended to include a non-minimally coupled scalar field, has been shown to lead precisely to the required modified entropy law, again for the same value of β.

[b] As long as one identifies the horizon degrees of freedom in the same way as has been done here, counting the j labels as distinguishing horizon states. See[63,64] for a discussion.

6. Summary and Discussion

We have reviewed in detail the mathematical foundations of IHs in classical and quantum gravity. Let us briefly summarize the main features of this framework.

An IH is defined to be a null hypersurface with compact spatial cross-sections, whose outgoing null normal is non-expanding, and on which certain components of the Levi–Civita connection are Lie dragged by the null normal. We have seen that a well-defined first-order action principle can be given for space-times possessing an IH as an inner boundary, with the IH boundary conditions ensuring that the action is functionally differentiable. From second variations of this action, a conserved symplectic current can be identified on the *covariant phase space* of solutions to the field equations. Using this, we have seen how one can construct a well-defined canonical framework. In this framework, local rotations and time translations at the horizon are generated by clear notions of angular momentum and energy of the IH. These quantities are independent of the ADM charges at infinity, and satisfy a quasi-local version of the first law of black hole mechanics. Furthermore, the expression for the horizon angular momentum can be naturally generalized to give a quasi-local, diffeomorphism-invariant notion of angular momentum and mass multipoles.

When the symplectic structure of this canonical framework is recast in terms of real $SU(2)$ Ashtekar connections and triads, the boundary term in the symplectic structure at the horizon becomes that of $SU(2)$ *Chern–Simons theory.* This leads one to quantize the bulk degrees of freedom using *loop quantum gravity* methods, and the horizon degrees of freedom using a well-understood quantization of $SU(2)$ Chern–Simons theory. The bulk and surface degrees of freedom are then coupled through a quantum version of a 'horizon boundary condition' embodying the fact that the inner boundary is an isolated horizon. The statistical entropy of an ensemble of states with area in a small window of values is found to be equal to $1/4$ times the horizon area for a single, universal value of the Barbero–Immirzi parameter, a quantization ambiguity present in the loop quantization of gravity.

There are some key features of the IH framework that we would like to highlight at this point. First, we stress that, in the IH framework, there is *no need to make reference to asymptotic infinity at all.* This paradigm shift is essential in order that we may understand situations in which a black hole is in equilibrium with dynamical fields in the exterior region that are in an arbitrarily small neighbourhood of the horizon. Moreover, we note that the isolated horizon definition places a restriction only on geometric structures *intrinsic to the horizon.* Not only on does this make the definition simpler to check in practice, but it ensures that the *full* complement of degrees of freedom outside the horizon remains intact. This is important in quantum theory, and allows the methods of loop quantum gravity to be used outside the horizon without change.

Lastly, perhaps the most notable feature of the IH framework is that it provides a unified mathematical construction for understanding equilibrium black holes in both classical and quantum gravity. Indeed, the IH framework can be used to study the geometry, topology, supersymmetry, mechanics and quantum statistical mechanics of quasi-local black holes in general relativity (and beyond).

Acknowledgements

We are grateful to Christopher Beetle for a careful reading of the manuscript for this chapter. This work was supported in part by the Natural Sciences and Engineering Research Council of Canada (TL), United States NSF grant PHY-1237510 (JE), and by NASA through the University of Central Florida's NASA-Florida Space Grant Consortium (JE).

References

1. A. Ashtekar, C. Beetle, and S. Fairhurst, "Isolated horizons: A Generalization of black hole mechanics," *Class. Quant. Grav.*, vol. 16, pp. L1–L7, 1999.
2. A. Ashtekar, C. Beetle, and S. Fairhurst, "Mechanics of isolated horizons," *Class. Quant. Grav.*, vol. 17, pp. 253–298, 2000.
3. A. Ashtekar, A. Corichi, and K. Krasnov, "Isolated horizons: The classical phase space," *Adv. Theor. Math. Phys.*, vol. 3, pp. 419–478, 1999.
4. A. Ashtekar and A. Corichi, "Laws governing isolated horizons: Inclusion of dilaton couplings," *Class. Quant. Grav.*, vol. 17, pp. 1317–1332, 2000.
5. A. Ashtekar, C. Beetle, O. Dreyer, S. Fairhurst, B. Krishnan, J. Lewandowski, and J. Wisniewski, "Isolated horizons and their applications," *Phys. Rev. Lett.*, vol. 85, pp. 3564–3567, 2000.
6. A. Ashtekar, S. Fairhurst, and B. Krishnan, "Isolated horizons: Hamiltonian evolution and the first law," *Phys. Rev.*, vol. D62, p. 104025, 2000.
7. A. Ashtekar, C. Beetle, and J. Lewandowski, "Mechanics of rotating isolated horizons," *Phys. Rev.*, vol. D64, p. 044016, 2001.
8. A. Ashtekar, C. Beetle, and J. Lewandowski, "Geometry of generic isolated horizons," *Class. Quant. Grav.*, vol. 19, pp. 1195–1225, 2002.
9. A. Ashtekar, J. Wisniewski, and O. Dreyer, "Isolated horizons in (2+1) gravity," *Adv. Theor. Math. Phys.*, vol. 6, pp. 507–555, 2003.
10. A. Ashtekar, A. Corichi, and D. Sudarsky, "Nonminimally coupled scalar fields and isolated horizons," *Class. Quant. Grav.*, vol. 20, pp. 3413–3426, 2003.
11. A. Ashtekar, J. Engle, T. Pawlowski, and C. Van Den Broeck, "Multipole moments of isolated horizons," *Class. Quant. Grav.*, vol. 21, pp. 2549–2570, 2004.
12. T. Liko and I. Booth, "Supersymmetric isolated horizons," *Class. Quant. Grav.*, vol. 25, p. 105020, 2008.
13. I. Booth and T. Liko, "Supersymmetric isolated horizons in ads spacetime," *Phys. Lett.*, vol. B670, p. 61, 2009.
14. J. Engle, K. Noui, A. Perez, and D. Pranzetti, "Black hole entropy from an SU(2)-invariant formulation of type I isolated horizons," *Phys. Rev. D*, vol. 82, p. 044050, 2010.
15. S. Hayward, "Energy and entropy conservation for dynamical black holes," *Phys. Rev. D*, vol. 70, p. 104027, 2004.
16. S. Hayward, "General laws of black -hole dynamics," *Phys. Rev. D*, vol. 49, pp. 6467–6474, 1994.

17. S. Hayward, "Spin-coefficient form of the new laws of black hole dynamics," *Class. Quant. Grav.*, vol. 11, pp. 3025–3036, 1994.
18. A. Ashtekar and B. Krishnan, "Dynamical horizons and their properties," *Phys. Rev.*, vol. D68, p. 104030, 2003.
19. A. Ashtekar and B. Krishnan, "Isolated and dynamical horizons and their applications," *Living Rev. Rel.*, vol. 7, p. 10, 2004.
20. A. Ashtekar and J. Lewandowski, "Background independent quantum gravity: A status report," *Class. Quant. Grav.*, vol. 21, 2004.
21. C. Rovelli, *Quantum Gravity.* Cambridge: Cambridge UP, 2004.
22. T. Thiemann, *Modern Canonical Quantum General Relativity.* Cambridge: Cambridge UP, 2007.
23. S. Hawking and G. Ellis, *The large scale structure of space-time.* Cambridge: Cambridge, 1973.
24. R. Wald, *Quantum Field Theory in Curved Space-time and Black Hole Thermodynamics.* Chicago: The University of Chicago Press, 1994.
25. R. M. Wald, *General Relativity.* Chicago: Chicago University Press, 1984.
26. W. Israel, "Event horizons in static electrovac space-times," *Commun. Math. Phys.*, vol. 8, pp. 245–260, 1968.
27. B. Carter, "Axisymmetric black hole has only two degrees of freedom," *Phys. Rev. Lett.*, vol. 26, pp. 331–332, 1971.
28. B. Carter, "Black hole equilibrium states," in *Black Holes* (C. DeWitt and B. DeWitt, eds.), New York: Gordon & Breach, 1973.
29. D. Robinson, "Uniqueness of the kerr black hole," *Phys. Rev. Lett.*, vol. 34, pp. 905–906, 1975.
30. P. Mazur, "Proof of uniqueness of the Kerr–Newman black hole solution," *J. Phys. A*, vol. 15, pp. 3173–3180, 1982.
31. W. Israel, "Third law of black hole dynamics: A formulation and proof," *Phys. Rev. Lett.*, vol. 57, pp. 397–399, Jul 1986.
32. J. D. Bekenstein, "Black holes and entropy," *Phys. Rev.*, vol. D7, pp. 2333–2346, 1973.
33. J. D. Bekenstein, "Generalized second law of thermodynamics in black hole physics," *Phys. Rev.*, vol. D9, pp. 3292–3300, 1974.
34. S. Hawking, "Black holes in general relativity," *Commun. Math. Phys.*, vol. 25, pp. 152–166, 1972.
35. J. M. Bardeen, B. Carter, and S. Hawking, "The Four laws of black hole mechanics," *Commun. Math. Phys.*, vol. 31, pp. 161–170, 1973.
36. R. M. Wald, "Black hole entropy is the Noether charge," *Phys. Rev.*, vol. D48, pp. 3427–3431, 1993.
37. V. Iyer and R. M. Wald, "Some properties of Noether charge and a proposal for dynamical black hole entropy," *Phys. Rev.*, vol. D50, pp. 846–864, 1994.
38. T. Jacobson, G. Kang, and R. C. Myers, "On black hole entropy," *Phys. Rev.*, vol. D49, pp. 6587–6598, 1994.
39. T. Jacobson, G. Kang, and R. C. Myers, "Increase of black hole entropy in higher curvature gravity," *Phys. Rev.*, vol. D52, pp. 3518–3528, 1995.
40. T. Jacobson and R. C. Myers, "Black hole entropy and higher curvature interactions," *Phys. Rev. Lett.*, vol. 70, pp. 3684–3687, 1993.
41. T. Liko, "Topological deformation of isolated horizons," *Phys. Rev.*, vol. D77, p. 064004, 2008.
42. I. Booth and S. Fairhurst, "Extremality conditions for isolated and dynamical horizons," *Phys. Rev.*, vol. D77, p. 084005, 2008.

43. G. Gibbons and C. Hull, "A Bogomolny Bound for General Relativity and Solitons in N=2 Supergravity," *Phys. Lett.*, vol. B109, p. 190, 1982.
44. A. Ashtekar, J. Engle, and D. Sloan, "Asymptotics and hamiltonians in a first order formalism," *Class. Quant. Grav.*, vol. 25, p. 095020, 2008.
45. A. Ashtekar and D. Sloan, "Action and Hamiltonians in higher dimensional general relativity: First order framework," *Class. Quant. Grav.*, vol. 25, p. 225025, 2008.
46. T. Liko and D. Sloan, "First-order action and Euclidean quantum gravity," *Class. Quant. Grav.*, vol. 26, p. 145004, 2009.
47. A. Ashtekar, L. Bombelli, and O. Reula, "The covariant phase space of asymptotically flat gravitational fields," in *Analysis, Geometry and Mechanics: 200 Years after Lagrange*, Amsterdam: North-Holland, 1991.
48. J. Lee and R. M. Wald, "Local symmetries and constraints," *J. Math. Phys.*, vol. 31, pp. 725–743, 1990.
49. E. Schnetter, B. Krishnan, and F. Beyer, "Introduction to dynamical horizons in numerical relativity," *Phys. Rev. D*, vol. 74, p. 024028, 2006.
50. R. Owen, "The final remnant of binary black hole mergers: Multipolar analysis," *Phys. Rev. D*, vol. 80, p. 084012, 2009.
51. M. Saijo, "Dynamic black holes through gravitational collapse: Analysis of multipole moment of the curvatures on the horizon," *Phys. Rev.*, vol. D83, p. 124031, 2011.
52. A. Ashtekar, J. Engle, and C. Van Den Broeck, "Quantum horizons and black hole entropy: Inclusion of distortion and rotation," *Class. Quant. Grav.*, vol. 22, pp. L27–L34, 2005.
53. J. Engle, K. Noui, and A. Perez, "Black hole entropy and SU(2) Chern–Simons theory," *Phys. Rev. Lett.*, vol. 105, p. 031302, 2010.
54. A. Ashtekar, J. Baez, A. Corichi, and K. Krasnov, "Quantum geometry and black hole entropy," *Phys. Rev. Lett.*, vol. 80, pp. 904–907, 1998.
55. A. Ashtekar, J. C. Baez, and K. Krasnov, "Quantum geometry of isolated horizons and black hole entropy," *Adv. Theor. Math. Phys.*, vol. 4, pp. 1–94, 2000.
56. C. Beetle and J. Engle, "Generic isolated horizons and loop quantum gravity," *Class. Quant. Grav.*, vol. 27, p. 235024, 2010.
57. A. Perez and D. Pranzetti, "Static isolated horizons: SU(2) invariant phase space, quantization, and black hole entropy," *Entropy*, vol. 13, pp. 744–777, 2011.
58. A. Ashtekar and A. Corichi, "Nonminimal couplings, quantum geometry and black hole entropy," *Class. Quant. Grav.*, vol. 20, pp. 4473–4484, 2003.
59. J. F. Barbero G, "Real Ashtekar variables for Lorentzian signature space-times," *Phys. Rev. D*, vol. 51, pp. 5507–5510, 1995.
60. E. Witten, "Quantum Field Theory and the Jones Polynomial," *Commun. Math. Phys.*, vol. 121, p. 351, 1989.
61. I. Agullo, J. F. Barbero, E. Borja, J. Diaz-Polo, and E. Villasenor, "The combinatorics of the SU(2) black hole entropy in loop quantum gravity," *Phys. Rev. D*, vol. 80, p. 084006, 2009.
62. J. Engle, K. Noui, A. Perez, and D. Pranzetti, "The SU(2) black hole entropy revisited," *J. High Energy Phys.*, vol. 2011, p. 016, 2011.
63. A. Ghosh and P. Mitra, "A Comment on black hole state counting in loop quantum gravity," *arXiv:0805.4302*, 2008.
64. A. Ghosh and P. Mitra, "Black hole state counting in loop quantum gravity," *Mod. Phys. Lett.*, vol. A26, pp. 1817–1823, 2011.

Chapter 6

Quantum Thermometers in Stationary Space-Times with Horizons

Sergio Zerbini

Dipartimento di Fisica, Università di Trento
and Istituto Nazionale di Fisica Nucleare — Gruppo Collegato di Trento
Via Sommarive 14, 38123 Povo, Italia
zerbini@science.unitn.it

A quantum field theoretical approach, in which a quantum probe is used to unveil the "intrinsic" temperature associated with stationary space-times with horizons is discussed. The probe is identified with a conformally coupled massless scalar field defined on a space-time with horizon and the procedure to measure the local temperature is realised by the use of Unruh–DeWitt detector. Another proposal to determine local temperature, due to Buchholz, is also considered.

1. Introduction

Hawking radiation[1] is considered one of the most important predictions of quantum field theory in curved space-time. Several derivations of this effect have been proposed[2–6] and recently the search for "experimental" verification making use of analogue models has been pursued by many investigators (see for example[7,8]).

In 2000 Parikh and Wilczek[9] (see also[10,11]) introduced a further approach, the so-called tunnelling method, for investigating Hawking radiation. A variant of their method has been also introduced and called Hamilton–Jacobi tunnelling method.[12–15] This method is covariant and enjoys a peculiar feature to admit the generalisation to the dynamical case.[16–19] It has also been used in the study of decay of massive particles and particle creation by naked singularities.[20] For a recent review, see[21] and the references therein.

To start with, we recall that within the tunnelling method, the semiclassical emission rate reads

$$\Gamma \propto |\text{Amplitude}|^2 \propto e^{-2\frac{\Im I}{\hbar}} . \tag{1.1}$$

with $\Im$ standing for the imaginary part, the appearance of the imaginary part due to the presence of the horizon. The leading term in the WKB approximation of the tunnelling probability reads

$$\Gamma \propto e^{-\frac{2\pi}{\kappa_H}\omega} , \tag{1.2}$$

in which the energy ω of the particle and the surface gravity evaluated at horizon appear. From this asymptotics, one obtains the Hawking temperature by the identification $T_H = \frac{\kappa_H}{2\pi}$. The method is quite general, and works for a generic stationary black hole, and the above formula is still valid in the spherically symmetric dynamical case, as shown in.[16–19] With regard to this issue, it should be important to have a quantum field theoretical confirmation of the tunnelling results.

As a first step toward this aim, in this chapter, we shall discuss a quantum field theoretical approach, in which a quantum probe is used to unveil the "intrinsic" temperature associated with stationary space-times with horizons, for which, as we will show, an exact treatment is possible. The probe which we are going to make use of is the simplest one and it is identified with a conformally coupled massless scalar field defined on a stationary space-time with horizon (see[22–24] and references therein). For a rigorous quantum theoretical approach, see also.[25]

We restrict our analysis to the spherically symmetric (dynamical) case, and, in this case, the use of invariant quantities plays a crucial role.[18, 19, 26, 23] For example, among the other things, this approach will permit the introduction of Kodama observers. For the sake of completeness, here we review the general formalism.

To begin with, let us recall that any spherically symmetric metric can locally be expressed in the form

$$ds^2 = \gamma_{ij}(x^i)dx^i dx^j + R^2(x^i)d\Omega^2 , \qquad i,j \in \{0,1\} , \tag{1.3}$$

where the two-dimensional metric

$$d\gamma^2 = \gamma_{ij}(x^i)dx^i dx^j \tag{1.4}$$

is referred to as the normal metric, $\{x^i\}$ are associated coordinates and $R(x^i)$ is the areal radius, considered as a scalar field in the two-dimensional normal space. A relevant scalar quantity in the reduced normal space is

$$\chi(x) = \gamma^{ij}(x)\partial_i R(x)\partial_j R(x) , \tag{1.5}$$

since the dynamical trapping horizon, if it exists, is located in correspondence of

$$\chi(x)\Big|_H = 0 , \tag{1.6}$$

provided that $\partial_i\chi|_H \neq 0$. Another important normal-space scalar is the Hayward surface gravity associated with this dynamical horizon, given by

$$\kappa_H = \frac{1}{2}\Box_\gamma R\Big|_H . \tag{1.7}$$

Recall that, in the spherical symmetric dynamical case, it is possible to introduce the Kodama vector field $\mathcal{K}$. Given the metric (1.3), the Kodama vector components are

$$\mathcal{K}^i(x) = \frac{1}{\sqrt{-\gamma}}\varepsilon^{ij}\partial_j R , \qquad \mathcal{K}^\theta = 0 = \mathcal{K}^\varphi . \tag{1.8}$$

We may introduce the Kodama trajectories, and related Kodama observer, by means of integral lines of Kodama vector

$$\frac{dx^i}{d\lambda} = \mathcal{K}^i = \frac{1}{\sqrt{-\gamma}}\varepsilon^{ij}\partial_j R\,. \tag{1.9}$$

As a result,

$$\frac{dR(x(\lambda))}{d\lambda} = \partial_i R\frac{dx^i}{d\lambda} = \frac{1}{\sqrt{-\gamma}}\varepsilon^{ij}\partial_j R\partial_i R = 0\,. \tag{1.10}$$

Thus, in a generic spherically symmetric space-times, the areal radius R is conserved along Kodama trajectories. In other word, a Kodama observer is characterised by the condition $R = R_0$.

The paper is organised as follows. In the next Section, after a brief survey of a scalar quantisation in a conformally flat space-time, we present the main formula of Unruh–De Witt response function detector. In Section 3, the formalism is applied to Rindler space and a generic static black hole space-time, and the de Sitter case is discussed in detail. In Section 4, the Buchholz proposal is presented and applied and in Section 5, the conclusions are reported.

2. Quantum Field Theory in Conformally Flat Space-Times

In the following, for the sake of completeness, we review the quantisation of a conformally coupled massless scalar field in a conformally flat space-time.

2.1. *The conformal quantum probe*

Since it will be sufficient for our aims to work on conformally FRW flat space-time, making use of the conformal time η by means $d\eta = \frac{dt}{a}$, we have

$$ds^2 = a^2(\eta)(-d\eta^2 + d\vec{x}^2)\,, \qquad x = (\eta, \vec{x})\,. \tag{2.1}$$

It is convenient to consider a free massless scalar field which is conformally coupled to gravity, since the related Wightman function $W(x, x')$ can be computed in an exact way. In fact, one has

$$W(x, x') = \sum_{\vec{k}} f_{\vec{k}}(x) f^*_{\vec{k}}(x')\,, \tag{2.2}$$

where the modes functions $f_{\vec{k}}(x)$ satisfy the conformally invariant equation ($\mathcal{R}$ being the curvature scalar)

$$\left(\Box - \frac{\mathcal{R}}{6}\right) f_{\vec{k}}(x) = 0\,. \tag{2.3}$$

The solution of this equation, which corresponds to the choice of the vacuum, is given by

$$f_{\vec{k}}(x) = \frac{e^{-i\eta k}}{2\sqrt{k}a(\eta)} e^{-i\vec{k}\cdot\vec{x}}\,, \tag{2.4}$$

with $k = |\vec{k}|$. As a consequence, one has[3]

$$W(x,x') = \frac{1}{4\pi^2 a(\eta)a(\eta')} \frac{1}{|\vec{x}-\vec{x}'|^2 - |\eta - \eta' - i\epsilon|^2} \,. \tag{2.5}$$

As is usual in distribution theory we shall leave understood the limit as $\epsilon \to 0^+$. However, it has been shown by Takagi[27] and Schlicht[28] that this prescription is manifestly non-covariant. Since one is dealing with distributions, the limit $\epsilon \to 0^+$ has to be taken in the weak sense, and it may lead to unphysical results with regard to instantaneous proper-time rate in Minkowski space-time. We adapt Schlicht's proposal to our conformally flat case, namely

$$W(x,x') = \frac{1}{4\pi^2 a(\eta)a(\eta')} \frac{1}{[(x-x') - i\epsilon(\dot{x}+\dot{x}')]^2} \,. \tag{2.6}$$

where an over dot stands for derivative with respect to proper time. In the flat case, this result has been generalised by Milgrom and Obadia, who made use of an analytical proper-time regularisation.[29,30] The appearance of Minkowski contribution, as a function of the conformal time η, should be noted.

2.2. *The Unruh–De Witt detector*

The Unruh–De Witt detector approach is a well known and used technique for exploring quantum field theoretical aspects in curved space-time. For a recent review see.[31] Here, we review the basic formula following Ref.[23]

The transition probability per unit proper time of the detector depends on the response function per unit proper time which, for radial trajectories, at finite time τ may be written as[32]

$$\frac{dF}{d\tau} = \frac{1}{2\pi^2}\mathrm{Re}\int_0^{\tau-\tau_0} \frac{ds}{a(\tau)a(\tau-s)} \frac{e^{-iEs}}{[x(\tau)-x(\tau-s) - i\epsilon(\dot{x}(\tau)+\dot{x}(\tau-s))]^2} \tag{2.7}$$

where τ_0 is the detector's proper time at which we turn on the detector, and E is the energy associated with the excited detector state (we are considering $E > 0$). Although the covariant $i\epsilon$-prescription is necessary in order to deal with the second order pole at $s = 0$, one may try to avoid the awkward limit $\epsilon \to 0^+$ by omitting the ϵ-terms but subtracting the leading pole at $s = 0$ (see[32] for details). In fact, the normalisation condition

$$g_{\mu\nu}\dot{x}^\mu \dot{x}^\nu \equiv [a(\tau)\dot{x}(\tau)]^2 = -1 \,, \tag{2.8}$$

characteristic of time-like four-velocities, has to be imposed. Thus, introducing the notation

$$\sigma^2(\tau,s) \equiv a(\tau)a(\tau-s)[x(\tau)-x(\tau-s)]^2 \,, \tag{2.9}$$

due to (2.8), for small s, one has

$$\sigma^2(\tau,s) = -s^2[1 + s^2 d(\tau,s)] \,. \tag{2.10}$$

As a consequence, for $\Delta\tau = \tau - \tau_0 > 0$, one can present the detector transition probability per unit time in the form

$$\frac{dF}{d\tau}(\tau) = \frac{1}{2\pi^2}\int_0^\infty ds\,\cos(Es)\left(\frac{1}{\sigma^2(\tau,s)} + \frac{1}{s^2}\right) + J_\tau\,, \tag{2.11}$$

where the "tail" or finite time fluctuating term is given by

$$J_\tau := -\frac{1}{2\pi^2}\int_{\Delta\tau}^\infty ds\,\frac{\cos(E\,s)}{\sigma^2(\tau,s)}\,. \tag{2.12}$$

In the important stationary cases (examples are the static black hole, the FRW de Sitter space, and Rindler space), one has $\sigma(\tau,s)^2 = \sigma^2(s) = \sigma^2(-s)$, and Eq. (2.11) simply becomes

$$\frac{dF}{d\tau}(\tau) = \frac{1}{4\pi^2}\int_{-\infty}^\infty ds\,e^{-iEs}\left(\frac{1}{\sigma^2(s)} + \frac{1}{s^2}\right) + J_\tau = \frac{dF}{d\tau} + J_\tau\,. \tag{2.13}$$

The first term is independent of τ, and all the time dependence is contained only in the fluctuating tail.

3. Quantum Thermometers in Static and Stationary Spaces

As an application of the formalism previously developed, first we are going to revisit the Rindler space and the case of the generic static black hole. The important case of the de Sitter space-time will be investigated in detail and the independence of the coordinate choice (gauge independence) will be explicitly verified.

3.1. *The Unruh effect*

Let us start with a short review of the well known Unruh Effect. The derivation is standard and it is reported here only for completeness.

Let us denote by (T, X, Y, Z) a set of Minkowski coordinates and by (t, x, y, z) a set of coordinates related to an accelerated observer along the Z direction, namely

$$T = \frac{e^{gz}}{g}\sinh gt\,, \tag{3.1}$$

$$Z = \frac{e^{gz}}{g}\cosh gt\,, \tag{3.2}$$

$$X = x\,, \quad Y = y\,, \tag{3.3}$$

where g is a real number. Making use of the new set of coordinates (t, x, y, z), one has that the Minkowski metric takes the form

$$ds^2 = -e^{2gz}dt^2 + e^{2gz}dz^2 + dx^2 + dy^2\,. \tag{3.4}$$

Furthermore, introducing the new "radial" coordinate

$$\rho = \frac{1}{g} e^{gz} , \tag{3.5}$$

one has the more familiar form of the Rindler space-time, namely

$$ds^2 = -g^2 \rho^2 dt^2 + d\rho^2 + dx^2 + dy^2 . \tag{3.6}$$

If one computes the 4-norm of the four acceleration of the Rindler observer at $z = z_0$ or $\rho = \rho_0$ constant, and noting that the proper time is

$$\tau = e^{gz_0} t , \tag{3.7}$$

one has

$$a_0^2 = g^2 e^{-2gz_0} . \tag{3.8}$$

In particular, if $z_0 = 0$, Rindler observer's origin, one has $a = g$, which gives the physical meaning of parameter g.

The response function depends on the geodesic distance, which in this case reads

$$\sigma^2(\tau, \tau - s) = -(T - T')^2 + (Z - Z')^2 = -\frac{4}{a_0^2} \sinh^2(\frac{a_0 s}{2}) . \tag{3.9}$$

Since $\sigma^2(\tau, s) = \sigma^2(s) = \sigma^2(-s)$, we can use equation (2.13), and the first time independent contribution reads

$$\frac{dF}{d\tau} = \frac{a}{4\pi^2} \int_{-\infty}^{\infty} dx e^{-\frac{2iEx}{a}} \left[-\frac{1}{\sinh^2 x} + \frac{1}{x^2} \right] . \tag{3.10}$$

The integral can be evaluated by the theorem of residues. The final result is well known and reads

$$\frac{dF}{d\tau} = \frac{1}{2\pi} \frac{E}{\exp\left(\frac{2\pi E}{a_0}\right) - 1} . \tag{3.11}$$

Since the transition rate exhibits the characteristic Planck distribution, it means that the Unruh–DeWitt thermometer detects a quantum system in thermal equilibrium at the local Unruh temperature

$$T_0 = \frac{a_0}{2\pi} = \frac{g e^{-gz_0}}{2\pi} . \tag{3.12}$$

3.2. *The generic static black hole*

The general metric for a static black hole reads

$$ds^2 = -V(r) dt^2 + \frac{dr^2}{V(r)} + r^2 d\Omega^2 , \tag{3.13}$$

with $V(r)$ having just simple zeroes. Let r_H be the (greatest) solution of $V(r) = 0$, and the horizon is located at $r = r_H$; the Killing vector is $(1, \vec{0})$, and the surface gravity, is $\kappa_H = \kappa = V'_H/2$.

First, we may make use of the standard argument of the near horizon approximation, thus $V(r) \simeq V'_H(r - r_H)$. Introducing the new radial co-ordinate

$$\rho = \frac{2}{\sqrt{V'_H}}\sqrt{r - r_H}\,, \tag{3.14}$$

and, noting also that $\kappa_H \rho = e^{\kappa_H z}$, one has a Rindler-like metric

$$ds^2 = -\kappa_H^2 \rho^2 dt^2 + d\rho^2 + r_H^2 d\Omega^2 = -e^{2\kappa_H z}dt^2 + e^{2\kappa_H z}dz^2 + r_H^2 d\Omega^2\,, \tag{3.15}$$

which becomes a Rindler metric for r_H very large. As a result, an observer at fixed distance r_0, but near the horizon, behaves as an Rindler accelerating observer with $g = \kappa_H$ and the related detector sees a Planckian thermal spectrum with a local temperature given by

$$T_0 = \frac{\kappa_H e^{-\kappa_H z_0}}{2\pi} = \frac{\kappa_H}{2\pi\sqrt{V(r_0)}}\,. \tag{3.16}$$

The same result can be obtained without any approximation in the following way. The first step consists in introducing the tortoise coordinate

$$dr^*(r) = \frac{dr}{V(r)}\,. \tag{3.17}$$

One has $-\infty < r^* < \infty$ and

$$ds^2 = V(r^*)[-dt^2 + (dr^*)^2] + r^2(r^*)d\Omega^2_{(2)}\,. \tag{3.18}$$

Then, one introduces Kruskal-like coordinates, defined by

$$R = \frac{1}{\kappa_H}e^{\kappa r^*}\cosh(\kappa_H t)\,, \quad T = \frac{1}{\kappa_H}e^{\kappa_H r^*}\sinh(\kappa_H t)\,, \tag{3.19}$$

so that

$$-T^2 + R^2 = \frac{1}{\kappa_H^2}e^{2\kappa_H r^*}\,, \tag{3.20}$$

and the original static metric becomes

$$ds^2 = e^{-2\kappa_H r^*}\,V(r^*)[-dT^2 + dR^2] + r^2(T,R)d\Omega^2\,, \tag{3.21}$$

where now the coordinates are T and R, $r^* = r^*(T,R)$, and the normal metric turns out to be conformally flat.

The key point to recall here is that in the Kruskal gauge (3.21) the normal metric, the important one for radial trajectories, is conformally related to two dimensional Minkowski space-time. The second observation is that Kodama observers are defined by the integral curves associated with the Kodama vector, thus the areal radius $r(T,R)$ and r^* are *constant*, say $r = r_0$. As a consequence, the proper time along Kodama trajectories reads

$$d\tau^2 = V_0\,dt^2 == V_0 e^{-2\kappa r_0^*}(dT^2 - dR^2) = a^2(r_0)(dT^2 - dR^2)\,. \tag{3.22}$$

Thus $t = \tau/\sqrt{V_0}$ and

$$R(\tau) = \frac{1}{\kappa_H} e^{\kappa_H r_0^*} \cosh\left(\kappa_H \frac{\tau}{\sqrt{V_0}}\right),$$
$$T(\tau) = \frac{1}{\kappa_H} e^{\kappa_H r_0^*} \sinh\left(\kappa_H \frac{\tau}{\sqrt{V_0}}\right). \tag{3.23}$$

The geodesic distance reads

$$\sigma^2(\tau, s) = V_0 e^{-2\kappa r_0^*} \left[-\left(T(\tau) - T(\tau - s)\right)^2 + \left(R(\tau) - R(\tau - s)\right)^2 \right], \tag{3.24}$$

and one gets, using (3.23),

$$\sigma^2(\tau, s) = -\frac{4V_0}{\kappa_H^2} \sinh^2\left(\frac{\kappa_H\, s}{2\sqrt{V_0}}\right). \tag{3.25}$$

Since $\sigma^2(\tau, s) = \sigma^2(s) = \sigma^2(-s)$, we can again use equation (2.13) and the final result is

$$\frac{dF}{d\tau} = \frac{1}{2\pi} \frac{E}{\exp\left(\frac{2\pi\sqrt{V_0}E}{\kappa_H}\right) - 1}. \tag{3.26}$$

Again, the transition rate exhibits the characteristic Planck distribution, and the Unruh–DeWitt detector in the generic spherically symmetric black hole space-time detects a quantum system in thermal equilibrium at the local temperature

$$T_0 = \frac{\kappa_H}{2\pi\sqrt{V_0}}. \tag{3.27}$$

With regard to the factor $\sqrt{V_0} = \sqrt{-g_{00}}$, recall the Tolman's theorem which states that, for a gravitational system at thermal equilibrium, $T\sqrt{-g_{00}} =$ constant. Thus, for a generic static black hole space-times, as well as for the Rindler space-times, one obtains the "intrinsic" constant temperature, the Hawking temperature, i.e.

$$T_H = \frac{\kappa}{2\pi} = \frac{V_H'}{4\pi}. \tag{3.28}$$

Furthermore, the local expression for the temperature given by the Tolmann form (3.2) may be written in the following form. First, we observe that in a static space-time, the Killing observers with $r = r_0$ constant, have an invariant acceleration

$$a_0^2 = \frac{V_0'^2}{4V_0}, \tag{3.29}$$

where $a^\mu = u^\nu \nabla_\nu u^\mu$, u^μ being the observer's four-velocity, that is the (normalised) tangent vector to the integral curves of the Killing vector field. Thus, the local equilibrium temperature can be rewritten in the form

$$T_0 = \frac{\kappa}{2\pi} \frac{2a_0}{V_0'} = T_H \frac{2a_0}{V_0'}. \tag{3.30}$$

In principle, V_0' can be expressed as a function only of a_0 and T_H. In the following, we will discuss two examples.

The first one is static patch of de Sitter space, with a metric defined by

$$V(r) = 1 - H_0^2 r^2\,, \qquad H_0^2 = \frac{\Lambda}{3}\,. \tag{3.31}$$

The unique horizon is located at $r_H = H_0^{-1}$ and the Gibbons–Hawking temperature is[33]

$$T_{GH} = H_0/2\pi. \tag{3.32}$$

The acceleration at fixed r_0 reads

$$a_0^2 = \frac{H_0^4 r_0^2}{1 - H_0^2 r_0^2}\,. \tag{3.33}$$

Thus

$$a_0^2 + H_0^2 = \frac{H_0^2}{1 - H_0^2 r_0^2}\,. \tag{3.34}$$

Since the de Sitter local temperature felt by the Unruh detector is

$$T_{dS} = \frac{H_0}{2\pi} \frac{1}{\sqrt{1 - H_0^2 r_0^2}} \tag{3.35}$$

one has[34,35]

$$T_{dS}(r_0) = \frac{1}{2\pi}\sqrt{a_0^2 + H_0^2} = T_{GH}\sqrt{1 + 4\pi^2 \frac{a_0^2}{T_{GH}^2}}\,. \tag{3.36}$$

In the case of four dimensional Schwarzschild black hole, one has

$$V(r) = 1 - \frac{2MG}{r}\,, \quad V'(r) = \frac{2MG}{r^2}\,. \tag{3.37}$$

The acceleration at fixed r_0 reads

$$a_0^2 = \frac{M^2 G^2}{r_0^4 - 2 r_H r_0^3}\,. \tag{3.38}$$

As a result, one has

$$4 r_0^4 a_0^2 - 4 r_0^3 r_H a_0^2 - r_H^2 = 0\,, \tag{3.39}$$

where $r_H = 2MG = \frac{1}{4\pi T_H}$. In principle, this algebraic equation of fourth order can be solved in the unknown r_0, but the final expression is quite involved. In the limit r_0 very large, one has

$$4 r_0^4 a_0^2 \simeq r_H^2\,, \tag{3.40}$$

and the local black hole temperature reads

$$T_{Schw}(r_0) \simeq T_H \left(1 + \frac{1}{2}\sqrt{\frac{a_0}{2\pi T_H}}\right)\,. \tag{3.41}$$

3.3. *The de Sitter space in FRW form*

In the previous subsection, as an example we have considered the de Sitter space-time in the static gauge. It may be instructive to investigate the de Sitter space-time in the cosmological gauge, due to the physical relevance of it with regard to inflation.

Recall that in a generic flat FRW space-time, the Kodama observer is given by

$$r(t) = \frac{R_0}{a(t)}, \tag{3.42}$$

with constant R_0. For a radial trajectory, the proper time in the flat FRW is

$$d\tau^2 = a^2(\eta)(d\eta^2 - dr^2) \,. \tag{3.43}$$

As a function of the proper time, the conformal time along a Kodama trajectory is

$$\begin{aligned} \eta(\tau) &= -\int d\tau \, \frac{1}{a(\eta)\sqrt{1 - R_0^2 H^2(\tau)}} d\eta \\ &\equiv -\int d\tau \, \frac{1}{a(\tau)\sqrt{V(\tau)}} \,, \end{aligned} \tag{3.44}$$

$H(\tau)$ being the Hubble parameter as a function of proper time. In the case of de Sitter, $a(t) = e^{H_0 t}$. Thus,

$$ds^2 = -dt^2 + e^{2H_0 t}(dr^2 + r^2 d\Omega^2) \,. \tag{3.45}$$

Here $H(t) = H_0$ is constant as well as $V = V_0 = 1 - H_0^2 R_0$. For Kodama observers

$$\tau = \sqrt{V_0}\, t \,, \qquad a(\tau) = e^{\frac{H_0}{\sqrt{V_0}}\tau} \,, \tag{3.46}$$

and

$$\eta(\tau) = -\frac{1}{H_0} e^{-\frac{H_0}{\sqrt{V_0}}\tau} \,, \quad r(\tau) = R_0\, e^{-\frac{H_0}{\sqrt{V_0}}\tau} \,, \tag{3.47}$$

so, the geodesic distance is

$$\sigma^2(\tau, s) = -\frac{4\,V_0}{H_0^2} \sinh^2\left(\frac{H_0\, s}{2\sqrt{V_0}}\right) \,. \tag{3.48}$$

This result is formally equal to the one obtained for the generic static black hole (3.25). Since again $\sigma^2(\tau, s) = \sigma^2(s) = \sigma^2(-s)$, we may use (2.13) and obtain, for $E > 0$

$$\frac{dF}{d\tau} = \frac{1}{2\pi} \frac{E}{e^{\frac{2\pi\sqrt{V_0}E}{H_0}} - 1} \,, \tag{3.49}$$

which shows that the Unruh–DeWitt thermometer in the FRW de Sitter space detects a quantum system in thermal equilibrium at a temperature $T_0 = \frac{H_0}{2\pi\sqrt{V_0}}$. Here, the Tolman factor is substituted by a Lorentz factor, which represents, as we already know, the Unruh acceleration part. In fact we recall that 4-acceleration of a Kodama observer in a FRW space-time turns out to be

$$\mathcal{A}^2 = \mathcal{A}^\mu \mathcal{A}_\mu = R_0^2 \left[\frac{\dot{H}(t) + (1 - H^2(t)R_0^2)H^2(t)}{(1 - H^2(t)R_0^2)^{\frac{3}{2}}} \right]^2 \tag{3.50}$$

where $\mathcal{A}^\mu := u^\nu \nabla_\nu u^\mu$, u^μ being the 4-velocity of the detector, that is the (normalised) tangent vector to the integral curves of the Kodama vector field. As a result, for dS space in a time dependent patch we have

$$\mathcal{A}^2 = \frac{R_0^2 H_0^4}{1 - R_0^2 H_0^2}, \tag{3.51}$$

showing that

$$\frac{H_0}{\sqrt{1 - H_0^2 R_0^2}} = \sqrt{H_0^2 + \mathcal{A}^2}$$

in agreement with the dS static calculation. When $R_0 = 0$, one has $V_0 = 1$ and the classical Gibbons–Hawking result $T_{dS} = \frac{H_0}{2\pi}$ is recovered. This is an important check of the approach, since it shows the coordinate independence of the result for the important case of de Sitter space.

The response function for unit proper time, in the stationary cases we have considered, gives information about the equilibrium temperature via the Planckian distribution. One may note that in the limit $E \to 0$, one has

$$\lim_{E \to 0} \frac{dF}{d\tau} = \frac{T_0}{2\pi}, \tag{3.52}$$

which shows in which sense a Unruh–De Witt detector is a quantum thermometer.

Finally, we recall that for stationary space-times, the fluctuating tail can be computed (see, for example[23]).

4. Buchholz Quantum Thermometer

Another proposal to detect local temperature associated with stationary space-time admitting an event horizon has been put forward by Buchholz and collaborators[41] (see also[42]). The idea may be substantiated by the following argument.

Let us start with a free massless quantum scalar field $\phi(x)$ in thermal equilibrium at temperature T in flat space-time. It is well known that finite temperature field theory effects of this kind may be investigated by considering the scalar field defined in the Euclidean manifold $S_1 \times R^3$, where one has introduced the imaginary time $\tau = -it$, compactified in the circle S_1, with period $\beta = \frac{1}{T}$ (see for example[43]).

Let us consider the local quantity $\langle \phi(x)^2 \rangle$. Formally, this is a divergent quantity, due to the operator-valued distribution in the same point x, and a regularisation and renormalisation are necessary. A simple and powerful way to deal with a regularized quantity is to make use of zeta-function regularisation see for example,[43–45] and references therein. Within zeta-function regularisation, one has

$$\langle \phi(x)^2 \rangle = \zeta(1|L_\beta)(x)\,, \tag{4.1}$$

where $\zeta(z|L_\beta)(x)$ is the analytic continuation of the local zeta-function associated with the operator L_β

$$L_\beta = -\partial_\tau^2 - \nabla^2\,, \tag{4.2}$$

defined on $S_1 \times R^3$. The local zeta-function is defined with $\mathrm{Re}\, z$ sufficiently large by means

$$\zeta(z|L_\beta)(x) = \frac{1}{\Gamma(z)} \int_0^\infty dt t^{z-1} K_t(x,x)\,, \tag{4.3}$$

where the heat-kernel on the diagonal is given by

$$K_t(x,x) = \langle x|e^{-tL_\beta}|x\rangle = \frac{1}{\beta(4\pi t)^{3/2}} \sum_n e^{-\frac{4\pi^2}{\beta^2} n^2}\,. \tag{4.4}$$

In (4.1), appears the analytic continuation of the local $\zeta(z|L_\beta)(x)$ and it is assumed this analytical continuation is regular at $z = 1$, which, as we shall see, is our case. If the analytic continuation has a simple pole in $z = 1$, the prescription has to be modified (see[46, 42]).

A standard computation, which makes use of the Jacobi–Poisson formula, leads to

$$K_t(x,x) = \frac{1}{(4\pi t)^2} \sum_n e^{-\frac{n^2\beta^2}{4t}}\,. \tag{4.5}$$

Let us plug this expression in (4.3). The term $n = 0$ leads to a formally divergent integral $\int_0^\infty dt t^{z-3}$, but this is zero in the sense of Gelfand analytic continuation, and it can be neglected. Thus, a direct computation gives the analytic continuation of the local zeta-function

$$\zeta(z|L_\beta)(x) = \frac{\Gamma(2-z)}{8\pi^2\Gamma(z)} \left(\frac{\beta^2}{4}\right)^{z-2} \zeta_R(4-2z)\,, \tag{4.6}$$

where $\zeta_R(z)$ is the Riemann zeta-function. It is easy to see that the analytic continuation of the local zeta-function is regular at $z = 1$, and from (4.1), recalling that $\zeta_R(2) = \frac{\pi^2}{6}$, one has

$$\langle \phi(x)^2 \rangle = \frac{1}{12\beta^2} = \frac{T^2}{12}\,. \tag{4.7}$$

Thus, the regularised vacuum expectation value of the observable ϕ^2 gives the temperature of the quantum field in thermal equilibrium, namely one is dealing with a quantum thermometer.

Motivated by this argument, let us consider again a conformal coupled scalar field in a FRW conformally flat space-time. We have seen that the off-diagonal Wigthmann function is

$$W(x,x') = \langle \phi(x)\phi(x') \rangle = \frac{1}{4\pi^2} \frac{1}{\sigma^2(x,x')}\,, \tag{4.8}$$

where

$$\sigma^2(x,x') = a(x)a(x')(x-x')^2\,, \tag{4.9}$$

with $a(x)$ being the conformal factor. In the limit $x \to x'$, formally one has

$$\langle \phi(x)^2 \rangle = W(x,x)\,, \tag{4.10}$$

but $W(x,x)$ is ill defined, and one has to regularise and then renormalize this object. In our case, we may make use of the simple point splitting procedure,[3] namely we consider $W(x,x')$ and evaluate the limit $x' \to x$.

Thus, the proposal is to consider the observable $\langle \phi^2 \rangle$ as a quantum thermometer in a stationary space-time with horizons. More precisely, the local temperature is given by

$$\frac{T^2}{12} = \langle \phi(x)^2 \rangle_R = F.P. \lim_{x' \to x} W(x,x')\,, \tag{4.11}$$

where $F.P.$ stands for finite part prescription, and the two points are joined by a Kodama trajectory parametrized by the proper time.

In the stationary space-times we have considered, the relevant Wightman functions associated with our conformally coupled massless scalar probe field, as a function of the proper time, all have the specific form

$$W(s, s+\varepsilon) = -\frac{1}{4\pi^2} \frac{\alpha_0^2}{4\sinh^2(\varepsilon\,\frac{\alpha_0}{2})}\,. \tag{4.12}$$

In fact, for the Rindler space

$$\alpha_0 = g e^{-g z_0}\,, \tag{4.13}$$

for the generic static black hole

$$\alpha_0 = \frac{\kappa_H}{\sqrt{V_0}}\,, \quad \kappa_H = \frac{V'_H}{2}\,, \tag{4.14}$$

and for the de Sitter space-time in flat FRW form

$$\alpha_0 = \frac{H_0}{\sqrt{1 - R_0^2 H_0^2}}\,. \tag{4.15}$$

Thus, one has

$$W(s, s+\varepsilon) = -\frac{1}{4\pi^2} \frac{\alpha_0^2}{4\sinh^2(\varepsilon\,\frac{\alpha_0)}{2}} = -\frac{1}{4\pi^2\varepsilon^2} + \frac{1}{12}\left(\frac{\alpha_0}{2\pi}\right)^2 + O(\varepsilon)\,. \tag{4.16}$$

As a consequence, the finite part prescription (4.11) gives

$$T_0 = \frac{\alpha_0}{2\pi}\,, \tag{4.17}$$

in agreement with the Unruh–DeWitt detector technique. Again, the local dependence comes from the localisation of the Kodama observer, expressed by constant $R = R_0$, where R is the areal radius.

5. Conclusion

In this paper, with the aim to better understand the temperature-versus-surface gravity paradigm, the asymptotic results obtained by the tunnelling semiclassical method have been tested with quantum field theory techniques as the Unruh–DeWitt detector and the Buchholz proposal. We have shown that for Rindler space and static black holes the two approaches are consistent and lead to well known results obtained by other standard QFT methods.[2,3]

However, we would like to recall that the tunnelling in its Hamilton–Jacobi covariant version leads to a very simple result for a spherically symmetric dynamical black hole, namely, the semiclassical result for the tunnelling probability reads

$$\Gamma \simeq e^{-2\Im I} \simeq e^{-\frac{2\pi}{\kappa_H}\omega_H}\,, \tag{5.1}$$

where k_H is the dynamical surface gravity (1.7) introduced by Hayward, and it should be very interesting to obtained a quantum field theoretical confirmation of this asymptotic semi-classical result.

With regard to this issue, a first attempt has been put forward in,[23] where a realistic cosmological model asymptotic to the de Sitter space has been investigated by means of the Unruh–de Witt detector technique, and there it has been shown that for a comoving detector, namely associated with $R_0 = 0$, the detector clicks close to a de Sitter response function and reaches thermalisation (possibly, through decaying oscillations) for sufficiently large proper time, and it should be interesting to generalise this result to a generic Kodama trajectory.

With regard to Buchholz approach, for a generic FRW flat space-time, but restricted to Kodama observers with $R_0 = 0$, a direct calculation gives

$$T^2(t) = \frac{H^2 + 2\dot{H}}{4\pi^2}\,, \tag{5.2}$$

which reduces to the Gibbon–Hawking temperature for the de Sitter case. It is still an unsolved issue to reconcile this result with the tunneling result.

Acknowledgement

We thank S. Hayward, L.Vanzo, G. Cognola, R. Di Criscienzo a G. Acquaviva for several discussions.

References

1. S W Hawking, Nature **248**, 30 (1974); Commun. Math. Phys.**43**, 199 (1975) [Erratum—ibid. **46**, 206 (1976)]
2. B. S. DeWitt, Phys. Rept. **19** (1975) 295.
3. N D Birrell and P C W Davies, *Quantum fields in curved space* (Cambridge University Press 1982). **43**, 199 (1975) [Erratum-ibid. **46**, 206 (1976)]
4. R. M. Wald, *Quantum Field Theory in Curved Spacetime and Black Hole Thermodynamics* (Chicago Lectures in Physics, Chicago University Press 1994).
5. S. A. Fulling, *Aspects of Quantum Field Theory in Curved Space-time* (Cambridge University Press 1996).
6. V. P. Frolov and I. D. Novikov, *Black hole physics*, Kluwer Academic Publisher, 2007.
7. W. G. Unruh, Phil. Trans. Roy. Soc. Lond. A **366**, 2905 (2008).
8. C. Barcelo, S. Liberati and M. Visser, Living Rev. Rel. **8**, 12 (2005). [gr-qc/0505065].
9. M. K. Parikh and F. Wilczek, Phys. Rev. Lett. **85**, 5042 (2000).
10. M. Visser, Int. J. Mod. Phys. **D12**, 649 (2003); A. B. Nielsen and M. Visser, Class. Quant. Grav. **23**, 4637 (2006).
11. L. Vanzo, Europhys. Lett. **95**, 20001 (2011) [arXiv:1104.1569 [gr-qc]].
12. M. Angheben, M. Nadalini, L. Vanzo and S. Zerbini, JHEP **0505**, 014 (2005); M. Nadalini, L. Vanzo and S. Zerbini, J. Physics A: Math. Gen. **39**, 6601 (2006).
13. K. Srinivasan and T. Padmanabhan, Phys. Rev. **D 60**, 24007 (1999).
14. R. Kerner and R. B. Mann, Phys. Rev. D **73**, 104010 (2006)
15. A. J. M. Medved and E. C. Vagenas, Mod. Phys. Lett. A **20**, 2449 (2005); M. Arzano, A. J. M. Medved and E. C. Vagenas, JHEP **0509**, 037 (2005); R. Banerjee and B. R. Majhi, Phys. Lett. B **662**, 62 (2008).
16. R. Di Criscienzo, M. Nadalini, L. Vanzo, S. Zerbini and G. Zoccatelli, Phys. Lett. **B657**, 107 (2007).
17. R. Di Criscienzo and L. Vanzo, Europhys. Lett. **82**, 60001 (2008).
18. S. A. Hayward, R. Di Criscienzo, L. Vanzo, M. Nadalini and S. Zerbini, Class. Quant. Grav. **26** , 062001 (2009).
19. R. Di Criscienzo, S. A. Hayward, M. Nadalini, L. Vanzo and S. Zerbini, Class. Quant. Grav. **27**, 015006 (2010).
20. R. Di Criscienzo, L. Vanzo and S. Zerbini, JHEP **1005**, 092 (2010).
21. L. Vanzo, G. Acquaviva and R. Di Criscienzo, Class. Quant. Grav. **28**, 183001 (2011). [arXiv:1106.4153 [gr-qc]].
22. N. Obadia, Phys. Rev. D **78**, 083532 (2008).
23. G. Acquaviva, R. Di Criscienzo, L. Vanzo and S. Zerbini, "Unruh–DeWitt detector and the interpretation of the horizon temperature in spherically symmetric dynamical space-times, [arXiv:1101.5254 [hep-th]].
24. R. Casadio, S. Chiodini, A. Orlandi, G. Acquaviva, R. Di Criscienzo and L. Vanzo, Mod. Phys. Lett. **A26**, 2149-2158 (2011). [arXiv:1011.3336 [gr-qc]].
25. V. Moretti and N. Pinamonti, "State independence for tunneling processes through black hole horizons and Hawking radiation", [arXiv:1011.2994 [gr-qc]].
26. H. Kodama, Prog. Theor. Phys. **63**, 1217 (1980).
27. S. Takagi, Prog. Theoretical Phys. Supp Grav. **88** (2004) 1.
28. S. Schlicht, Class. Quant. Grav. **21** (2004) 4647.
29. P. Langlois, Annals Phys. **321** (2006) 2027.
30. N. Obadia and M. Milgrom, Phys. Rev. D **75** (2007) 065006.
31. L. C. B. Crispino, A. Higuchi and G. E. A. Matsas, Rev. Mod. Phys. **80** (2008) 787.

32. J. Louko and A. Satz, Class. Quant. Grav. **23** (2006) 6321; J. Louko and A. Satz, Class. Quant. Grav. **25**, 055012 (2008).
33. G. W. Gibbons and S. W Hawking, Phys. Rev. D **14**, 2738 (1977).
34. H. Narnhofer, I. Peter and W. E. Thirring, Int. J. Mod. Phys. B **10**, 1507 Int. (1996).
35. S. Deser and O. Levin, Class. Quant. Grav. **14** (1997) L163.
36. S. Åminneborg, I. Bengtsson, S. Holst and P. Peldán, Class. Quantum Grav. 13 (1996) 2707.
37. R.B. Mann, Class. Quantum Grav. 14 (1997) L109.
38. D.R. Brill, Helv. Phys. Acta 69 (1996) 249; D.R. Brill, J. Louko and P. Peldán, Phys. Rev. D.56 (1997) 3600.
39. L. Vanzo, Phys. Rev. D56 (1997) 6475.
40. B. Garbrecht and T. Prokopec, Class. Quant. Grav. **21**, 4993 (2004).
41. D. Buchholz and J. Schlemmer, Class. Quant. Grav. **24**, F25–F31 (2007); C. Solveen, Class. Quant. Grav. **27**, 235002 (2010). [arXiv:1005.3696 [gr-qc]].
42. D. Binosi, S. Zerbini, J. Math. Phys. **40**, 5106-5116 (1999). [gr-qc/9901036].
43. A. A. Bytsenko, G. Cognola, L. Vanzo and S. Zerbini, Phys. Rept. **266**, 1–126 (1996). [arXiv:hep-th/9505061 [hep-th]].
44. S. W. Hawking, Commun. Math. Phys. **55**, 133 (1977).
45. E. Elizalde, S. D. Odintsov, A. Romeo, A. A. Bytsenko and S. Zerbini, Singapore, Singapore: World Scientific (1994) 319 p.
46. D. Iellici and V. Moretti, Phys. Lett. **B425**, 33-40 (1998). [gr-qc/9705077].

Chapter 7

Relativistic Thermodynamics

Sean A. Hayward

Center for Astrophysics, Shanghai Normal University, 100 Guilin Road, Shanghai 200234, China,
sean_a_hayward@yahoo.co.uk

A generally relativistic theory of thermodynamics is developed, based on four main physical principles: heat is a local form of energy, therefore described by a thermal energy tensor; conservation of mass, equivalent to conservation of heat, or the local first law; entropy is a local current; and non-destruction of entropy, or the local second law. A fluid is defined by the thermostatic energy tensor being isotropic. The entropy current is related to the other fields by certain equations, including a generalized Gibbs equation for the thermostatic entropy, followed by linear and quadratic terms in the dissipative (thermal minus thermostatic) energy tensor. Then the second law suggests certain equations for the dissipative energy tensor, generalizing the Israel–Stewart dissipative relations, which describe heat conduction and viscosity including relativistic effects and relaxation effects. In the thermostatic case, the perfect-fluid model is recovered. In the linear approximation for entropy, the Eckart theory is recovered. In the quadratic approximation for entropy, the theory is similar to that of Israel and Stewart, but involving neither state-space differentials, nor a non-equilibrium Gibbs equation, nor non-material frames. Also, unlike conventional thermodynamics, the thermal energy density is not assumed to be purely thermostatic, though this is derived in the linear approximation. Otherwise, the theory reduces in the non-relativistic limit to the extended thermodynamics of irreversible processes due to Müller. The dissipative energy density seems to be a new thermodynamical field, but also exists in relativistic kinetic theory of gases.

1. Introduction

Thermodynamics is perhaps best known to many relativists by its analogies with black-hole dynamics, made concrete by the famous result of Hawking that stationary black holes radiate quantum fields with a thermal spectrum. Another link has been obtained: a unified first law which contains first laws of both black-hole dynamics and relativistic thermodynamics.[1] This result requires little thermodynamics other than the general form of the energy tensor given long ago by Eckart.[2] However, the literature does not seem to contain a fully satisfactory general theory of relativistic thermodynamics, in the classical sense of a general macroscopic theory independent of (but consistent with) microscopic theories. This remarkable situation has prompted the development of the theory described in this paper. Such a theory naturally has widespread applications in astrophysics and cosmology. For simplicity, only a single material will be considered here, ignoring mixtures, phase changes, chemical reactions and external forces.

Perhaps the greatest progress in relativistic thermodynamics was made by Eckart,[2] who: identified the basic fields, formulated in this paper as the material current, the thermal energy tensor, the entropy current and the temperature; gave the basic conservation equations, of mass and energy-momentum; related the entropy density for fluids by a relativistic Gibbs equation; and gave a relativistic second law, showing that it could be satisfied by simple dissipative relations which generalize the Fourier and Navier–Stokes equations for heat conduction and viscosity. Thus the thermodynamic system is described by a finite number of macroscopic fields which may be measured by standard techniques.

Although Eckart was concerned with special relativity, the theory generalizes immediately to general relativity. The only unacceptable part of the Eckart theory is that the dissipative relations entail non-causal propagation of disturbances in temperature and velocity. The problem of infinite propagation speeds existed even in non-relativistic thermodynamics for longer still, until Müller[3] showed how to resolve it by allowing the entropy current to contain terms which are quadratic (rather than just linear) in the dissipative fields, namely thermal flux and viscous stress. Physically, this means assuming that the dissipative fields are small enough to justify the approximation: thermostatic, linear or quadratic. This incidentally means that entropy is generally not a state function, as claimed in classical thermodynamics, though this is retained in the linear approximation.

One might expect to obtain an acceptable theory of relativistic thermodynamics simply by applying a Müller extension to the Eckart theory. This was attempted by Israel and Stewart,[4] who obtained modified dissipative relations which are consistent with the Müller theory. However, they gave a type of Gibbs equation which differs from that of Eckart and does not reduce appropriately to the Gibbs equation of non-relativistic local thermodynamics, as described for example by de Groot and Mazur.[5] Also, Israel and Stewart gave a theory intended to hold not only in the material frame of Eckart, but also in the frame of Landau and Lifshitz.[6] This seems to involve some further approximation which is unclear, at least to this author. In any case, this paper will use the material frame only, for reasons explained below.

The theory described in this article was developed from first principles, but includes Müller-type quadratic dissipative effects, without which the Eckart theory is recovered. The theory is based on simple physical principles which are consistent with local non-relativistic thermodynamics, including local first and second laws. These principles serve to define what is meant by thermodynamic matter, which is intended to describe real fluids and solids in circumstances where a macroscopic description is adequate. The most basic principle is that *heat is a local form of energy*. In relativity, heat is therefore described by an energy tensor, called the thermal energy tensor, whose various components are thermal energy density, thermal flux and thermal stress. Thermal flux is what is normally called heat flux only in the material frame, thereby fixing this frame as the physically natural one. Thermal stress is what is normally called material stress. Thermal energy is what is normally called internal energy, but with fixed zero.

This principle entails one further generalization: the thermal energy density need not be purely thermostatic, allowing a further dissipative field, namely dissipative energy density. Classical thermodynamics excludes this by the claim that internal energy is a state function. However, this is inconsistent with relativistic kinetic theory of gases,[7] where thermal energy density is defined in terms of the molecular distribution and is generally not purely thermostatic. In fact, dissipative energy density emerged unrecognized even in non-relativistic kinetic theory as essentially the 14th moment of the 14-moment approximation of Grad.[8] It turns out that relativistic causality requires dissipative energy density to vanish in the linear approximation for entropy, thereby recovering the Eckart theory. However, this need not hold in the quadratic approximation, thereby generalizing the Israel–Stewart dissipative relations.

This article is intended for relativists who may have no previous experience of thermodynamics. (Thermodynamicists with a nodding acquaintance with relativity should also find it accessible). Therefore brief reviews are given of Müller's extended thermodynamics, of the previously standard local (or field) thermodynamics, and of relevant classical thermodynamics. This seems necessary because most textbooks on thermodynamics, even today, insist on applying thermostatic concepts in ways which are quite inapplicable in genuine thermodynamics, even in the formulation of the first and second laws. Such confusion often survive in work on relativistic thermodynamics.

The article is organized as follows. Sections 2–6 respectively review classical thermodynamics, non-relativistic hydrodynamics, non-relativistic thermodynamics, extended thermodynamics and basic general relativity. Section 7 introduces the relativistic concept of heat and the local first law, or conservation of heat, equivalent to conservation of mass. Section 8 checks the Newtonian limit. Section 9 divides the thermal energy tensor into thermostatic and dissipative parts and defines a fluid by the thermostatic energy tensor being isotropic. Section 10 introduces the relativistic concept of entropy and the local second law, and describes the thermostatic case using a generalized Gibbs equation. Section 11 describes the linear approximation for entropy, showing that it reduces to a minor generalization of the Eckart theory. Section 12 describes the quadratic approximation for entropy, obtaining dissipative relations which generalize the Israel–Stewart relations (in the material frame). Section 13 concerns integral quantities and corresponding laws. Section 14 concludes. A point of style is that equations in the text are sometimes used for temporary arguments, whereas displayed equations are intended to carry more weight.

2. Classical Thermodynamics

In classical thermodynamics, the basic quantities are temperature ϑ, heat supply Q, work W, internal energy H and entropy S. The classical first law is said to be $dH = \delta Q + \delta W$ and the second law $\vartheta dS \geq \delta Q$, where d is said to be a differential and δ not, neither being actually defined except in thermostatics, where the second law becomes an equality, or in the limit where they become difference operators between two equilibrium states separated in time, corresponding to the classical experimental situation. Clearly this is nonsense, as has been emphasized eloquently by Truesdell.[9]

The resolution is simply that these operators are really derivatives with respect to time. In fluid mechanics the relevant derivative is that along the fluid flow, called the material or comoving derivative, henceforth denoted by a dot. Similarly, in solid mechanics the relevant derivative is that in the centre-of-momentum frame, which may also be called the material derivative. Then the first law is

$$\dot{H} = \dot{Q} + \dot{W} \tag{2.1}$$

where $\dot{Q}$ and $\dot{W}$ should be taken as the basic quantities, Q and W themselves not being uniquely defined. For instance, for an inviscid fluid, the work is given by $\dot{W} = -p\dot{V}$, where V is the volume and p the pressure of the fluid, so that the first law reads

$$\dot{H} = \dot{Q} - p\dot{V}. \tag{2.2}$$

Similarly, the second law is

$$\vartheta\dot{S} \geq \dot{Q}. \tag{2.3}$$

Both laws now make sense, though it turns out that they respectively require the pressure and temperature to be uniform, again reflecting the classical experimental situation. Here and henceforth, uniform means spatially constant. In general, the laws need to be formulated locally, as described subsequently.

The classical zeroth law states that temperature ϑ is constant in thermal equilibrium. The classical third law, expressed in its loosest form, is that absolute zero temperature cannot be physically attained. Thus $\vartheta > 0$ in practice.

The entropy may be divided into entropy supply $S_\circ$, given by

$$\vartheta\dot{S}_\circ = \dot{Q} \tag{2.4}$$

and entropy production $S - S_\circ$. Then the second law may be written

$$\dot{S} \geq \dot{S}_\circ \tag{2.5}$$

which expresses entropy production. In words, S is the entropy of the system, where system means a comoving volume of material, and $S_\circ$ is the entropy supplied to the system. Thus the second law implies that the total entropy of the universe cannot decrease.

Equality in the second law,

$$\dot{S} = \dot{S}_\circ \tag{2.6}$$

holds in thermostatics, often called equilibrium thermodynamics or reversible thermodynamics. In the thermostatic case, the first and second laws for an inviscid fluid imply

$$\vartheta\dot{S} = \dot{H} + p\dot{V} \tag{2.7}$$

which is the Gibbs equation.

It was found experimentally that the thermostatic state of a gas could be described by just two quantities, such as density ρ and temperature. For uniform density, one may use the material volume V instead of density, since mass $M = V\rho$ is conserved:

$$\dot{M} = 0. \tag{2.8}$$

In particular, H and p are said to be state functions, meaning functions of (V, ϑ). If S is also a state function, then the Gibbs equation can be rewritten in terms of the differential d in state space as $\vartheta dS = dH + pdV$, as Gibbs originally did.[10] This follows because the state functions f are now uniform by assumption, so that $df = \dot{f}dt$. However, in the non-uniform case, the last relation does not hold and the two types of Gibbs equation cannot both be generally correct. It turns out that the material Gibbs equation (2.7), localized, forms part of the thermodynamic field equations,[5] but that the state-space Gibbs equation is unnecessary.[a] This point deserves emphasis because the Gibbs equation is often given in state-space form in introductory textbooks on thermodynamics, or on mechanics including thermodynamics, as well as in much work on relativistic thermodynamics. Indeed, this misunderstanding seems to be the root of the meaningless first and second laws mentioned above. The thermodynamic theory developed in this article will involve the material derivative rather than state-space differentials.

3. Non-Relativistic Hydrodynamics

In non-relativistic hydrodynamics the basic quantities are the velocity vector v, density ρ and stress tensor τ of the fluid. The stress tensor is divided into thermostatic pressure p and viscous stress σ by

$$\tau = \sigma + ph \tag{3.1}$$

where h is the flat spatial metric. Some authors use the opposite sign convention for stress, but the above convention is standard in relativity. The viscous stress σ

[a] For instance, in a static, spherically symmetric case with radius r, one has $df = f'dr$ and the state-space Gibbs equation implies $\vartheta S' = H' + pV'$. For liquid helium II at absolute zero, this implies an unphysical equation of state $H = -pV$, whereas the material Gibbs equation (4.8) would relate incompressibility, $\dot{V} = 0$, to conserved internal energy, $\dot{H} = 0$.

may be further divided into trace and traceless parts

$$\varpi = \tfrac{1}{3}h : \sigma \tag{3.2}$$
$$\varsigma = \sigma - \tfrac{1}{3}(h : \sigma)h \tag{3.3}$$

where the colon denotes double symmetric contraction. Then ϖ is the viscous pressure (or bulk viscous stress) and ς is the shear (or deviatoric) stress, satisfying

$$\sigma = \varsigma + \varpi h. \tag{3.4}$$

The inviscid case is defined by $\sigma = 0$.

Denoting the spatial volume form by $*$ and a spatial region by Σ, the volume is recovered as

$$V = \int_\Sigma *1 \tag{3.5}$$

and the mass as

$$M = \int_\Sigma *\rho. \tag{3.6}$$

Note the kinematic relation

$$\dot{*} = *D \cdot v \tag{3.7}$$

where D is the spatial gradient and the centred dot denotes contraction. Then conservation of mass (2.8) may be written in a local form as the continuity equation

$$\dot{\rho} + \rho D \cdot v = 0. \tag{3.8}$$

Similarly, conservation of momentum, or Newton's first law for the fluid, yields the Euler–Cauchy equation

$$\rho(\dot{v} + D\Phi) = -D \cdot \tau \tag{3.9}$$

where Φ is the Newtonian gravitational potential. Conservation of mass and momentum are the basic hydrodynamic field equations.

4. Non-Relativistic Thermodynamics

The classical theory of thermodynamics must be reformulated locally to be compatible with hydrodynamics. This can be done by introducing local quantities which replace the classical integral quantities. Firstly there are the internal energy density ε and entropy density s, in terms of which

$$H = \int_\Sigma *\varepsilon \tag{4.1}$$
$$S = \int_\Sigma *s. \tag{4.2}$$

Actually it is more traditional to use the specific internal energy $\epsilon_* = \varepsilon/\rho$ and the specific entropy $s_* = s/\rho$, due to frequent occurrences of combinations of the form

$$\rho \dot{f}_* = \dot{f} + f D \cdot v \tag{4.3}$$

where $f_* = f/\rho$ for a function f; the identity follows from conservation of mass (3.8).

Heat supply is replaced locally by a heat flux vector q, in terms of which

$$\dot{Q} = -\oint_{\partial\Sigma} \cdot q = -\int_{\Sigma} *D \cdot q \tag{4.4}$$

where the second expression follows from the Gauss divergence theorem. Similarly, defining the entropy flux vector $\varphi = q/\vartheta$ allows the entropy supply to be recovered as

$$\dot{S}_\circ = -\oint_{\partial\Sigma} \cdot \varphi = -\int_{\Sigma} *D \cdot \varphi. \tag{4.5}$$

This relates to $\dot{Q}$ as in the classical definition (2.4) if the temperature is uniform, $D\vartheta = 0$.

The local first law is

$$\dot{\varepsilon} + \varepsilon D \cdot v = -D \cdot q - \tau : (D \otimes v) \tag{4.6}$$

where $\otimes$ denotes the symmetric tensor product. This local first law integrates to the classical form (2.1) for an inviscid fluid with uniform pressure, $Dp = 0$. Similarly, the local second law is

$$\dot{s} + sD \cdot v + D \cdot \varphi \geq 0 \tag{4.7}$$

which integrates to the classical second law (2.5). Lastly, the local form of the Gibbs equation is

$$\vartheta \dot{s}_* = \dot{\epsilon}_* - p\dot{\rho}/\rho^2. \tag{4.8}$$

For uniform pressure and temperature, this integrates to the classical Gibbs equation (2.7).

It was found experimentally that the thermostatic state of a gas could be described by the density and temperature only. Often this is generalized by giving two equations relating $(\rho, \vartheta, \epsilon, p)$ which specify a two-dimensional subspace. However, this article will maintain (ρ, ϑ) as the state variables for definiteness. In particular, p and ϵ_* are functions of (ρ, ϑ), these relations being called, respectively, thermal and caloric equations of state. The simplest equations of state are those of an ideal gas:

$$p = R_0 \rho \vartheta \tag{4.9}$$

$$\epsilon_* = c_0 \vartheta \tag{4.10}$$

where R_0 is the gas constant and c_0 is the specific heat capacity at constant density, usually called constant volume. The local Gibbs equation then integrates to give the specific entropy, up to a time-independent function, as

$$s_* = c_0 \log \vartheta - R_0 \log \rho. \tag{4.11}$$

In the inviscid case, the first law yields

$$-D \cdot q = c_0 \rho \dot{\vartheta} - R_0 \vartheta \dot{\rho}. \tag{4.12}$$

Comparing with the classical definition of heat capacity $\dot{Q}/\dot{\vartheta}$, implicitly assuming uniform temperature, it can be seen that the specific heat capacities at constant density and pressure are c_0 and $c_0 + R_0$ respectively. Experimentally, R_0 and c_0 are nearly constant at low pressure, for a wide range of density and temperature.[11] Moreover, it is found that there is a universal gas constant $R_0 m_0$ independent of the type of gas, where m_0 is the molecular mass, and that monatomic gases satisfy

$$c_0 = \tfrac{3}{2} R_0. \tag{4.13}$$

This relation and the ideal gas laws (4.9)–(4.10) can be derived in the kinetic theory of gases [5], with $R_0 m_0$ being the Boltzmann constant.

Completing the system of equations requires further equations for (q, σ) which are consistent with the second law. This is a tight restriction, as may be seen by eliminating the internal energy between the first law (4.6) and Gibbs equation (4.8), yielding

$$\dot{s} + sD \cdot v + D \cdot \varphi = -\frac{q \cdot D\vartheta}{\vartheta^2} - \frac{\sigma : (D \otimes v)}{\vartheta}. \tag{4.14}$$

The simplest way to ensure compliance with the second law (4.7) is for the right-hand side to be a sum of squares, known as Onsager's principle, which leads to

$$q = -\kappa_0 D\vartheta \tag{4.15}$$

$$\varpi = -\lambda_0 D \cdot v \tag{4.16}$$

$$\varsigma = -2\mu_0 (D \otimes v - \tfrac{1}{3}(D \cdot v) h) \tag{4.17}$$

with

$$\kappa_0 \geq 0, \qquad \lambda_0 \geq 0, \qquad \mu_0 \geq 0. \tag{4.18}$$

If these coefficients are non-zero, the entropy production is given explicitly by

$$\dot{s} + sD \cdot v + D \cdot \varphi = \frac{q \cdot q}{\kappa_0 \vartheta^2} + \frac{\varpi^2}{\lambda_0 \vartheta} + \frac{\varsigma : \varsigma}{2\mu_0 \vartheta} \geq 0 \tag{4.19}$$

which is manifestly non-negative. This makes it clear that (q, σ) cause entropy production, i.e. they are thermally dissipative in nature. In this article, (q, σ) will be called the *dissipative fields* and the equations (4.15)–(4.17) determining them the *dissipative relations*. The dissipative relations and the equations of state are usually described collectively as the constitutive relations.

One may recognize the dissipative relation (4.15) for q as the Fourier equation and the dissipative relations (4.16)–(4.17) for σ as the definition of a Newtonian fluid; substituting the latter into conservation of momentum (3.9) yields

$$\rho(\dot{v} + D\Phi) = -Dp + \mu_0 D^2 v + (\lambda_0 + \tfrac{1}{3}\mu_0)D(D \cdot v) \tag{4.20}$$

which is the Navier–Stokes equation. These are the classical equations describing heat conduction and viscosity respectively, with κ_0 being the thermal conductivity, μ_0 the shear viscosity and λ_0 the bulk viscosity. Originally, these dissipative relations were discovered experimentally, with $\lambda_0 = 0$, but the above method shows that they may be loosely derived from the local second law. This seems to have been originally noticed by Eckart,[12] who immediately generalized it to relativity.[2]

Note that one of the dynamical equations is the local Gibbs equation (4.8), even though this was originally concerned only with thermostatics. Thus it has been implicitly assumed that, in the general thermodynamic case, the Gibbs equation, as well as the thermal and caloric equations of state (4.9)–(4.10), continue to take their thermostatic forms. Also, note that there is no need for a state-space Gibbs equation $\vartheta ds_* = d\epsilon_* + pd(1/\rho)$; indeed it would be generally inconsistent with the material Gibbs equation (4.8). In this article, the material Gibbs equation and the relation $\varphi = q/\vartheta$ will be called *entropic relations*, since they effectively define the entropic fields (s, φ), up to the usual ambiguity in s. In the kinetic theory of gases, the entropic fields are defined in terms of the molecular distribution, allowing the entropic relations to be derived.[5]

In summary, the basic thermohydrodynamic fields are $(\rho, v, p, \sigma, \varepsilon, q, \vartheta, s, \varphi)$, which constitutes twenty functions. There are five conservation equations, namely the first law and conservation of mass and momentum. (The first law may be regarded as conservation of heat, as will become clear in the relativistic theory). The system is completed by two equations of state for (p, ϵ_*), four components of the entropic relations for (s, φ) and nine dissipative relations for (q, σ), the latter being chosen to imply the second law.

The fields pair up neatly with the equations expressing them or their material derivatives in terms of the basic fields and their spatial derivatives, except for temperature and the first law; even this becomes a temperature propagation equation on substituting other relations. For instance, for an inviscid ideal gas at constant density $\rho = \rho_0$, the local first law reads

$$c_0 \rho_0 \dot{\vartheta} = \kappa_0 D^2 \vartheta \tag{4.21}$$

which is the diffusion equation. This describes infinitely fast propagation of temperature and is therefore physically implausible. There is a similar problem of infinitely fast propagation implied by the Navier–Stokes equations.[13] Thus the standard local thermodynamics described so far, e.g. by de Groot and Mazur,[5] needs to be modified.

5. Extended Thermodynamics

The problem of infinite propagation speeds stems from the assumption that the entropy density s satisfies the Gibbs equation (4.8), originally intended to apply only to the thermostatic case. A generalisation due to Müller[3] involves allowing the entropy to depend on quadratic terms in the dissipative fields (q, σ). This reduces to the foregoing theory in an approximation where the quadratic terms may be neglected, i.e. the new theory is a quadratic rather than linear approximation to non-equilibrium. Applying the second law as before yields modified dissipative relations containing relaxation terms that provide finite propagation speeds. Moreover, dissipative relations of exactly this type may be obtained in the kinetic theory of gases by taking sufficiently many moments of the Boltzmann equation, as originally shown by Grad.[8] Müller and Ruggeri[13] now describe this as extended thermodynamics of irreversible processes, reserving extended thermodynamics to describe a more systematic theory which assumes a further divergence-type balance law.

Since a distinction between entropy and thermostatic entropy is now required, it is convenient to use s_* for the thermostatic specific entropy henceforth, with s being the entropy density as before. The equations displayed in the preceding subsection have been written with this distinction in mind, e.g. the second law (4.7) contains s and the Gibbs equation (4.8) contains s_*. The thermostatic relation $s = \rho s_*$ is now modified to include quadratic terms in (q, σ), giving the general form

$$s = \rho s_* - \tfrac{1}{2}\rho(b_q q \cdot q + b_\varpi \varpi^2 + b_\varsigma \varsigma : \varsigma). \tag{5.1}$$

Here the coefficients b are arbitrary apart from being non-negative, since entropy must reach a maximum in the thermostatic case:

$$b_{\{q,\varpi,\varsigma\}} \geq 0. \tag{5.2}$$

Similarly, the entropy flux φ may differ from the thermostatic entropy flux q/ϑ by quadratic terms in (q, σ), giving the general form

$$\varphi = \frac{q}{\vartheta} - k_\varpi \varpi q - k_\varsigma \varsigma \cdot q. \tag{5.3}$$

Then the entropy production may be written explicitly as

$$\begin{aligned}\dot{s} + sD\cdot v + D\cdot\varphi = &-q\cdot\left(\frac{D\vartheta}{\vartheta^2} + b_q\rho\dot{q} + k_\varsigma D\cdot\varsigma + k_\varpi D\varpi\right)\\ &-\varpi\left(\frac{D\cdot v}{\vartheta} + b_\varpi\rho\dot{\varpi} + k_\varpi D\cdot q\right)\\ &-\varsigma:\left(\frac{D\otimes v}{\vartheta} + b_\varsigma\rho\dot{\varsigma} + k_\varsigma D\otimes q\right)\end{aligned} \tag{5.4}$$

where the coefficients (b, k) have been assumed constant for the sake of simplicity.

So the second law suggests modified dissipative relations

$$q = -\kappa_0(D\vartheta + b_q\rho\vartheta^2\dot{q} + k_\varsigma\vartheta^2 D\cdot\varsigma + k_\varpi\vartheta^2 D\varpi) \tag{5.5}$$

$$\varpi = -\lambda_0(D\cdot v + b_\varpi\rho\vartheta\dot{\varpi} + k_\varpi\vartheta D\cdot q) \tag{5.6}$$

$$\varsigma = -2\mu_0(D\otimes v - \tfrac{1}{3}(D\cdot v)h + b_\varsigma\rho\vartheta\dot{\varsigma} + k_\varsigma\vartheta(D\otimes q - \tfrac{1}{3}(D\cdot q)h)) \tag{5.7}$$

which yield the entropy production with the same quadratic form as (4.19). The key point here is that the relaxation coefficients b_f introduce the time derivatives of the dissipative fields $f = (q, \varpi, \varsigma)$ into the dissipative relations, giving equations for $f + t_f\dot{f}$, where the relaxation timescales are

$$t_q = \kappa_0 b_q\rho\vartheta^2, \tag{5.8}$$

$$t_\varpi = \lambda_0 b_\varpi\rho\vartheta, \tag{5.9}$$

$$t_\varsigma = 2\mu_0 b_\varsigma\rho\vartheta. \tag{5.10}$$

This leads to finite propagation speeds, at least for linear perturbations.[13] The temperature propagation equation for an inviscid ideal gas at constant density now becomes

$$c_0\rho_0(\dot{\vartheta} + t_q\ddot{\vartheta}) = \kappa_0 D^2\vartheta \tag{5.11}$$

for constant t_q. This is the telegraph equation, with finite propagation speed $\sqrt{\kappa_0/c_0\rho_0 t_q}$.

In summary, Müller[3] obtained a causal theory of thermodynamics simply by allowing the entropic fields (s, φ) to depend on quadratic terms in the dissipative fields (q, σ). There has been no tampering with the conservation laws, the second law, the equations of state or even the Gibbs equation, provided the latter is taken in thermostatic form, i.e. in terms of the thermostatic specific entropy s_*. This is often explained misleadingly as a generalization of the Gibbs equation; in reality, the thermostatic Gibbs equation is still assumed but the entropy density is no longer assumed to be purely thermostatic. One might continue to suppose that the thermostatic s_* is a state function, but the entropy density s can no longer be a state function, even for an ideal gas, thereby toppling what many textbooks erect as a principle of thermodynamics.

6. General Relativity

General relativity is based on a four-dimensional manifold, space-time, with a symmetric bilinear form g, the metric. Taking units such that the Newtonian gravitational constant and the speed of light are unity, the Einstein equation is

$$G = 8\pi T \tag{6.1}$$

where G is the Einstein tensor of g and T is the energy tensor of the matter, which will be taken in their contravariant forms. The energy tensor is more fully described as the energy-momentum-stress tensor, due to the physical interpretation of its various projections in a given frame as the energy density, momentum density or energy flux, and stress.

The contracted Bianchi identity, a purely geometrical identity, reads

$$\nabla \cdot G = 0 \tag{6.2}$$

where ∇ is the covariant derivative operator of g. Therefore

$$\nabla \cdot T = 0 \tag{6.3}$$

which expresses energy-momentum conservation. A material model is specified by giving the general form of T in terms of the material fields, supplemented by any required equations other than energy-momentum conservation. This paper gives such a procedure as a definition of thermodynamic matter. Any such procedure requires at least a first law of thermodynamics. Physically, the first law is an energy conservation equation, but it differs from conservation of energy in the sense of a time-component of energy-momentum conservation (6.3). The difference turns out to be conservation of mass. In the Newtonian limit, conservation of energy yields conservation of mass to leading order and the first law as the next correction. Thus relativistic thermodynamics requires a concept of mass. In a microscopic description, this would be just the mass of the particles.

In relativity, mass is described by an energy-momentum vector J which will be called the *material current*. Its magnitude is the density ρ and its direction is the velocity vector u, assumed timelike:

$$J = \rho u, \tag{6.4}$$

$$u \cdot u^\flat = -1. \tag{6.5}$$

Here the sign convention is that spatial metrics are positive definite and $\flat$ denotes the covariant dual with respect to g (index lowering). Similarly, $\sharp$ will denote the contravariant dual (index raising). These accents are included for book-keeping and are irrelevant for many purposes. In relativistic kinetic theory of gases,[7] J is the first moment of the molecular distribution, with T being the second moment. Conservation of mass and energy-momentum may then be derived from the relativistic Boltzmann equation.

Conservation of mass is expressed simply by

$$\nabla \cdot J = 0. \tag{6.6}$$

Written explicitly,

$$0 = \nabla \cdot J = \dot{\rho} + \rho \nabla \cdot u \tag{6.7}$$

where the overdot henceforth denotes the covariant derivative along u, $\dot{f} = u \cdot \nabla f$, which is the relativistic material derivative, or comoving derivative along the flow. This has a similar form to non-relativistic conservation of mass (3.8).

Mass is a local form of energy in relativity, so it has an energy tensor T_M which will be called the *material energy tensor.* This takes the standard form

$$T_M = \rho u \otimes u \tag{6.8}$$

so that the relation

$$\nabla \cdot J = -u \cdot (\nabla \cdot T_M)^\flat \tag{6.9}$$

shows that conservation of mass is consistent with energy-momentum conservation for this type of matter, known as dust.

7. Heat

The proposed theory of relativistic thermodynamics is based on four main physical principles. The first principle is that (i) *heat is a local form of energy.* In general relativity, heat is therefore described by an energy tensor T_H which will be called the *thermal energy tensor. Thermodynamic matter* will be taken to mean a form of matter for which mass and heat are the only local forms of energy:

$$T = T_M + T_H. \tag{7.1}$$

Comparing with a microscopic description, T_H is the effective energy tensor produced by the random motion of particles around the average flow given by u, as can be made precise in relativistic kinetic theory of gases.[7]

The physical meaning of the various components of the thermal energy tensor are now determined:

$$\varepsilon = u \cdot T_H^\flat \cdot u \tag{7.2}$$

is the *thermal energy density,*

$$q = -\perp(T_H \cdot u^\flat) \tag{7.3}$$

is the *thermal flux* and

$$\tau = \perp T_H \tag{7.4}$$

is the *thermal stress,* where $\perp$ denotes projection by the spatial metric

$$h^{-1} = g + u^\flat \otimes u^\flat \tag{7.5}$$

orthogonal to the fluid flow. Note that h denotes the contravariant form and h^{-1} the covariant form. Therefore

$$T_H = \varepsilon u \otimes u + 2u \otimes q + \tau. \tag{7.6}$$

As the notation indicates, thermal energy density, flux and stress will be identified respectively with the internal energy density, heat flux and stress of the non-relativistic theory. Therefore the energy tensor of the thermodynamic matter is

$$T = (\rho + \varepsilon) u \otimes u + 2u \otimes q + \tau. \tag{7.7}$$

This suffices to determine the quasi-local first law of relativistic thermodynamics in spherical symmetry.[1] Written in terms of the effective density $\varrho = \rho + \varepsilon$, the energy tensor has the same form as that of Eckart,[2] who defined specific internal energy as ϱ/ρ plus an undetermined constant. This is consistent, since specific thermal energy is ε/ρ.

The above principle implies that (ε, q, τ) are different aspects of heat. In particular, one may say that thermal energy density or heat density ε represents heat at rest, and q heat in motion, relative to the average flow. This furnishes the common-sense explanation for such simple physical processes as the heating, insulation and cooling of a body, namely that heat flows in, remains at rest, then flows out. This intuitive explanation is often mocked in thermodynamics textbooks, accompanied by the claim that heat flux is the only form of heat. In relativity, thermal flux is just one projection of a thermal energy tensor.

Similarly, what is traditionally called internal energy is just thermal energy, or simply heat, as the term is used in everyday language. Moreover, the ambiguous zero of internal energy is fixed for thermal energy. Recall that the original meaning of internal energy derived from the fact that the measurable energies, due to heat flux and work, were not conserved. According to conservation of energy, there had to be some other form of energy, internal to the system. The experiments of Joule established that this internal energy satisfied simple laws for an ideal gas. This stop-gap concept of internal energy differs intrinsically from that of thermal energy, yet they turn out to agree. That internal energy is really thermal energy should anyway be clear even from non-relativistic kinetic theory of gases,[5] where internal energy is defined as the average kinetic energy of the particles in the centre-of-momentum frame.

The second physical principle is (ii) *conservation of mass* (6.6). As energy-momentum conservation (6.3) is automatic in general relativity, conservation of mass is equivalent by (6.9) to *conservation of heat*

$$u \cdot (\nabla \cdot T_H)^\flat = 0. \tag{7.8}$$

Written explicitly,

$$0 = -u \cdot (\nabla \cdot T_H)^\flat = \dot{\varepsilon} + \varepsilon \nabla \cdot u + \nabla \cdot q + q \cdot \dot{u}^\flat + \tau : (\nabla \otimes u^\flat). \tag{7.9}$$

This reduces to the non-relativistic first law (4.6) in the Newtonian limit, since the term in the acceleration $\dot{u}$ disappears, as shown in the next section. Thus principle (ii) is equivalent to the relativistic *first law*. This reveals that the first law is conservation of energy minus conservation of mass, cf. Eckart.[2] One may say that the first law is conservation of heat for the thermodynamic matter. Note that this could be modified in the presence of other matter, which would also contribute to the total energy tensor T.

Several points perhaps need to be stressed. Firstly, although kinetic theory of gases has been mentioned for comparison, the above theory of heat is manifestly independent of such microscopic theories. Secondly, conservation of mass is required by the first law of thermodynamics. In the perspective of general relativity, this is the primary distinction between thermodynamic matter and any other energy tensor T to be used in the Einstein equation. Thirdly, conservation of mass involves a material current which determines a preferred time direction u, the material frame of Eckart. It is only in this frame that thermal flux—meaning the energy-flux component of the thermal energy tensor—agrees with what is normally called heat flux, even in the limit of non-relativistic thermodynamics. Finally, the basic conservation equations (6.3) and (6.6) are standard in relativistic thermodynamics and date back to Eckart.[2] What seems to have been lacking is the physical interpretation in terms of mass and heat. In particular, conservation of heat should not be confused with the old caloric theory, as discussed further in the Conclusion.

8. Newtonian Limit

The Newtonian limit may be described in terms of a vector λ which plays the role of Newtonian time. The function Φ defined by

$$e^{2\Phi} = -\lambda \cdot \lambda^\flat \tag{8.1}$$

turns out to reduce to the Newtonian gravitational potential. Similarly, the component of u orthogonal to λ turns out to reduce to the Newtonian velocity:

$$v = u + e^{-2\Phi}(u \cdot \lambda^\flat)\lambda. \tag{8.2}$$

Inverting,

$$\lambda = \frac{e^\Phi (u - v)}{\sqrt{1 + |v|^2}}. \tag{8.3}$$

Taking units of length, factors of the speed c of light may be introduced by the formal replacements

$$h \mapsto h, \tag{8.4}$$

$$(u, \lambda, v) \mapsto c^{-1}(u, \lambda, v), \tag{8.5}$$

$$(\Phi, \rho) \mapsto c^{-2}(\Phi, \rho), \tag{8.6}$$

$$(\varepsilon, \tau) \mapsto c^{-4}(\varepsilon, \tau), \tag{8.7}$$

$$q \mapsto c^{-5} q. \tag{8.8}$$

These factors are determined by the desired physical interpretation of the fields in Newtonian theory. This implicitly includes consequences like $D \mapsto D$ for the covariant derivative of h^{-1} and $\dot{f} \mapsto c^{-1}\dot{f}$ (where $f \mapsto f$) for the material derivative $\dot{f} = u \cdot \nabla f$. These replacements will be assumed in any subsequent equation involving c.

For λ to qualify as a candidate for Newtonian time, it should satisfy

$$\perp L_\lambda h = O(c^{-2}) \tag{8.9}$$

where L_λ is the Lie derivative along λ. This condition reflects the absoluteness of Newtonian space. Then

$$\bot(\nabla \otimes u^\flat) = D \otimes v^\flat + O(c^{-2}) \tag{8.10}$$

and in particular

$$\nabla \cdot u = D \cdot v + O(c^{-2}). \tag{8.11}$$

So conservation of mass (6.6) and heat (7.8) read

$$\dot{\rho} + \rho D \cdot v = O(c^{-2}) \tag{8.12}$$

$$\dot{\varepsilon} + \varepsilon D \cdot v + D \cdot q + \tau : (D \otimes v^\flat) = O(c^{-2}) \tag{8.13}$$

which manifestly reduce to non-relativistic conservation of mass (3.8) and heat (4.6).

Conservation of momentum may be written explicitly as

$$0 = \bot(\nabla \cdot T) = (\rho + \varepsilon)\bot\dot{u} + \bot(q \cdot \nabla)u + q(\nabla \cdot u) + \bot\dot{q} + \bot(\nabla \cdot \tau). \tag{8.14}$$

Inserting factors of c as in (8.4)–(8.8), all terms except the first and last disappear in the Newtonian limit. A longer argument, skipped here, confirms that this reduces to non-relativistic conservation of momentum (3.9). Thus all the conservation equations of non-relativistic thermodynamics have been generalized to general relativity, as conservation of energy-momentum and heat or mass. Similarly, it is straightforward to check the Newtonian limits of the second law, entropic relations and constitutive relations yet to be given.

9. Thermostatics and Thermodynamics

The next step, giving equations of state, requires a distinction between thermostatic and dissipative quantities, e.g. between thermostatic pressure and viscous pressure. Thus the thermal energy tensor T_H is divided into a thermostatic part T_0 and a dissipative part T_1 by

$$T_H = T_0 + T_1 \tag{9.1}$$

with $T_1 = 0$ in the thermostatic case. The two are to be distinguished as in the non-relativistic theory by the type of equations determining them: T_0 is to be given by equations of state and T_1 by dissipative relations consistent with the second law, the latter being described in the next section.

A *fluid* may be defined by T_0 being isotropic, therefore taking the general form

$$T_0 = \epsilon u \otimes u + ph. \tag{9.2}$$

Then p is the *thermostatic pressure* and ϵ the *thermostatic energy density*. This isotropic form is appropriate for a fluid but not a solid; for instance, an elastic solid generally has anisotropic thermostatic stress, i.e. a general tensor replacing ph in T_0.

The desired thermal and caloric equations of state may then be combined into an equation of state for T_0 in terms of (ρ, ϑ). For instance, an ideal gas may be defined by

$$T_0 = (c_0 u \otimes u + R_0 h)\rho\vartheta \tag{9.3}$$

which gives thermal and caloric equations of state with the same form as for the non-relativistic ideal gas (4.9)–(4.10). Generally, the equations determining T_0 may be called the *thermostatic relations*.

In the thermostatic case $T_1 = 0$, this recovers the relativistic model of a perfect fluid,[14] consisting of the energy tensor

$$T_M + T_0 = (\rho + \epsilon)u \otimes u + ph \tag{9.4}$$

together with equations of state for (ϵ, p) as functions of ρ and conservation of mass or heat (7.8), which yields

$$\dot{\epsilon}_* = p\dot{\rho}/\rho^2 \tag{9.5}$$

in terms of the specific thermostatic energy

$$\epsilon_* = \epsilon/\rho. \tag{9.6}$$

For an ideal gas, this gives $\vartheta = f_0 \rho^{R_0/c_0}$ for a time-independent function f_0. Thus

$$\epsilon = c_0 f_0 \rho^{1+R_0/c_0}, \tag{9.7}$$

$$p = R_0 f_0 \rho^{1+R_0/c_0}. \tag{9.8}$$

These equations of state specify a thermostatic ideal gas. Actually, it is more common to use the effective density $\varrho = \rho + \epsilon$ instead of ρ and ϵ, forget mass conservation and give a single equation of state for p as a function of ϱ. Indeed, it is common to ignore ϵ and replace ρ with ϱ in the last equation, giving, for constant f_0, the polytropic equation of state. This is a reasonable approximation if $\epsilon \ll \rho$, as in Newtonian theory, but the distinction between ρ and ϱ is important in general.

Returning to the general thermodynamic case, defining the *dissipative stress* or viscous stress

$$\sigma = \tau - ph \tag{9.9}$$

and the *dissipative energy density*

$$\eta = \varepsilon - \epsilon \tag{9.10}$$

allows the dissipative energy tensor to be composed as

$$T_1 = \eta u \otimes u + 2u \otimes q + \sigma \tag{9.11}$$

which is the general form of such a tensor. It will turn out that η vanishes in the linear approximation for entropy, allowing consistency with standard non-relativistic thermodynamics.[5] Nevertheless, η will be retained on the grounds that, as a matter of principle, thermal energy need not be purely thermostatic. The same distinction between thermostatic and dissipative energy density could be drawn in the non-relativistic case, leaving ambiguities to be fixed in equations where η might occur; these were fixed in equations displayed in Section 4 by using ε and ϵ_* with appropriate foresight.

Exhibiting all terms in the energy tensor,

$$T = T_M + T_0 + T_1 = (\rho + \epsilon + \eta)u \otimes u + 2u \otimes q + ph + \sigma. \tag{9.12}$$

This shows that T incorporates all the basic thermohydrodynamic fields of the non-relativistic theory described previously, apart from temperature ϑ and entropy (s, φ), but adding dissipative energy density η.

10. Entropy

The third physical principle is that (iii) *entropy is a local current.* In relativity, it is therefore described by an *entropy current* vector Ψ whose components are *entropy density*

$$s = -u^\flat \cdot \Psi \tag{10.1}$$

and *entropy flux*

$$\varphi = \perp \Psi. \tag{10.2}$$

As their names indicate, entropy density and entropy flux will be identified with their non-relativistic versions. Therefore

$$\Psi = su + \varphi. \tag{10.3}$$

The fourth physical principle is (iv) *non-destruction of entropy.* In relativity, this is therefore expressed by

$$\nabla \cdot \Psi \geq 0. \tag{10.4}$$

Explicitly,

$$\dot{s} + s\nabla \cdot u + \nabla \cdot \varphi \geq 0 \tag{10.5}$$

which reduces to the non-relativistic second law (4.7) in the Newtonian limit. Thus principle (iv) is the relativistic *second law.* These two principles are standard in relativistic thermodynamics. Again it seems to be Eckart[2] who first gave a version of (10.4), assuming $\varphi = q/\vartheta$.

The final stage consists of introducing entropic relations for Ψ in terms of the other fields, then finding dissipative relations for T_1 which imply the second law. Successive approximations for Ψ will be taken: $\Psi = \Psi_0$ for the thermostatic case, $\Psi = \Psi_0 + \Psi_1$ for the linear approximation and $\Psi = \Psi_0 + \Psi_1 + \Psi_2$ for the quadratic approximation for entropy.

The thermostatic entropy current Ψ_0 may be assumed to satisfy

$$\perp\Psi_0 = 0, \tag{10.6}$$

$$\nabla \cdot \Psi_0 = -\frac{u}{\vartheta} \cdot (\nabla \cdot T_0)^\flat. \tag{10.7}$$

The first equation states that the thermostatic entropy current is comoving with the material. Therefore Ψ_0 may be written in terms of the specific thermostatic entropy

$$s_* = -u^\flat \cdot \Psi_0/\rho. \tag{10.8}$$

For a fluid (9.2), the second equation may be written as

$$\vartheta\nabla \cdot \Psi_0 = \dot{\epsilon} + \epsilon\nabla \cdot u + p\nabla \cdot u \tag{10.9}$$

or

$$\vartheta\dot{s}_* = \dot{\epsilon}_* - p\dot{\rho}/\rho^2. \tag{10.10}$$

This form was given by Eckart[2] and is identical to that of the non-relativistic local Gibbs equation (4.8).[b] This is usually regarded as defining thermostatic temperature. For an ideal gas (9.3), it integrates to

$$s_* = c_0 \log\vartheta - R_0 \log\rho \tag{10.11}$$

which has the same form as in the non-relativistic case (4.11).

The relations (10.6)–(10.7) therefore constitute a generalized Gibbs equation, relativistically unified in terms of Ψ_0 and T_0. This relativistic Gibbs equation has the desired property that entropy is conserved in the thermostatic case:

$$T_1 = 0 \quad \Rightarrow \quad \nabla \cdot \Psi_0 = \rho\dot{s}_* = 0 \tag{10.12}$$

as follows from conservation of heat (7.8). In other words, thermostatics requires equality in the second law. Indeed, one might propose a stronger version of the second law, to the effect that non-thermostatic processes necessarily produce entropy. One could then define thermostatics, or equilibrium thermodynamics, or reversible thermodynamics, by equality in the second law (10.4).

[b]Most references on relativistic thermodynamics give a different Gibbs equation. Sometimes the material derivative is replaced with the covariant derivative in the Gibbs equation,[15,16] giving an equation whose spatial components are quite unnecessary and generally overdetermine the fields. This equation has even been described as the first law. In other cases, the material derivative is replaced with a state-space differential in the Gibbs equation,[4,6] which generally does not imply the material Gibbs equation (10.10). It has even been claimed that the thermostatic (Ψ_0, T_0) should be replaced with (Ψ, T) in this state-space equation.[4]

11. Linear Approximation

The linear correction Ψ_1 to entropy current may be assumed as

$$\Psi_1 = -\frac{u^\flat}{\vartheta} \cdot T_1 \tag{11.1}$$

which may be regarded as defining temperature ϑ away from equilibrium. Then the entropy flux $\perp\Psi_1 = q/\vartheta$ has the same form as in non-relativistic thermodynamics. The component of Ψ_1 along the flow, η/ϑ, is not so determined, except that the above linear form is the simplest. However, it turns out to yield a useful cancellation in the entropy production, which, using conservation of heat (7.8), becomes simply

$$\nabla \cdot (\Psi_0 + \Psi_1) = -T_1 : \left(\nabla \otimes \frac{u^\flat}{\vartheta}\right). \tag{11.2}$$

Thus the entropy production is a contraction of the dissipative energy tensor T_1 with the gradient of the inverse temperature vector u/ϑ, thereby unifying what are sometimes called, respectively, the thermodynamic fluxes and thermodynamic forces.[5] Writing the components explicitly,

$$\nabla \cdot (\Psi_0 + \Psi_1) = -\frac{q \cdot (\nabla\vartheta + \vartheta\dot{u}^\flat)}{\vartheta^2} - \frac{\sigma : (\nabla \otimes u^\flat)}{\vartheta} - \frac{\eta\dot{\vartheta}}{\vartheta^2}. \tag{11.3}$$

So the second law (10.4) suggests the dissipative relations

$$q = -\kappa_0 \perp(\nabla^\sharp\vartheta + \vartheta\dot{u}) \tag{11.4}$$

$$\varpi = -\lambda_0 \nabla \cdot u \tag{11.5}$$

$$\varsigma = -2\mu_0 \perp(\nabla^\sharp \otimes u - \tfrac{1}{3}(\nabla \cdot u)h) \tag{11.6}$$

$$\eta = -\nu_0\dot{\vartheta} \tag{11.7}$$

with

$$\kappa_0 \geq 0, \qquad \lambda_0 \geq 0, \qquad \mu_0 \geq 0, \qquad \nu_0 \geq 0. \tag{11.8}$$

If these coefficients are non-zero, the entropy production is given explicitly by

$$\nabla \cdot \Psi = \frac{q \cdot q^\flat}{\kappa_0\vartheta^2} + \frac{\varpi^2}{\lambda_0\vartheta} + \frac{\varsigma : \varsigma^\flat}{2\mu_0\vartheta} + \frac{\eta^2}{\nu_0\vartheta^2} \geq 0. \tag{11.9}$$

The dissipative relation for q is a relativistic modification of the Fourier equation, including a term in the acceleration $\dot{u}$ which disappears in the Newtonian limit. The dissipative relations for σ are relativistic versions of the viscosity relations for a Newtonian fluid, yielding a relativistic Navier–Stokes equation. With $\varpi = 0$, these were originally derived by Eckart.[2] The last dissipative relation for η appears to be new and modifies the temperature propagation equation: taking the example of an inviscid ideal gas at constant density with vanishing acceleration, the first law (7.8) becomes

$$\rho_0 c_0 \dot{\vartheta} - \nu_0 \ddot{\vartheta} = \kappa_0 D^2 \vartheta. \tag{11.10}$$

Thus the parabolic (diffusion) equation for $\nu_0 = 0$ becomes a hyperbolic (telegraph) equation for $\nu_0 < 0$, as desired, but an elliptic equation for $\nu_0 > 0$. Since ellipticity is even worse than parabolicity on causal grounds, this leads to $\nu_0 = 0$ and therefore $\eta = 0$ as promised. Thus the linear approximation for entropy recovers and generalizes the Eckart theory,[2] which concerned the case of a fluid without bulk viscosity.

12. Quadratic Approximation

For the quadratic correction Ψ_2 to the entropy current, the most general form which is quadratic in the dissipative energy tensor T_1 is

$$\Psi_2 = -\frac{1}{2}(b_q q \cdot q^\flat + b_\varpi \varpi^2 + b_\varsigma \varsigma : \varsigma^\flat + b_\eta \eta^2)\rho u - k_\varpi \varpi q - k_\varsigma \varsigma \cdot q^\flat - k_\eta \eta q \quad (12.1)$$

where

$$b_{\{q,\varpi,\varsigma,\eta\}} \geq 0 \quad (12.2)$$

so that entropy is maximized in equilibrium. The coefficients b lead to relaxation effects as in the non-relativistic case. Explicitly, the entropy production is

$$\begin{aligned}\nabla \cdot (\Psi_0 + \Psi_1 + \Psi_2) = &-q \cdot \left(\frac{\nabla\vartheta + \vartheta\dot{u}^\flat}{\vartheta^2} + b_q\rho\dot{q}^\flat + k_\varsigma(\nabla\cdot\varsigma)^\flat + k_\varpi\nabla\varpi + k_\eta\nabla\eta\right)\\ &-\varpi\left(\frac{\nabla\cdot u}{\vartheta} + b_\varpi\rho\dot{\varpi} + k_\varpi\nabla\cdot q\right)\\ &-\varsigma : \left(\frac{\nabla\otimes u^\flat}{\vartheta} + b_\varsigma\rho\dot{\varsigma}^\flat + k_\varsigma\nabla\otimes q^\flat\right)\\ &-\eta\left(\frac{\dot{\vartheta}}{\vartheta^2} + b_\eta\rho\dot{\eta} + k_\eta\nabla\cdot q\right)\end{aligned} \quad (12.3)$$

where the coefficients (b, k) have been assumed constant for the sake of simplicity, the generalization being straightforward. The entropy production takes the same quadratic form (11.9) as in the linear approximation if the dissipative relations are taken as

$$\begin{aligned} q &= -\kappa_0\bot(\nabla^\sharp\vartheta + \vartheta\dot{u} + b_q\rho\vartheta^2\dot{q} + k_\varsigma\vartheta^2\nabla\cdot\varsigma + k_\varpi\vartheta^2\nabla^\sharp\varpi \\ &\quad + k_\eta\vartheta^2\nabla^\sharp\eta), && (12.4)\\ \varpi &= -\lambda_0(\nabla\cdot u + b_\varpi\rho\vartheta\dot{\varpi} + k_\varpi\vartheta\nabla\cdot q), && (12.5)\\ \varsigma &= -2\mu_0\bot(\nabla^\sharp\otimes u - \tfrac{1}{3}(\nabla\cdot u)h + b_\varsigma\rho\vartheta\dot{\varsigma} + k_\varsigma\vartheta(\nabla^\sharp\otimes q \\ &\quad -\tfrac{1}{3}h : (\nabla\otimes q^\flat)h)), && (12.6)\\ \eta &= -\nu_0(\dot{\vartheta} + b_\eta\rho\vartheta^2\dot{\eta} + k_\eta\vartheta^2\nabla\cdot q). && (12.7)\end{aligned}$$

This completes the system of thermodynamic field equations for specified coefficients (b, k). In the case $\eta = 0$, these dissipative relations reduce to those of Israel and

Stewart,[4] also given by Jou et al.[17] If the coefficients (b, k) are not constant, additional derivatives appear, as noted by Hiscock and Lindblom.[18]

Relativistic kinetic theory of gases[7] confirms the existence of dissipative energy density η; indeed it emerged unrecognised in the non-relativistic theory of Grad[8] as essentially the 14th moment of the molecular distribution, the previous 13 moments being $(\rho, v, \epsilon, \varsigma, q)$, with (ϑ, p) being given by equations of state and ϖ vanishing. For instance, for a monatomic (4.13) ideal (9.3) gas with $\varpi = 0$, setting

$$b_q = \frac{2}{5p^2\vartheta} \qquad k_\varsigma = \frac{2}{5p\vartheta} \qquad b_\varsigma = \frac{1}{2p\rho\vartheta} \qquad k_\eta = \frac{8\rho}{15p^2\vartheta} \qquad b_\eta = \frac{8\rho}{15p^3\vartheta} \tag{12.8}$$

and

$$\kappa_0 = \frac{5p^2}{2B_0\rho^2\vartheta} \qquad \mu_0 = \frac{2p}{3B_0\rho} \qquad \nu_0 = \frac{15p^3}{8B_0\rho^3\vartheta} \tag{12.9}$$

where B_0 is a constant, the dissipative relations in the non-relativistic limit become

$$0 = \dot{q} + B_0\rho q + \tfrac{5}{2}R_0^2\rho\vartheta D\vartheta + R_0\vartheta D\cdot\varsigma + \tfrac{4}{3}D\eta \tag{12.10}$$

$$0 = \dot{\varsigma} + \tfrac{3}{2}B_0\rho\varsigma + 2R_0\rho\vartheta(D\otimes v - \tfrac{1}{3}(D\cdot v)h) + \tfrac{4}{5}(D\otimes q - \tfrac{1}{3}(D\cdot q)h) \tag{12.11}$$

$$0 = \dot{\eta} + B_0\rho\eta + R_0\vartheta D\cdot q \tag{12.12}$$

where the speed of light has been introduced as in (8.4–8.8), with $R_0 \mapsto c^2R_0$, $B_0 \mapsto cB_0$, $\vartheta \mapsto c^{-4}\vartheta$ and $\eta \mapsto c^{-6}\eta$. Together with the non-relativistic conservation laws, these are the 14-moment equations (1.42) of Müller and Ruggeri,[13] their 14th moment being 8η, with the factor determined in relativistic kinetic theory.[7] This confirms the existence of dissipative energy density and thereby topples another pillar of textbook thermodynamics, that internal energy necessarily be a state function.

13. Integral Quantities and Laws

There exist integral forms of the second law and conservation of mass, because the local forms are current (semi-) conservation laws, allowing the Gauss divergence theorem to be applied. In particular, the mass of a region Σ of a spatial hypersurface may be defined by

$$M(\Sigma) = -\int_\Sigma J \tag{13.1}$$

where the volume form and contraction with unit normal are implicit. Consider a space-time region Ω bounded by Σ, a later Σ' and a hypersurface region $\hat{\Sigma}$ generated by flowlines of u. Applying the Gauss theorem,

$$M(\Sigma') - M(\Sigma) = \int_\Omega \nabla\cdot J = 0 \tag{13.2}$$

which is the integral form of conservation of mass. Similarly, the entropy of Σ may be defined by

$$S(\Sigma) = -\int_{\Sigma} \Psi \tag{13.3}$$

and the entropy supply through $\hat{\Sigma}$ by

$$S_{\circ}(\hat{\Sigma}) = -\int_{\hat{\Sigma}} \Psi. \tag{13.4}$$

Then the Gauss theorem yields

$$S(\Sigma') - S(\Sigma) - S_{\circ}(\hat{\Sigma}) = \int_{\Omega} \nabla \cdot \Psi \geq 0 \tag{13.5}$$

which is the integral second law. Thus there is an integral second law of thermodynamics even in general relativity. An integral first law may seem awkward in similar generality, since the local first law (7.8) is not a current conservation law, but in spherical symmetry a simple integral (or quasi-local) first law exists.[1]

The integral laws become particularly simple if the vorticity vanishes,

$$\perp(\nabla \wedge u^{\flat}) = 0 \tag{13.6}$$

where $\wedge$ denotes the antisymmetric tensor product. Then there exist spatial hypersurfaces orthogonal to u which may be used to define preferred volume integrals. For instance, the mass is

$$M = \int_{\Sigma} *\rho \tag{13.7}$$

where Σ henceforth denotes a region of a hypersurface orthogonal to u, with volume form $*$. The induced metric of Σ is just h^{-1}, which defines $*$. One may assume that Σ has finite volume, or infinite volume with all relevant integrals existing.

Concerning time derivatives: in the relativistic case, $\dot{M}$ is generally meaningless, since proper time along u generally does not label the hypersurfaces orthogonal to u. Labelling these hypersurfaces by t, choosing vanishing shift vectors and denoting $\Delta f = \partial f/\partial t$ for functions f, the relationship

$$\dot{f} = e^{-\phi}\Delta f \tag{13.8}$$

defines the function ϕ, up to relabellings of t. Then ΔM is meaningful and the Gauss theorem yields

$$\int_{\Omega} \nabla \cdot J = \int_{\Sigma'} *\rho - \int_{\Sigma} *\rho \tag{13.9}$$

which differentiates to

$$\int_{\Sigma} *e^{\phi}\nabla \cdot J = \Delta M. \tag{13.10}$$

Therefore local conservation of mass (6.6) implies the integral form

$$\Delta M = 0 \tag{13.11}$$

which has the same form as classical conservation of mass (2.8).

Similarly, the entropy is

$$S = \int_\Sigma *s \tag{13.12}$$

and the entropy supply through $\hat{\Sigma}$ is

$$S_\circ = -\int_{\hat{\Sigma}} \hat{*} \cdot \varphi \tag{13.13}$$

where $\hat{*}$ is the volume form of $\hat{\Sigma}$ times the unit outward normal covector tangent to Σ. Then

$$\Delta S_\circ = -\oint_{\partial\Sigma} \cdot e^\phi \varphi \tag{13.14}$$

where the surface $\partial\Sigma$ is the intersection of Σ and $\hat{\Sigma}$, with area form and unit normal implicit in $\oint$. This has the same form as the classical definition of entropy supply (4.5), apart from the use of Δ and the corresponding factor e^ϕ, which tends to one in the Newtonian limit. Applying the Gauss theorem again,

$$\int_\Omega \nabla \cdot \Psi = \int_{\Sigma'} *s - \int_\Sigma *s + \int_{\hat{\Sigma}} \hat{*} \cdot \varphi \tag{13.15}$$

which differentiates to

$$\int_\Sigma *e^\phi \nabla \cdot \Psi = \Delta S - \Delta S_\circ. \tag{13.16}$$

Therefore $\nabla \cdot \Psi$ is the entropy density production rate and the local second law (10.4) integrates to

$$\Delta S \geq \Delta S_\circ \tag{13.17}$$

which has the same form as the classical second law (2.5).

Similarly, the heat supply through $\hat{\Sigma}$ may be defined by

$$Q = -\int_{\hat{\Sigma}} \hat{*} \cdot q \tag{13.18}$$

which yields

$$\Delta Q = -\oint_{\partial\Sigma} \cdot e^\phi q \tag{13.19}$$

again with the same form as the classical definition of heat supply (4.4). Finally, the thermal energy may be defined by

$$H = \int_\Sigma *\varepsilon. \tag{13.20}$$

The forms

$$M = \int_\Sigma *u \cdot T^\flat_M \cdot u \tag{13.21}$$

$$H = \int_\Sigma *u \cdot T^\flat_H \cdot u \tag{13.22}$$

explain the notation adopted earlier: the material and thermal energy tensors T_M and T_H are the energy tensors of mass and heat respectively.

Thermal energy H is what thermodynamicists call internal energy and what ordinary people call heat. In particular, this defines the heat of a body or volume of fluid. Thermodynamics textbooks often claim that there is no such thing as the heat of a body, in bizarre conflict with common experience. This seems to involve the belief that heat flux is the only form of heat, i.e. what was really meant is that there is no conserved quantity corresponding to the flux of heat, as supposed by the old caloric theory. A nice analogy,[11] intended to support the above claim but actually revealing the opposite, is that there is no such thing as the amount of rain in a lake. True, because rain is not the only form of water, but the amount of water in the lake does make sense.

14. Conclusion

A general macroscopic theory of generally relativistic thermodynamics has been developed, based on clear principles and sufficient to include relaxation effects. There is a full set of dynamical equations, easily applied in practice, with no meaningless derivatives or other common thermostatic confusions. The basic fields are the temperature ϑ, the material, thermostatic and dissipative energy tensors (T_M, T_0, T_1) and the entropy current Ψ. These fields may be paired respectively with the thermodynamic field equations: conservation of mass (6.6) or heat (7.8), conservation of energy-momentum (6.3), the thermostatic relations for T_0, e.g. the equations of state (9.3) for an ideal gas, the dissipative relations for T_1 and the entropic relations for Ψ: thermostatic (10.6)–(10.7), linear (11.1) and quadratic (12.1) approximations. The dissipative relations (11.4)–(11.7) or (12.4)–(12.7) are chosen to yield the second law (10.4), subject to relativistic causality. Having ensured these conditions, the dynamical system may be reduced by eliminating Ψ and the entropic relations, since Ψ does not appear in the remaining equations. Similarly, the thermostatic relations may be used to eliminate further fields, numbering two in the case of a fluid.

Four physical principles have been emphasized, which may be regarded as minimal requirements for a definition of thermodynamic matter. In principle this includes solids, necessarily imperfect in relativity, though the form of the thermostatic relations has been specified only for fluids. Three of the principles are standard. However, the first principle, that heat is a local form of energy, seems not to have been clearly stated before. This principle unifies various thermodynamic quantities in an essentially relativistic way as the energy, flux and stress components of a thermal energy tensor. This unification is analogous to that of electricity and magnetism in relativistic electromagnetism. Moreover, this agrees with non-relativistic thermodynamics in the physical interpretation of thermal flux only in the material frame. The principle thereby resolves the controversy of Eckart[2] versus Landau–Lifshitz[6] frame.

A further consequence of this formulation of basic principles is that it is unnatural to insist that thermal energy be purely thermostatic, despite the traditional dogma that internal energy be a state function. The similarly traditional dogma that entropy be a state function had already disappeared in Müller's extended thermodynamics.[3] Thus we have an apparently new thermodynamical quantity, dissipative energy, which nevertheless occurred unrecognised in the kinetic theory of gases.[7] These former assumptions may now be derived in the linear approximation to non-equilibrium, but need not hold in general, e.g. in the quadratic approximation.

One artificial feature of this theory, shared with the original Müller[3] and Israel-Stewart[4] theories, is that the quadratic correction to entropy current involves certain coefficients which are left unspecified. However, these coefficients can be determined for an ideal gas by kinetic theory of gases.[13] Kinetic theory also automatically implies conservation of mass and energy-momentum, and higher moments can be used to obtain a hierarchy of divergence equations for which there is a standard mathematical theory of local existence, uniqueness and causality.[19] However, this type of relativistic extended thermodynamics has so far been worked out under the assumption of vanishing dissipative energy density.[13] The main physical problem with this type of theory is that it is not known how to directly measure the higher moments.

It seems necessary to stress that conservation of heat for thermodynamic matter has been derived from conservation of mass and energy-momentum, and definitely differs from the old caloric theory of Carnot.[20] A relativistic version of the caloric theory would be to describe heat by an energy-momentum vector, the caloric current $J_H = \varepsilon u + q$, and conservation of heat by $\nabla \cdot J_H = 0$, as for conservation of mass. This would integrate by (13.18) and (13.20) to $\Delta H = \Delta Q$. Carnot's theory of heat can be interpreted as involving this equation and the claim that the heat H is a state function, which contradicts experiment. Unfortunately the discrepancy between this and the classical first law (2.1) has led to a historical over-reaction against the intuitive notion of heat, specifically the widespread insistence that the internal energy of Clausius differs from the heat of Carnot, whereas Carnot's mistake was instead in the formulation of conservation of heat. The relativistic theory reveals that the caloric theory failed only because the mathematical nature of energy, as an energy-momentum-stress tensor T_H rather than an energy-momentum vector J_H, was not understood in pre-relativistic physics. In particular, material stress appears in the form of work in the first law, because it is also a form of heat. If the thermodynamic matter is not isolated, heat and other forms of energy may be exchanged.

The relativistic theory similarly resolves the ancient philosophical question of whether heat is a substance. Put simply, both mass and heat are forms of energy, but with different natures: mass is described by a current J which satisfies a vectorial conservation law (6.6), conservation of mass, whereas heat is a general form of energy T_H satisfying a tensorial conservation law (7.8), the first law of thermodynamics.

Acknowledgements

Research supported by the National Natural Science Foundation of China under grant 11075107. Thanks are due to Ingo Müller and Jürghen Ehlers for discussions during the initial development, many years ago.

References

1. S. A. Hayward, *Class. Qu. Grav.* **15**, 3147 (1998).
2. C. Eckart, *Phys. Rev.* **58**, 919 (1940).
3. I. Müller, *Z. Phys.* **198**, 329 (1967).
4. W. Israel and J. M. Stewart, *Ann. Phys.* **118**, 341 (1979).
5. S. R. de Groot and P. Mazur, *Non-equilibrium Thermodynamics.* (North-Holland 1962).
6. L. Landau and E. M. Lifshitz, *Fluid Mechanics.* (Addison-Wesley 1987).
7. S. A. Hayward, *Relativistic kinetic theory of gases* (in preparation).
8. H. Grad, *Comm. Pure App. Math.* **2**, 331 (1949).
9. C. Truesdell, *Rational Thermodynamics.* (Springer-Verlag 1984).
10. J. W. Gibbs, *Elementary Principles in Statistical Mechanics.* (Dover 1960).
11. M. W. Zemansky, *Heat and Thermodynamics.* (McGraw-Hill 1968).
12. C. Eckart, *Phys. Rev.* **58**, 267 (1940).
13. I. Müller and T. Ruggeri, *Extended Thermodynamics.* (Springer-Verlag 1993).
14. B. F. Schutz, *A First Course in General Relativity.* (Cambridge University Press 1985).
15. I. D. Novikov and K. S. Thorne, in *Black Holes*, ed. C. DeWitt and B. S. DeWitt. (Gordon and Breach 1973).
16. C. W. Misner, K. S. Thorne and J. A. Wheeler, *Gravitation.* (Freeman 1973).
17. D. Jou, J. Casas-Vázquez and G. Lebon, *Extended Irreversible Thermodynamics.* (Springer-Verlag 1996).
18. W. A. Hiscock and L. Lindblom, Ann. Phys. **151**, 466 (1983).
19. R. Geroch and L. Lindblom, Phys Rev. **D41**, 1855 (1990).
20. S. Carnot, *Reflections on the motive power of fire*, ed. E. Mendoza. (Dover 1960).

Chapter 8

Trapped Surfaces

José M. M. Senovilla
Física Teórica, Universidad del País Vasco,
Apartado 644, 48080 Bilbao, Spain
josemm.senovilla@ehu.es

I review the definition and types of (closed) trapped surfaces. Surprising global properties are pointed out, such as their "clairvoyance" and the possibility that they enter into flat portions of the spacetime. Several results on the interplay of trapped surfaces with vector fields and with spatial hypersurfaces are presented. Applications to the quasi-local definition of Black Holes are analyzed, with particular emphasis set onto marginally trapped tubes, trapping horizons and the boundary of the region with closed trapped surfaces. Finally, the *core* of a trapped region is introduced, and its importance briefly discussed.

1. Introduction

Black Holes (BH) are fundamental physical objects predicted classically in General Relativity (GR) which show a very deep relation between Gravitation, Thermodynamics, and Quantum Theory.[50] Classically, the characteristic feature of a BH is its event horizon EH: the boundary of the region from where one can send signals to a faraway asymptotic external region. This EH is usually identified as the surface of the BH and its area to the entropy. Unfortunately, the EH is essentially a global object, as it depends on the *whole* future evolution of the spacetime. Thus, EHs are determined by future causes, they are *teleological*, see[6,7,10,11,17,20] and references therein.

However, it is important to recognize the presence of a BH locally, for instance in numerical GR,[47] in the 3+1 or Cauchy-problem perspective of GR[25] (see Jaramillo's chapter in this volume), or in Astrophysics. In the former two, one needs to pinpoint when the BH region has been entered, while in the latter there are so many candidates to real BHs that a precise meaning of the sentence "there is a BH in the region such and such" is required. Of course, this meaning cannot rely on the existence of an EH as the real BHs undergo evolutionary processes and are usually dynamical.

Over the recent years, there has been a number of efforts to give a general quasi-local description of a dynamical black hole. In particular, the case has been made for quasi-local objects called Marginally Trapped Tubes (MTT) or Trapping Horizons (TH),[6,7,11,23,24] and their particular cases called *dynamical horizons* (DH)[6,7] or Future Outer Trapping Horizons (FOTH).[23,24] The underlying ideas were discussed in the 90s by Hayward.[23,24] MTTs are hypersurfaces foliated by closed (compact without boundary) marginally trapped surfaces. It is accepted that closed (marginally) trapped surfaces constitute the most important ingredient in the formation of BHs, so that the idea of using MTTs as the surface of BHs, and as viable replacements for the EH, looked very promising. This is one of the main reasons to study trapped surfaces, the subject of this chapter.

Unfortunately, MTTs have an important problem: they are highly non-unique.[7] This manifests itself because the 2-dimensional Apparent Horizons[21,49] depend on the choice of a reference foliation of spacelike hypersurfaces. Hence, another reasonable alternative, which is manifestly independent of any foliation, has been pursued and investigated more recently:[1,9,10,45] the boundary $\mathscr{B}$ of the future-trapped region $\mathscr{T}$ in spacetime (this is the region through which future-trapped surfaces pass). This is also a very natural candidate for the surface of a BH, as it defines a frontier beyond which MTTs and general trapped surfaces cannot be placed. One can show that $\mathscr{B}$ cannot be an MTT itself, and that it suffers from problems similar to that of EHs: it is unreasonably global. This seems to be an intrinsic problem linked to a fundamental property of closed trapped surfaces: they are *clairvoyant*.[1,9,10,45]

Recently, a novel idea[10] has been put forward to address all these issues: the *core* of a trapped region, and its boundary. This is a minimal region which is indispensable to sustain the existence of closed trapped surfaces in the spacetime. It has some interesting features and it may help in solving, or better understanding, the difficulties associated to BHs.

2. Trapped Surfaces: Definition and Types

In 1965, Penrose[36] introduced a new important concept: *closed trapped surfaces*. These are closed spacelike surfaces (usually topological spheres) such that their area tends to decrease locally along any possible *future* direction. (There is a dual definition to the past). The traditional Black Hole solutions in GR, constituted by the Kerr–Newman family of metrics, have closed trapped surfaces in the region inside the Event Horizon.[21,49] Actually, the existence of closed trapped surfaces is a fundamental ingredient in the singularity theorems:[37,21,22,40] if they form, then singularities will surely develop.

2.1. *Co-dimension two surfaces: Notation*

Let $(\mathcal{V}, g)$ be an n-dimensional Lorentzian manifold with metric tensor g of signature $(-,+,\dots,+)$. A co-dimension two (dimension $n-2$) connected surface S can be represented by means of its embedding $\Phi : S \longrightarrow \mathcal{V}$ into the spacetime $\mathcal{V}$ via some parametric equations $x^\mu = \Phi^\mu(\lambda^A)$ where $\{x^\mu\}$ are local coordinates in $\mathcal{V}$ $(\mu, \nu, \dots =$

$0, 1 \ldots, n-1$), while $\{\lambda^A\}$ are local coordinates for S $(A, B, \cdots = 2, \ldots, n-1)$.

The tangent vectors ∂_{λ_A} are pushed forward to $\mathcal{V}$ to define the tangent vectors to S as seen on $\mathcal{V}$

$$\vec{e}_A \equiv \Phi'(\partial_{\lambda^A}) \Longleftrightarrow e^\mu_A = \frac{\partial \Phi^\mu}{\partial \lambda^A}$$

and the first fundamental form of S in $(\mathcal{V}, g)$ is defined as the pull-back of g:

$$\gamma = \Phi^* g \Longrightarrow \gamma_{AB}(\lambda) = g|_S(\vec{e}_A, \vec{e}_B) = g_{\mu\nu}(\Phi) e^\mu_A e^\nu_B.$$

From now on, S will be assumed to be *spacelike* which means that γ_{AB} is *positive definite*. Then, $\forall x \in S$ one can canonically decompose the tangent space $T_x\mathcal{V}$ as

$$T_x\mathcal{V} = T_x S \oplus T_x S^\perp$$

which are called the *tangent* and *normal* parts. In particular, we have[35, 29]

$$\nabla_{\vec{e}_A} \vec{e}_B = \overline{\Gamma}^C_{AB} \vec{e}_C - \vec{K}_{AB}$$

where ∇ is the covariant derivative in $(\mathcal{V}, g)$, $\overline{\Gamma}^C_{AB}$ is the Levi–Civita connection associated to the first fundamental form in S $(\bar{\nabla}_C \gamma_{AB} = 0)$ and $\vec{\boldsymbol{K}} : \mathfrak{X}(S) \times \mathfrak{X}(S) \longrightarrow \mathfrak{X}(S)^\perp$ is called the *shape tensor* or *second fundamental form vector* of S in $\mathcal{V}$. [Here $\mathfrak{X}(S)$ $(\mathfrak{X}(S)^\perp)$ is the set of vector fields tangent (orthogonal) to S.] Observe that $\vec{\boldsymbol{K}}$ is orthogonal to S. $\vec{\boldsymbol{K}}$ measures the difference between the pull-back to S of the covariant derivative of covariant tensor fields and the covariant derivative in S of the pull-back of those tensor fields according to the formula

$$\Phi^*(\nabla v) = \bar{\nabla}(\Phi^* v) + v(\vec{\boldsymbol{K}}) \quad \Longleftrightarrow \quad e^\mu_B e^\nu_A \nabla_\nu v_\mu = \overline{\nabla}_A \bar{v}_B + v_\mu K^\mu_{AB} \qquad (2.1)$$

where for all one-forms v_μ of $\mathcal{V}$ we write $\bar{v}_A \equiv v_\mu|_S \, e^\mu_A = v_\mu(\Phi) \, e^\mu_A$ for the components of its pull-back to S, $\bar{v} = \Phi^*(v)$.

A *second fundamental form* of S in $(\mathcal{V}, g)$ relative to any normal vector field $\vec{n} \in \mathfrak{X}(S)^\perp$ is defined as

$$K_{AB}(\vec{n}) \equiv n_\mu K^\mu_{AB}.$$

For each $\vec{n}$, these are 2-covariant symmetric tensor fields on S.

2.1.1. *Mean curvature vector. Null expansions*

For a spacelike surface S there are two *independent* normal vector fields, and one can choose them to be future-pointing and null everywhere, $\vec{k}^\pm \in \mathfrak{X}(S)^\perp$ with

$$k^+_\mu e^\mu_A = 0, \; k^-_\mu e^\mu_A = 0, \; k^+_\mu k^{+\mu} = 0, \; k^-_\mu k^{-\mu} = 0.$$

By adding the normalization condition $k_{+\mu} k^\mu_- = -1$, there still remains the freedom

$$\vec{k}^+ \longrightarrow \vec{k}'^+ = \sigma^2 \vec{k}, \qquad \vec{k}^- \longrightarrow \vec{k}'^- = \sigma^{-2} \vec{k}^- . \qquad (2.2)$$

One obviously has

$$\vec{K}_{AB} = -K_{AB}(\vec{k}^-) \, \vec{k}^+ - K_{AB}(\vec{k}^+) \, \vec{k}^- .$$

The *mean curvature vector* of S in $(\mathcal{V}, g)$ is the trace of the shape tensor:

$$\vec{H} \equiv \gamma^{AB}\vec{K}_{AB}, \qquad \vec{H} \in \mathfrak{X}(S)^{\perp}$$

and its decomposition in the null normal basis

$$\vec{H} \equiv \gamma^{AB}\vec{K}_{AB} = -\theta^{-}\vec{k}^{+} - \theta^{+}\vec{k}^{-}$$

defines the so-called *future null expansions:* $\theta^{\pm} \equiv \gamma^{AB}K_{AB}(\vec{k}^{\pm})$. Notice that, even though the null expansions are not invariant under the boost transformations (2.2), $\vec{H}$ is actually invariant.

2.2. *The trapped surface fauna: a useful symbolic notation*

The concept of trapped surfaces was originally formulated in terms of the signs or the vanishing of the null expansions,[36] and has remained as such for many years. This is obviously related to the causal orientation of the mean curvature vector. Thus, it has become clear over the recent years that the causal orientation of the mean curvature vector provides a better and powerful characterization[a] of the trapped surfaces.[29,35,33,41–44] Therefore, a symbolic notation for the causal orientation of $\vec{H}$ becomes very useful. Using an arrow to denote $\vec{H}$ and denoting the future as the upward direction and null vectors at 45° with respect to the vertical, the symbolic notation was introduced in:[44]

$\vec{H}$	Causal orientation
$\downarrow$	past-pointing timelike
$\swarrow$ or $\searrow$	past-pointing null ($\propto \vec{k}^{+}$ or $\vec{k}^{-}$)
$\leftarrow$ or $\rightarrow$	spacelike
$\cdot$	vanishes
$\nearrow$ or $\nwarrow$	future-pointing null ($\propto \vec{k}^{+}$ or $\vec{k}^{-}$)
$\uparrow$	future-pointing timelike

A surface is said to be *weakly future-trapped* (f-trapped from now on) if the mean curvature vector is future-pointing all over S, similarly for weakly past-trapped. The traditional f-trapped surfaces have a timelike (non-vanishing) future-pointing $\vec{H}$ all over S, while the marginally f-trapped surfaces have $\vec{H}$ future pointing and proportional to one of the null normal directions. Using the previously introduced notation — if $\vec{H}$ changes causal orientation over S the corresponding symbols are superposed[44]— the main cases are summarized in the next table, where the signs of the null expansions are also shown.

Sometimes, only the sign of one of the expansions is relevant. This may happen if there is a consistent or intrinsic way of selecting a particular null normal on S. Then, one can use the $\pm$-symbols to denote the preferred direction and define $\pm$-trapped

[a]For instance, the characterization by means of the mean curvature vector has permitted the extension of the classical singularity theorems to cases with trapped submanifolds of arbitrary co-dimension embedded in spacetimes of arbitrary dimension, see Ref.18.

Table 1. The main cases of future-trapped surfaces.

Symbol	Expansions	Type of surface
·	$\theta^+ = \theta^- = 0$	stationary or minimal
↑	$\theta^+ < 0, \theta^- < 0$	f-trapped
·↗	$\theta^+ = 0, \theta^- \leq 0$	marginally f-trapped
↖·	$\theta^+ \leq 0, \theta^- = 0$	marginally f-trapped
↖↑↗	$\theta^+ \leq 0, \theta^- \leq 0$	weakly f-trapped

surfaces. However, in the literature the preferred direction is usually declared to be "outer", and then the nomenclature speaks about "outer trapped", no matter whether or not this outer direction coincides with any particular outer or external part to the surface. Thus, (marginally) +-trapped surfaces are usually referred to as (marginally) outer trapped surfaces ((M)OTS) and similarly for the – case. The main possibilities are summarized in Table 2. Important studies concerning these

Table 2. The main cases of +-trapped, also "outer trapped", surfaces.

Symbol	Expansions	Type of surface
↖↑↙	$\theta^+ < 0$	half converging, or +-trapped (or OTS)
↙·↗	$\theta^+ = 0$	null dual or M+TS (or MOTS)
↖↑↗↙	$\theta^+ \leq 0$	weakly +-trapped (W+TS or WOTS)

surfaces, and in particular MOTS, have been carried out recently in,[2–4,13] with relevant results for black holes and the existence of MTTs. The reader is referred to the chapters by Carrasco and Mars in this volume for further information.

For completeness, I also present the characterization and symbols of some other distinguished types of surfaces, such as (weakly) untrapped surfaces or the null untrapped surfaces —recently also named "generalized apparent horizons" in Ref.12— in Table 3. The last type of surface shown in Table 3 was proposed as a viable replacement[12] for marginally trapped surfaces in a new version of the Penrose inequality.[38,32] However, this version cannot hold as a recent counterexample has been found in Ref.14. Nevertheless, it is known that there cannot be null untrapped closed surfaces embedded in spacelike hypersurfaces[b] of Minkowski spacetime.[26] An important question related with these issues and raised in Ref.33 is whether or not there can be any closed $*$-surfaces in Minkowski spacetime — or more generally, in stationary spacetimes.

[b]The question of whether or not there are null untrapped surfaces which are not embedded in any spacelike hypersurface in Minkowski space-time is still open, though. The existence of compact, embedded spacelike surfaces in Minkowski space-time which are not contained in any spacelike hypersurface follows, for instance, from Ref.28.

Table 3. Miscellaneous surfaces.

Symbol	Expansions	Type of surface
→ or ←	$\theta^+\theta^- < 0$	untrapped
↗→↘ or ↖←↙	$\theta^+ \geq 0, \theta^- \leq 0$ or $\theta^+ \leq 0, \theta^- \geq 0$	weakly untrapped
↓↘	$\theta^+ > 0$	half diverging or +-untrapped (or outer untrapped)
↗→↙↓↘	$\theta^+ \geq 0$	weakly +-untrapped (or weakly outer untrapped)
↖↑↗↙↓↘	$\theta^+\theta^- \geq 0$	$\ast$-surfaces[44]
↖↗↙↘	$\theta^+\theta^- = 0$	null $\ast$-surfaces
↗↘	$\theta^+\theta^- = 0$ and $\theta^+ \geq 0, \theta^- \leq 0$	null untrapped or generalized apparent horizon[12]

2.3. *A useful formula for the scalar $H_\mu H^\mu$.*

Now, I am going to present a formula for the norm of the mean curvature vector associated to a given family of co-dimension-2 surfaces.[41] This allows one to ascertain which surfaces within the family can be trapped, marginally trapped, etc. Assume you are given a family of $(n-2)$-dimensional spacelike surfaces S_{X^a} described by

$$\{x^a = X^a\}, \qquad a, b, \dots = 0, 1$$

where X^a are constants and $\{x^\alpha\}$ local coordinates in $(\mathcal{V}, g)$. Locally, the line-element can be written as

$$ds^2 = g_{ab}dx^a dx^b + 2g_{aA}dx^a dx^A + g_{AB}dx^A dx^B \tag{2.3}$$

where $g_{\mu\nu}(x^\alpha)$ and $\det g_{AB} > 0$. There remains the freedom

$$x^a \longrightarrow x'^a = f^a(x^b), \; x^A \longrightarrow x'^A = f^A(x^B, x^c) \tag{2.4}$$

keeping the form (2.3) and the chosen family of surfaces.

The imbedding Φ for the surfaces S_{X^a} is

$$x^a = \Phi^a(\lambda) = X^a = \text{const.}, \qquad x^A = \Phi^A(\lambda) = \lambda^A$$

from where the first fundamental form for each S_{X^a} reads

$$\gamma_{AB} = g_{AB}(X^a, \lambda^C)$$

while the future null normal one-forms become

$$\boldsymbol{k}^\pm = k_b^\pm dx^b|_{S_{X^a}}, \;\; g^{ab}k_a^\pm k_b^\pm = 0, \;\; g^{ab}k_a^+ k_b^- = -1.$$

Notice that g^{ab} are components of $g^{\mu\nu}$ and therefore (g^{ab}) is not necessarily the inverse of (g_{ab}) !

Set

$$G \equiv +\sqrt{\det g_{AB}} \equiv e^U, \;\; \boldsymbol{g}_a \equiv g_{aA}dx^A$$

where $\boldsymbol{g}_a$ are considered to be two one-forms, one for each $a = 0, 1$. When pull-backed to S_{X^a} they read $\bar{\boldsymbol{g}}_a = g_{aA}(X^a, \lambda^C)d\lambda^A$. A direct computation[41] provides the null expansions ($f_{,\mu} = \partial_\mu f$)

$$\theta^{\pm} = k^{\pm a}\left(\frac{G_{,a}}{G} - \frac{1}{G}(G\gamma^{AB}g_{aA})_{,B}\right)\Bigg|_{S_{X^a}} .$$

and thereby the mean curvature one-form

$$\boxed{H_\mu = \delta^a_\mu\left(U_{,a} - \text{div}\vec{g}_a\right)} \tag{2.5}$$

where div is the divergence operating on vector fields at each surface S_{X^a}. These surfaces are thus trapped if and only if the scalar

$$\boxed{\kappa = -\left. g^{bc}H_bH_c\right|_{S_{X^a}}}$$

is positive, and a necessary condition for them to be marginally trapped is that κ vanishes on the surface. Observe that one only needs to compute the norm of H_a *as if it were a one-form in the "2-dimensional" contravariant metric* g^{ab}.

3. Horizons: MTTs, FOTHs and Dynamical Horizons

3.1. $\{S_{X^a}\}$-*horizons*

In the construction of the previous section, in general the mean curvature vector $\vec{H}$ of the S_{X^a}-surfaces will change its causal character at different regions of the spacetime. The hypersurface(s) of separation

$$\mathcal{H} \equiv \{\kappa = 0\}, \qquad \text{“}S_{X^a}\text{ -horizon”}$$

is a fundamental place in $(\mathcal{V}, g)$ associated to the given family of surfaces S_{X^a} called the S_{X^a}-horizon. It contains the regions with marginally trapped, marginally outer trapped, null untrapped, and null $\ast$-surfaces S_{X^a} (as well as those parts of each S_{X^a} where one of the $\theta^{\pm}$ vanishes).

As an example and to show the applicability of the previous formulas, consider the 4-dimensional Kerr spacetime ($G = c = 1$, $n = 4$) in advanced (or Kerr) coordinates[21]

$$\begin{aligned} ds^2 = &-\left(1 - \frac{2Mr}{r^2 + a^2\cos^2\theta}\right)dv^2 + 2dvdr - 2a\sin^2\theta d\varphi dr - \frac{4Mar\sin^2\theta}{r^2 + a^2\cos^2\theta}d\varphi dv \\ &+ \left(r^2 + a^2\cos^2\theta\right)d\theta^2 + \left(r^2 + a^2 + \frac{2Mar\sin^2\theta}{r^2 + a^2\cos^2\theta}\right)\sin^2\theta d\varphi^2 \end{aligned} \tag{3.6}$$

where M and a are constants. A case of physical interest is given by the topological spheres defined by constant values of v and r, so that with the notation of the previous section one has $\{x^a\} = \{v, r\}$, $\{x^A\} = \{\theta, \varphi\}$. The two one-forms $\boldsymbol{g}_a$ are

$$\boldsymbol{g}_r = -a\sin^2\theta d\varphi, \qquad \boldsymbol{g}_v = -\frac{2Mar\sin^2\theta}{r^2 + a^2\cos^2\theta}d\varphi$$

so that div$\vec{g}_a = 0$. On the other hand

$$e^{2U} = \sin^2\theta[(r^2+a^2)(r^2+a^2\cos^2\theta)+2Mra^2\sin^2\theta],$$

so that the mean curvature one-form becomes

$$H_a dx^a = U_{,a}dx^a = \frac{r(r^2+a^2+r^2+a^2\cos^2\theta)+Ma^2\sin^2\theta}{(r^2+a^2)(r^2+a^2\cos^2\theta)+2Mra^2\sin^2\theta}dr$$

from which one easily obtains

$$\kappa = -\frac{(r^2-2Mr+a^2)}{(r^2+a^2\cos^2\theta)}\frac{\left[r(r^2+a^2+r^2+a^2\cos^2\theta)+Ma^2\sin^2\theta\right]^2}{\left[(r^2+a^2)(r^2+a^2\cos^2\theta)+2Mra^2\sin^2\theta\right]^2}$$

ergo (for $r>0$) sign $\kappa = -$sign $(r^2-2Mr+a^2)$. Thus $\mathcal{H}$ are the classical Cauchy and event horizons of the Kerr spacetime.[21]

As another important example, to be used repeatedly later on in this chapter, consider general spherically symmetric spacetimes, whose line-element can always be cast in the form

$$ds^2 = g_{ab}(x^c)dx^a dx^b + r^2(x^c)d\Omega^2_{n-2}$$

where $d\Omega^2_{n-2}$ is the round metric on the $(n-2)$-sphere, r is a function depending only on the x^b-coordinates called the area coordinate, and $\det g_{ab} < 0$ so that g_{ab} is a 2-dimensional Lorentzian metric. By taking the family of round spheres as the selected family of co-dimension two surfaces, so that they are given by constant values of x^b, the two one-forms $\boldsymbol{g}_a$ vanish so that

$$H_a = U_{,a} \propto \frac{r_{,a}}{r}$$

One can thus define the standard "mass function"

$$2m(x^a) \equiv r^{n-3}\left(1-g^{bc}r_{,b}r_{,c}\right)$$

so that

$$\kappa = -g^{ab}H_aH_b \propto -\left(1-\frac{2m}{r^{n-3}}\right)$$

Hence, the round $(n-2)$-spheres are (marginally) trapped if $2m/r^{n-3}$ is (equal to) greater than 1. The corresponding horizon $\mathcal{H}: r^{n-3} = 2m$ thus becomes the classical spherically symmetric apparent $(n-1)$-horizon AH[c].

[c]Observe that AH is considered here to be a hypersurface in spacetime, while the traditional "apparent horizons" are co-dimension two marginally trapped surfaces.[21,49] However, AH is foliated by such marginally trapped round spheres, so that it is a collection of apparent horizons

3.2. *MTTs, FOTHs, and dynamical horizons*

The apparent $(n-1)$-horizon AH in spherical symmetry, as well as some S_{X^a}-horizons, are examples of marginally trapped tubes (MTTs).

Definition 1. A marginally trapped tube is a hypersurface foliated by closed marginally f-trapped surfaces.

This corresponds to the concept of "future trapping horizons" as defined by Hayward,[23] who also introduced the concept of *future outer trapping horizon* by requiring that the vanishing null expansion $\theta^+ = 0$ becomes negative when perturbed along the other future null normal direction $\vec{k}^-$. In that case, the horizon is necessarily a spacelike hypersurface — unless the whole second fundamental form $K_{AB}(\vec{k}^+) = 0$ vanishes and also $G_{\mu\nu}k^{+\mu}k^{+\nu} = 0$, where $G_{\mu\nu}$ is the Einstein tensor of the spacetime, in which case the horizon is null and actually a non-expanding or isolated horizon.[6]

However, in many situations the "outer" condition in Hayward's definition is not required to obtain results about MTTs, and at the same time there are speculations that in dynamical situations describing the collapse to a realistic black hole the MTTs will eventually become spacelike. This led to the definition of *dynamical horizons* (DH), which are simply *spacelike* MTTs. Actually, the previously mentioned requirements $K_{AB}(\vec{k}^+) \neq 0 \neq G_{\mu\nu}k^{+\mu}k^{+\nu}$ are sometimes termed as "genericity" conditions for dynamical horizons.[7] They are known to fail in some very special situations, as there are explicit examples of such non-generic dynamical horizons in pp-wave sacetimes,[43] and actually they do not enclose any f-trapped surfaces in those explicit cases.

In spherical symmetry, the apparent $(n-1)$-horizons AH are the unique *spherically symmetric* MTTs. This does not only mean that they are the unique MTTs foliated by marginally trapped *round spheres*, but also that they are the only spherically symmetric hypersurfaces foliated by any kind of marginally trapped surfaces.[10] As seen before, these preferred MTTs can be invariantly defined by

$$\nabla_\mu r \nabla^\mu r = 0 \Longleftrightarrow r^{n-3} = 2m\,.$$

Despite this fact, it is known that MTTs, even in spherical symmetry, are not unique;[7] an explicit proof of this fact in generic spherically symmetric spacetimes can be found in Ref. 9, see Corollary 1 below.

4. The Future-Trapped Region $\mathscr{T}$ and its Boundary $\mathscr{B}$

The non-uniqueness of MTTs poses a difficult problem for the physics of black holes, and casts some doubts on whether MTTs provide a good quasi-local definition of the surface of a black hole. It is also a manifestation that the $(n-2)$-dimensional apparent horizons will depend on a chosen foliation in the spacetime. Thus, a reasonable alternative is to consider the boundary of the region containing f-trapped surfaces. Thus, following Refs.[23,10] we define the following.

Definition 2. The future-trapped region $\mathscr{T}$ is defined as the set of points $x \in \mathcal{V}$ such that x lies on a closed future-trapped surface. We denote by $\mathscr{B}$ the boundary of the future trapped region $\mathscr{T}$:

$$\mathscr{B} \equiv \partial \mathscr{T}.$$

$\mathscr{T}$ and $\mathscr{B}$ are invariant by the isometry group of the spacetime. More precisely[10]

Proposition 1. *If G is the group of isometries of the spacetime $(\mathcal{V}, g)$, then $\mathscr{T}$ is invariant under the action of G, and the transitivity surfaces of G, relative to points of $\mathscr{B}$, remain in $\mathscr{B}$.*

Therefore, in arbitrary spherically symmetric spacetimes, $\mathscr{T}$ and $\mathscr{B}$ have spherical symmetry. Actually the result is stronger.

Proposition 2. *In arbitrary spherically symmetric spacetimes, $\mathscr{B}$ (if not empty) is a spherically symmetric hypersurface without boundary.*

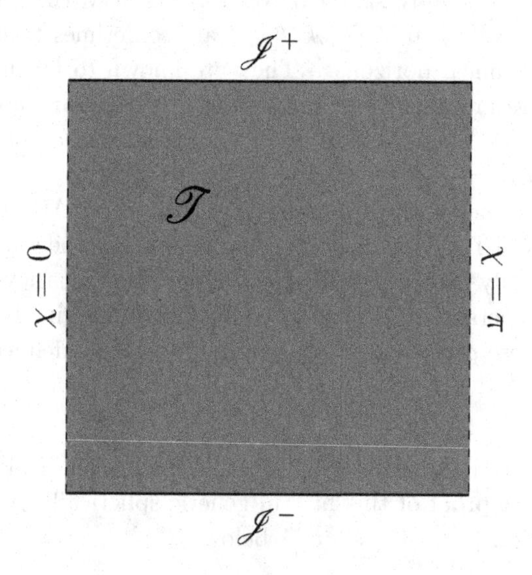

Figure 1. (Color online) A Penrose conformal diagram for the de Sitter spacetime. The whole spacetime is coloured in red because there are f-trapped spheres passing through every point, so that $\mathscr{T}$ is the entire de Sitter spacetime. Thus, there is no boundary $\mathscr{B}$ for the f-trapped region.

As simple examples of these concepts, consider de Sitter spacetime.[21,46] It is well known that there are f-trapped round spheres in such spacetime, but given that it is a homogeneous spacetime then there must be such a f-trapped sphere through every point. Therefore, $\mathscr{T}$ is the whole spacetime and $\mathscr{B} = \emptyset$. This is represented in Fig. 1.

An example with non-empty $\mathscr{B}$ is provided by the closed Robertson–Walker dust model with $\Lambda = 0$.[21,46] In this case, $\mathscr{T}$ covers only 'half' of the spacetime and the boundary $\mathscr{B}$ can be seen to correspond to the recollapsing time, which is a maximal hypersurface, as shown in Fig. 2, see Ref. 10.

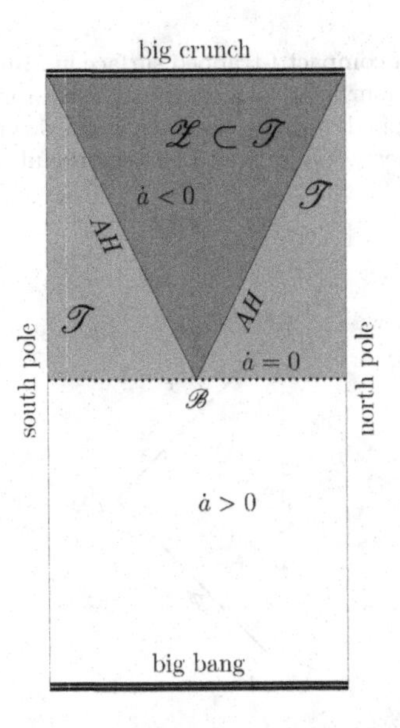

Figure 2. (Color online) Conformal diagram for the Robertson–Walker dust model. Closed f-trapped round spheres exist in the contracting phase, the coloured region, defined by $\dot{a} < 0$, where a is the scale factor. Thus, the boundary $\mathscr{B}$ is the hypersurface corresponding to the instant of recollapse, shown as a dotted horizontal line. A MTT corresponding to AH as defined in the text is also shown. This will be relevant for the definition of a core in Section 10.

As a final example with $\mathscr{B} = \emptyset$, consider the flat spacetime, with line-element

$$ds^2 = -dt^2 + dx^2 + dy^2 + dz^2$$

One can easily check[40] that the family of surfaces $S_{x_0,t_0} : \{x = x_0,\ e^{t_0 - t} = \cosh z\}$ are f-trapped. However, they are non-compact, see Fig. 3. It is a well-known result that there cannot be any *closed* trapped surface in flat spacetime. Actually, they are absent in arbitrary (globally) stationary spacetimes.[33] Therefore, in this case $\mathscr{T} = \emptyset$ and there is no boundary. There are no MTTs (such as AH) either. A conformal diagram is presented in Fig. 4.

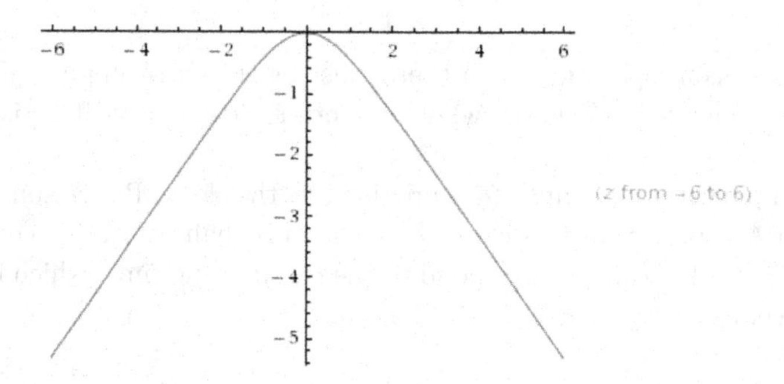

Figure 3. (Color online) A non-compact f-trapped surface in Minkowski spacetime. In spacetimes with selected foliations by hypersurfaces with some extrinsic properties — such as the static time t in flat spacetime — all f-trapped surfaces have to "bend down" in time t, as shown (t is the vertical coordinate). This property will prove to be very useful in some studies concerning black holes.

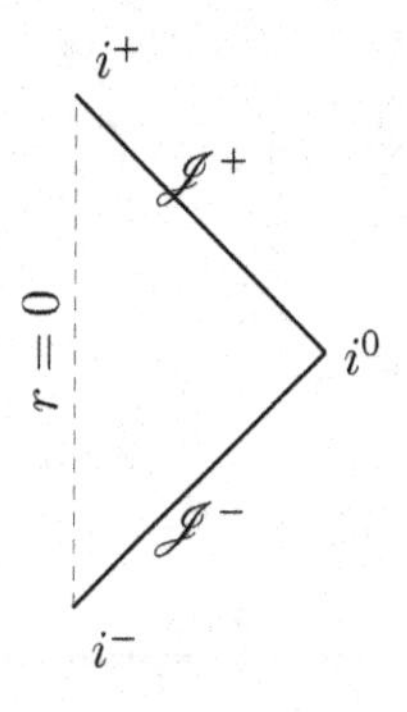

Figure 4. (Color online) Flat (Minkowski) spacetime conformal diagram. There is no red region now ($\mathscr{T} = \emptyset$) because there are no closed (marginally) trapped surfaces in flat spacetime. Observe that the structure of infinity is given by two null hypersurfaces, $\mathscr{J}^+$ and $\mathscr{J}^-$, and three points i^+, i^- and i^0. Spacetimes with a similar structure of infinity are called *asymptotically flat.* In flat spacetime all causal endless curves have an initial point at $\mathscr{J}^- \cup i^-$, and a final point at $\mathscr{J}^+ \cup i^+$. This will not happen in general.

5. The Event Horizon, and its Relation to $\mathscr{B}$ and AH.

Consider now the Schwarzschild solution ($n = 4$) in "Eddington–Finkelstein" advanced coordinates[21] (units with $G = c = 1$), given by

$$ds^2 = -\left(1 - \frac{2M}{r}\right) dv^2 + 2dvdr + r^2 d\Omega^2 \tag{5.7}$$

This is (locally) the only spherically symmetric solution of the vacuum Einstein field equations. The mass function is now a constant $m = M$ representing the total mass, and v is advanced (null) time. This metric is also the case $a = 0$ (no rotation) of the Kerr metric (3.6). From the above we know that the round spheres — defined by constant values of v and r — are trapped if and only if $r < 2M$. If $r = 2M$ they are marginally trapped. Thus, the unique spherically symmetric MTT is given by AH, defined as

$$\text{AH}: \quad r - 2M = 0.$$

One can actually prove that all possible closed f-trapped surfaces, be they round spheres or not, must lie completely inside the region $r < 2M$. Therefore we also have $\mathscr{T} = \{r < 2m\}$ and hence, $\mathscr{B}$ =AH. Even more, one can see, as shown in Fig. 5, that the spacetime is asymptotically flat (as remarked in Fig. 4, this is when the asymptotic region is "Minkowskian" with $\mathscr{J}^{\pm}$ and i^0), but that there are many future-endless causal curves that never reach future infinity, $\mathscr{J}^+ \cup i^+$. Therefore, there is a hypersurface separating those points which can send signals to infinity, from those which cannot. This separating membrane is called the *event*

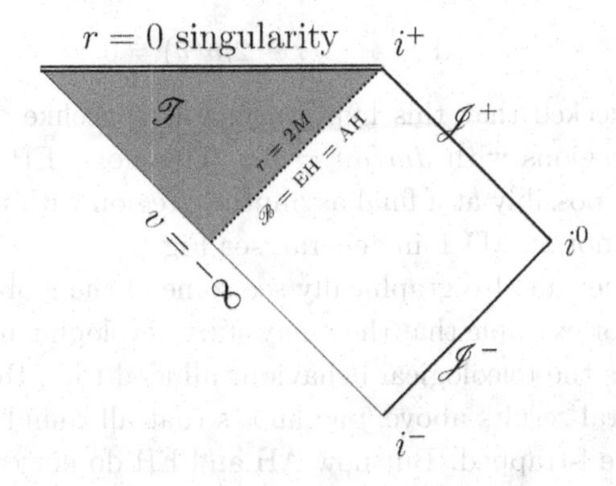

Figure 5. (Color online) Conformal diagram for the Schwarzschild spacetime (5.7). The red region is now confined by a boundary $\mathscr{B}$ which coincides with the unique spherically symmetric MTT, and with the event horizon EH.

horizon (EH) in general, and in this case it happens to coincide with the spherically symmetric MTT and with the boundary of the f-trapped region

$$\text{AH} = \text{EH} = \mathscr{B}: \qquad r - 2M = 0\,.$$

However, this is an exceptional case, and these coincidences do not hold in general, dynamical, situations.

For general dynamical but *asymptotically flat* spacetimes one can define the region from where $\mathscr{J}^+$ cannot be reached by any causal means. The boundary of this region is called the *Event Horizon* EH. (The formal definition is that EH is the boundary of the past of $\mathscr{J}^+$). By definition, this is always a null hypersurface. However, and contrary to what happens in the Schwarzschild metric, it is not an MTT in general.

To prove this claim, a simple example will suffice. Consider the imploding Vaidya spacetime,[48] given in advanced coordinates by the line-element ($n = 4$)

$$ds^2 = -\left(1 - \frac{2m(v)}{r}\right) dv^2 + 2dvdr + r^2 d\Omega^2. \tag{5.8}$$

Observe that this has exactly the same form as (5.7) but now the mass function depends on advanced null time v. The Einstein tensor of this metric is

$$G_{\mu\nu} = \frac{2}{r^2}\frac{dm}{dv}\ell_\mu \ell_\nu, \qquad \ell_\mu dx^\mu = -dv, \qquad \vec{\ell} = -\partial_r \quad (\ell^\mu \ell_\mu = 0)$$

which vanishes only if $dm/dv = 0$. The null convergence condition (or the weak energy condition)[21,49] requires that $dm/dv \geq 0$. For simplicity, and to illustrate some important points concerning EHs and BHs, consider the following restrictions on the mass function

$$m(v) = 0 \ \ \forall v < 0; \qquad m(v) \leq M < \infty \ \ \forall v > 0.$$

Then, the spacetime is flat for all $v < 0$, while M can be considered as the total mass. The unique spherically symmetric MTT is now given by

$$\text{AH:} \qquad r - 2m(v) = 0$$

and it is easily checked that this hypersurface is spacelike whenever $dm/dv > 0$ (and null at the regions with $dm/dv = 0$). Therefore, EH is different from AH everywhere except possibly at a final asymptotic region with $m = M = \text{const}$. This proves that EH is not an MTT in general, see Fig. 6.

In the figure one can also graphically see some of the global properties of event horizons, such as for example that they may start developing in regions whose whole past is flat. This is the teleological behaviour alluded to in the Introduction.

From the general results above, one knows that all round spheres in the region with $r < 2m(v)$ are f-trapped. But now AH and EH do not coincide. Can there be any other f-trapped surfaces which extend outside AH? Will they be able to extend all the way down to EH? Put another way, one wants to know if the boundary $\mathscr{B}$ is EH, or AH, or neither. Ben-Dov proved[8] that the event horizon cannot be the boundary of closed f-trapped surfaces for the particular case of a shell of null radiation. This proves that EH $\neq \mathscr{B}$ in general. Interestingly enough, he also proved that EH is the boundary for (marginally) *outer* f-trapped closed surfaces (MOTS) in the Vaidya spacetimes. That this must be the case is a general conjecture due to Eardley.[17] I refer the reader to the chapter by Bengtsson where this is explained in some detail and with illuminating pictures. The conclusion is that closed surfaces having essentially the same properties as the built MOTS but extending beyond EH can actually be constructed. However, they cannot be embedded into a suitable spacelike hypersurface if they cross EH, and then the notion of "outer" is not

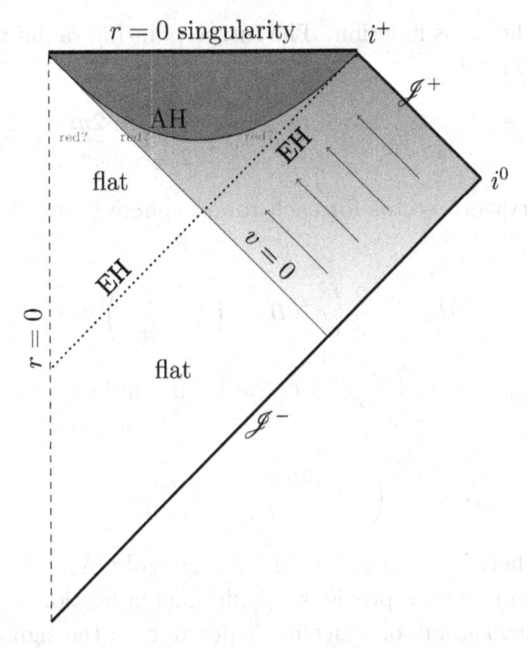

Figure 6. An imploding Vaidya spacetime. The spacetime is flat until null radiation flows in in spherical manner and produces a singularity when meeting at the center. This is "clothed" by an event horizon, as shown. However, this EH has a portion inside the flat zone of the spacetime, showing that it is "aware" of things that will happen in the future. The unique spherically symmetric MTT, denoted by AH, is a spacelike hypersurface (ergo a dynamical horizon) which never coincides with EH, and it approaches it asymptotically. One knows that there are closed f-trapped round spheres in the red region ($r < 2m(v)$), so the question arises of whether or not AH is the boundary $\mathscr{B}$ of the f-trapped region. One can also wonder if the boundary $\mathscr{B}$ will actually be EH.

meaningful any more. This is a nice and satisfactory explanation, and renders absolutely transparent the important differences between 'outer' f-trapped and f-trapped surfaces.

Coming back to genuine f-trapped surfaces, some numerical investigations[39] were not capable of finding closed f-trapped surfaces to the past of the apparent 3-horizon AH. Thus, a natural question arises: is AH $= \mathscr{B}$?

The answer is negative for general spherically symmetric spacetimes with inflowing matter and radiation, as discussed in the next section.

6. Imploding Spherically Symmetric Spacetimes in Advanced Coordinates: $\mathscr{B} \neq$ AH.

The general 4-dimensional spherically symmetric line-element can be locally given in advanced coordinates by

$$ds^2 = -e^{2\beta}\left(1 - \frac{2m(v,r)}{r}\right) dv^2 + 2e^{\beta} dv dr + r^2 d\Omega^2$$

where $m(v,r)$ is the mass function. The future-pointing radial null geodesic vector fields ($k_\mu \ell^\mu = -1$) read

$$\vec{\ell} = -e^{-\beta}\partial_r, \qquad \vec{k} = \partial_v + \frac{1}{2}\left(1 - \frac{2m}{r}\right)e^{\beta}\partial_r$$

and the mean curvature vector for each round sphere (defined by constant values of r and v) is:

$$\vec{H}_{sph} = \frac{2}{r}\left(e^{-\beta}\partial_v + \left(1 - \frac{2m}{r}\right)\partial_r\right) \tag{6.9}$$

so that setting $\vec{k}^+_{sph} = \vec{k}$ and $\vec{k}^-_{sph} = \vec{\ell}$, the future null expansions for these round spheres become

$$\theta^+_{sph} = \frac{e^{\beta}}{r}\left(1 - \frac{2m}{r}\right), \qquad \theta^-_{sph} = -\frac{2e^{-\beta}}{r}$$

and the unique spherically symmetric MTT is given by AH : $r - 2m(r,v) = 0$ ($\Longleftrightarrow$ $\theta^+_{sph} = 0$), in agreement with previous calculations in Section 3.1.

AH can be timelike, null or spacelike depending on the sign of

$$\left.\frac{\partial m}{\partial v}\left(1 - 2\frac{\partial m}{\partial r}\right)\right|_{AH}.$$

In particular, AH is null (in fact it is an *isolated horizon*[6]) on any open region where $m = m(r)$. This isolated horizon portion of AH, denoted by AH^{iso}, is characterized by:

$$\text{AH}^{iso} \equiv \text{AH} \cap \{G_{\mu\nu}k^\mu k^\nu = 0\}$$

An example (based on a Vaidya spacetime) is shown in the next Fig. 7.

6.1. *Perturbations on the spherical MTT*

In order to prove that there are closed f-trapped surfaces extending beyond AH one can use a perturbation argument, see Ref. 10.

Theorem 1. *In arbitrary spherically symmetric spacetimes, there are closed f-trapped surfaces (topological spheres) penetrating both sides of the apparent 3-horizon AH at any region where* $G_{\mu\nu}k^\mu k^\nu|_{AH} \neq 0$.

Proof. Recall that $\theta^-_{sph} = -e^{-\beta}\frac{2}{r} < 0$, $\theta^+_{sph} = 0$ on AH. Perturb any marginally f-trapped 2-sphere $\varsigma \in$ AH along a direction $f\vec{n}$ orthogonal to ς, where f is a function on ς and

$$\vec{n} = -\left.\vec{\ell} + \frac{n_\mu n^\mu}{2}\vec{k}\right|_{\varsigma} \qquad k_\mu n^\mu = 1.$$

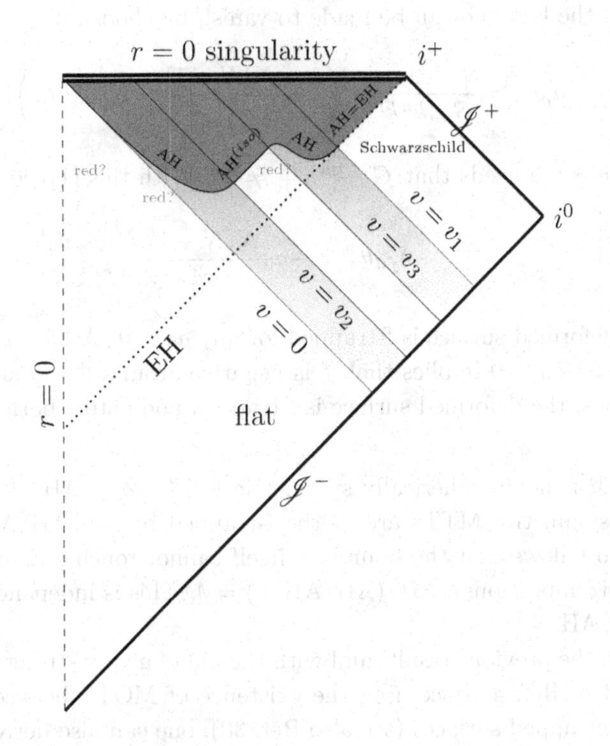

Figure 7. (Color online) A dynamical situation with non-empty $\mathrm{AH}^{iso}\setminus\mathrm{EH}$. Null radiation flows for $v \in (0, v_2)$, and then stops. There is no inflow of energy for $v \in (v_2, v_3)$ and then it comes in again from past null infinity for $v \in (v_3, v_1)$. From $v \geq v_1$ there is no more infalling energy and the spacetime settles down to a Schwarzschild black hole. The unique spherically symmetric MTT, denoted by AH, is spacelike for $v \in (0, v_2) \cup (v_3, v_1)$, but it also possesses two portions of isolated-horizon type: one given by the part of EH with $v \geq v_1$, and the other one for $v \in (v_2, v_3)$, represented here by AH^{iso}. The red portion is part of the f-trapped region $\mathscr{T}$, but one can prove (see main text) that there are closed f-trapped surfaces extending below $\mathrm{AH}\setminus\mathrm{AH}^{iso}$.

Observe that the causal character of $\vec{n}$ is unrestricted. The variation of the vanishing expansion $\theta^+ = 0$ reduces to[2,3,10]

$$\delta_{f\vec{n}}\theta^+ = -\Delta_\varsigma f + f\left(\frac{1}{r^2} - G_{\mu\nu}k^\mu \ell^\nu - \frac{n_\rho n^\rho}{2} G_{\mu\nu}k^\mu k^\nu\right)\Bigg|_\varsigma$$

where Δ_ς is the Laplacian on ς. Now, choose $f = a_0 + a_N P_N(\cos\theta)$, ($a_0, a_N =$ const.), where P_l are the Legendre polynomials. Using $\Delta_\varsigma P_l = -\frac{l(l+1)}{r^2_\varsigma} P_l$, the previous variation becomes

$$\delta_{f\vec{n}}\theta^+ = a_0\left(\frac{1}{r^2} - G_{\mu\nu}k^\mu \ell^\nu - \frac{n_\rho n^\rho}{2} G_{\mu\nu}k^\mu k^\nu\right)\Bigg|_\varsigma$$
$$+a_N\left(\frac{1+N(N+1)}{r^2} - G_{\mu\nu}k^\mu \ell^\nu - \frac{n_\rho n^\rho}{2} G_{\mu\nu}k^\mu k^\nu\right)\Bigg|_\varsigma P_N(\cos\theta).$$

The term in the last line can be made to vanish by choosing

$$n_\rho n^\rho = \frac{2}{G_{\mu\nu}k^\mu k^\nu|_\varsigma}\left(\frac{1+N(N+1)}{r^2} - G_{\mu\nu}k^\mu \ell^\nu\right)\Big|_\varsigma .$$

(Here is where one needs that $G_{\mu\nu}k^\mu k^\nu|_\varsigma \neq 0$). With this choice

$$\delta_{f\vec{n}}\theta^+ = -a_0\frac{N(N+1)}{r_\varsigma^2},$$

hence, the deformed surface is f-trapped for any $a_0 > 0$. As $f = a_0 + a_N P_N(\cos\theta)$, setting $a_N < -a_0 < 0$ implies that f is negative around $\theta = 0$ and positive where $P_N \leq 0$. Thus, the deformed surface is f-trapped and enters both sides of AH. □

Therefore, $\mathscr{B}$ is not a spherically symmetric MTT: $\mathscr{B} \neq$ AH. It follows that the spherically symmetric MTTs are in the f-trapped region, AH$\backslash$AH$^{iso} \subset \mathscr{T}$, from where it also follows that the boundary itself cannot touch AH (except perhaps in isolated-horizon portions): $\mathscr{B} \cap$ (AH$\backslash$AH^{iso}) $= \emptyset$. This is independent of the causal character of AH.

Based on the previous result, and with the aid of a very strong and fundamental result found in Ref. 4 concerning the existence of MOTS between outer trapped and outer untrapped surfaces (see also Ref. 30), one can also derive a result on the existence of non-spherical MTTs, see Ref. 10.

Corollary 1. *In arbitrary spherically symmetric spacetimes there exist MTTs penetrating both sides of the apparent 3-horizon AH : $\{r = 2m\}$ at any region where $G_{\mu\nu}k^\mu k^\nu|_{AH} \neq 0$.*

Actually, all spacelike MTTs (that is, DHs) other than AH must lie partly to the future of AH and partly to its past, as proven in Ref. 7.

Theorem 2. *No closed weakly f-trapped surface can be fully contained in the past domain of dependence $D^-(AH)$ of a spacelike AH.*

However, closed f-trapped surfaces may lie on D^-(AH) almost completely. In other words, closed f-trapped surfaces can intersect the region $\{r < 2m\}$ in just an arbitrarily tiny portion, as small as desired. This surprising result was obtained in Ref. 10 and will be of fundamental importance for the concept of "core" of a black hole, see section 10. More precisely:

Theorem 3. *In spherically symmetric spacetimes, there are closed f-trapped surfaces (topological spheres) penetrating both sides of the apparent 3-horizon AH$\backslash$AH^{iso} with arbitrarily small portions outside the region $\{r > 2m\}$.*

6.2. *Closed trapped surfaces are "clairvoyant"*

Now that we have learnt that closed f-trapped surfaces can extend beyond the spherically symmetric MTT, one wonders how far can they go. In particular, an important question is whether or not they may extend all the way 'down' to the flat regions of the spacetime, if these exist.

Surprisingly, the short answer is yes: sometimes they do extend and penetrate flat portions of the space-time, regions whose whole past is also flat. This was proven, via an explicit example, in Ref. 9, later refined with more elaborated examples in Ref. 1. A picture showing points of the surface in a Penrose diagram is shown in Fig. 8. A longer discussion of this and related explicit examples, with some interesting insights, is provided in the chapter by Bengtsson.

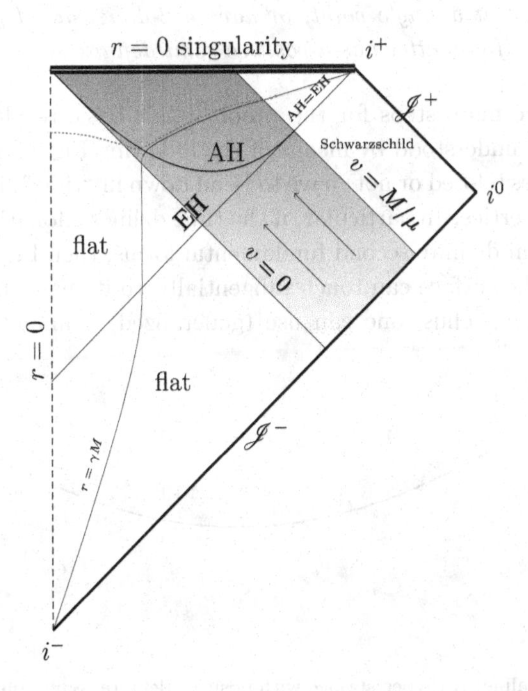

Figure 8. (Color online) A closed f-trapped surface penetrating the flat portion of a black hole spacetime constructed with the Vaidya solution. The surface has points on every sphere on the dotted line in red. It thus enters the flat portion of the spacetime.

In conclusion, closed trapped surfaces may penetrate flat regions of the spacetime. This implies that they are highly non-local: they can have portions in a flat region of spacetime whose *whole* past is also flat in clairvoyance of energy that crosses them elsewhere to make their compactness feasible. They are *clairvoyant*, that is to say, 'aware' of things that happen elsewhere, far away, with no causal connection.

Observe finally that this result also implies that the boundary $\mathscr{B}$ of the f-trapped region may penetrate the flat regions too.

7. Interplay of Surfaces and Generalized Symmetries

Hitherto, we have proven that the boundary $\mathscr{B}$ of the f-trapped region must be generically below the spherically symmetric MTT defined by AH: $\{r = 2m\}$, but it cannot extend all the way down to the event horizon EH. In order to try to put restrictions on the location of $\mathscr{B}$ and the extension of $\mathscr{T}$ we use the following fundamental property[10]

Proposition 3. *No f-trapped surface (closed or not) can touch a spacelike hypersurface to its past at a single point, or have a 2-dimensional portion contained in the hypersurface, if the latter has a positive semi-definite second fundamental form.*

Before giving the main steps for the proof of this basic result, the intuitive idea behind it can be understood by means of the following Fig. 9. As we saw in Fig. 3, f-trapped surfaces (closed or not) have to bend down in "time", if this time has some appropriate properties. In particular, if the time defines a foliation by hypersurfaces with positive semi-definite second fundamental forms, then Fig. 9 shows that there is no way that the surface can touch tangentially, to its past (in "time"), any such level hypersurface. Thus, one can use (generalized) symmetries, or equivalently

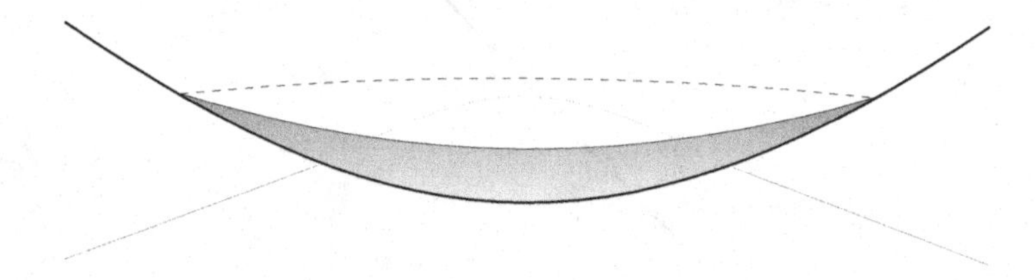

Figure 9. (Color online) A (hyper)surface with positive-definite second fundamental form (shown as a hyperboloid) and a f-trapped curve (shown in red) in 3-dimensional Minkowski spacetime.

some distinguished hypersurface-orthogonal future-pointing vector fields — such as (conformal, homothetic) Killing vectors, Kerr–Schild vector fields,[15] the Kodama vector field in spherically symmetric spacetimes, etc.— to constrain the possible existence of f-trapped surfaces. This interplay between surfaces and generalized symmetries has proven very useful in several investigations concerning the trapped-surface fauna.[10,13,33,42,43,45]

To prove the fundamental property given in Proposition 3 and related interesting results, start with the identity for the Lie derivative of the metric along a vector field $\vec{\xi}$

$$(\pounds_{\vec{\xi}} g)_{\mu\nu} = \nabla_\mu \xi_\nu + \nabla_\nu \xi_\mu .$$

Projecting to the surface S and using (2.1)

$$(\pounds_{\vec{\xi}} g|_S)_{\mu\nu}\, e^\mu_A e^\nu_B = \overline{\nabla}_A \overline{\xi}_B + \overline{\nabla}_B \overline{\xi}_A + 2\xi_\mu|_S K^\mu_{AB}$$

so that contracting now with γ^{AB} we arrive at the main formula

$$\boxed{\frac{1}{2} P^{\mu\nu} (\pounds_{\vec{\xi}} g|_S)_{\mu\nu} = \overline{\nabla}_C \overline{\xi}^C + \xi_\rho H^\rho} \tag{7.10}$$

where $P^{\mu\nu} \equiv \gamma^{AB} e^\mu_A e^\nu_B$ is the orthogonal projector of S.

This elementary formula is very useful and permits one to obtain many interesting, immediate, consequences. For instance:

(1) If S is minimal ($\vec{H} = \vec{0}$), integrating the formula for closed S

$$\oint_S P^{\mu\nu} (\pounds_{\vec{\xi}} g|_S)_{\mu\nu} = 0 .$$

This relation must be satisfied for *all* possible vector fields $\vec{\xi}$. Therefore, closed minimal surfaces are very rare.

(2) If $\vec{\xi}$ is a Killing vector, integrating again for closed S

$$\oint_S \xi_\rho H^\rho = 0 .$$

Therefore, if the Killing vector $\vec{\xi}$ is timelike on S, then S cannot be weakly trapped, unless it is minimal.[33]

More sophisticated and useful consequences can be derived, such as[10,33]

Lemma 1. *Let $\vec{\xi}$ be future-pointing on a region $\mathscr{R} \subset \mathcal{V}$ and let S be a closed surface contained in $\mathscr{R}$ with $P^{\mu\nu}(\pounds_{\vec{\xi}} g|_S)_{\mu\nu} \geq 0$. Then, S cannot be closed and weakly f-trapped (unless $\xi_\mu H^\mu = 0$ and $P^{\mu\nu}(\pounds_{\vec{\xi}} g|_S)_{\mu\nu} = 0$).*

Proof. Integrating the main formula on S, the divergence term integrates to zero so that

$$\oint_S \xi_\rho H^\rho = \frac{1}{2} \oint_S P^{\mu\nu} (\pounds_{\vec{\xi}} g|_S)_{\mu\nu} \geq 0$$

Hence, $\vec{H}$ cannot be future pointing on all S (unless $\xi_\mu H^\mu = P^{\mu\nu}(\pounds_{\vec{\xi}} g|_S)_{\mu\nu} = 0$). □

Stronger results can be obtained for *hypersurface-orthogonal* $\vec{\xi}$, that is, $\vec{\xi}$ satisfying

$$\xi_{[\mu} \nabla_\nu \xi_{\rho]} = 0 \iff \xi_\mu = -F \partial_\mu \tau$$

for some local functions $F > 0$ and τ. This means that $\vec{\xi}$ is orthogonal to the hypersurfaces τ =const. (called the level hypersurfaces.)

Theorem 4. *Let $\vec{\xi}$ be future-pointing and hypersurface-orthogonal on a region $\mathscr{R} \subset \mathcal{V}$ and let S be a f-trapped surface. Then, S cannot have a local minimum of τ at any point $q \in \mathscr{R}$ where $P^{\mu\nu}(\pounds_{\vec{\xi}}g)_{\mu\nu}|_q \geq 0$.*

Proof. Let $q \in S \cap \mathscr{R}$ be a point where S has a local extreme of τ. Noting that $\bar{\xi}_A = -\bar{F}\partial\bar{\tau}/\partial\lambda^A$ with $\bar{F} \equiv F|_S$ this means that

$$\bar{\xi}_A|_q = \left.\frac{\partial\bar{\tau}}{\partial\lambda^A}\right|_q = 0$$

where $\tau = \bar{\tau}(\lambda^A)$ is the local parametric expression of τ on S.

An elementary calculation leads then to:

$$\left.\overline{\nabla}_A\bar{\xi}^A\right|_q = \gamma^{AB}\overline{\nabla}_A\left(-\bar{F}\frac{\partial\bar{\tau}}{\partial\lambda^B}\right)\bigg|_q = -\bar{F}\gamma^{AB}\left.\frac{\partial^2\bar{\tau}}{\partial\lambda^A\partial\lambda^B}\right|_q.$$

Introducing this in the main formula (7.10)

$$\bar{F}\gamma^{AB}\left.\frac{\partial^2\bar{\tau}}{\partial\lambda^A\partial\lambda^B}\right|_q = -\frac{1}{2}P^{\mu\nu}(\pounds_{\vec{\xi}}g)_{\mu\nu}\Big|_q + \xi_\rho H^\rho\big|_q \leq \xi_\rho H^\rho\big|_q$$

hence $\partial^2\bar{\tau}/\partial\lambda^A\partial\lambda^B|_q$ cannot be positive (semi)-definite if $\vec{\xi}$ and $\vec{H}$ are both future-pointing. □

Remark 1.

(1) S does not need to be compact, nor contained in $\mathscr{R}$.
(2) It is enough to assume $P^{\mu\nu}(\pounds_{\vec{\xi}}g)_{\mu\nu}|_q \geq 0$ only at the local extremes of τ on S.
(3) A positive semi-definite $\partial^2\bar{\tau}/\partial\lambda^A\partial\lambda^B|_q$ is also excluded.
(4) The theorem holds true for weakly trapped surfaces with the only exception of

$$\left.\frac{\partial^2\bar{\tau}}{\partial\lambda^A\partial\lambda^B}\right|_q = 0 \quad \text{and} \quad P^{\mu\nu}(\pounds_{\vec{\xi}}g)_{\mu\nu}|_q = 0 \quad \text{and} \quad \xi_\rho H^\rho\big|_q = 0$$

If $\vec{\xi}|_q$ is timelike, the last of these implies that $\vec{H}|_q = \vec{0}$.
(5) Setting aside this exceptional possibility, τ always decreases at least along one tangent direction in T_qS. Starting from any point $x \in S \cap \mathscr{R}$ one can always follow a connected path along $S \cap \mathscr{R}$ with decreasing τ.

8. The Past Barrier Σ

The results above on the interplay of vector fields with special convexity properties and the trapped-surface fauna were essential in order to detect a general past barrier for closed f-trapped surfaces in general imploding, asymptotically flat, spherically symmetric spacetimes.[10] The mild assumptions used to get this past barrier are:[10]

(1) The total mass function is finite, and there is an initial flat region

$$m(v,r)=0 \quad \forall v<0; \qquad \forall v>0: \quad 0\leq m(v,r)\leq M<\infty \ (M>0)$$

and a regular future null infinity $\mathscr{J}^+$ with associated event horizon EH.[16] Let AH_1 denote the connected component of $\text{AH} \equiv \{r=2m\}$ associated to this EH. AH_1 separates the region $\mathscr{R}_1$, defined as the connected subset of $\{r>2m\}$ which contains the flat region of the spacetime, from a region containing f-trapped 2-spheres.

(2) The dominant energy condition holds, and furthermore the matter-energy is falling in

$$\frac{\partial m}{\partial v}\geq 0 \quad \text{on} \quad \{r\geq 2m\}\cap J^+(EH)$$

where $J^+(EH)$ is the causal future of the event horizon.

The connected component of $\mathscr{B}$ associated to AH_1 will be denoted by $\mathscr{B}_1$, in analogy with $\mathscr{R}_1$ which denotes the corresponding $\{r>2m\}$-region. The same notation is used for AH_1^{iso}, $\mathscr{T}_1$, etc. Some examples are provided in the Penrose diagrams of Fig. 10.

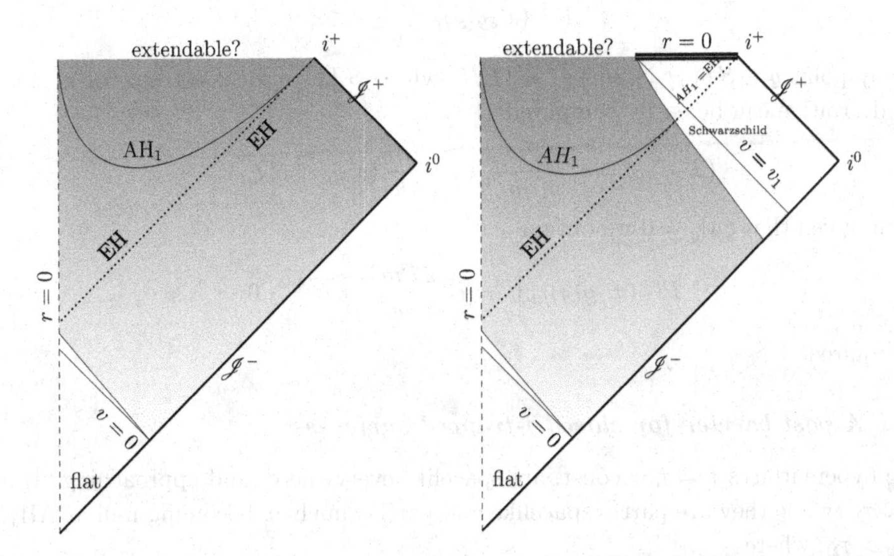

Figure 10. Two conformal diagrams that correspond to the possibilities $m(v,r)<M$ everywhere (left), and to $m(v,r)=M$ in some asymptotic region (right). In the second case the spacetime becomes eventually Schwarzschild spacetime. The shaded regions are non-flat spherically symmetric spacetimes with non-zero energy-momentum tensor. A connected component of AH, labeled AH_1, is shown. This may have timelike portions, and it merges with the EH either asymptotically (left) or at a finite value of v (right). Whether or not a singularity develops in the shaded region depends on the specific properties of the matter (of the mass function $m(v,r)$), and there can be cases where the spacetime can be continued towards the future, where some other connected parts of AH may appear.

8.1. *The Kodama vector field*

The Kodama vector field,[27] which in these coordinates takes the simple form

$$\vec{\xi} = e^{-\beta}\partial_v$$

characterizes the spherically symmetric directions tangent to the hypersurfaces $r =$ const. In other words, it points into the unique direction where the round spheres have vanishing expansion, as can be checked using (6.9): $\xi_\mu H^\mu_{sph} = 0$.

$\vec{\xi}$ is hypersurface orthogonal, with the level function τ defined by

$$\xi_\mu dx^\mu = -F d\tau = dr - e^\beta \left(1 - \frac{2m(v,r)}{r}\right) dv.$$

Furthermore

$$\xi_\mu \xi^\mu = -\left(1 - \frac{2m(v,r)}{r}\right), \qquad \ell_\mu \xi^\mu = -1$$

so that $\vec{\xi}$ is future-pointing timelike on the region $\{r > 2m\}$, and future-pointing null at AH : $\{r = 2m\}$. Thus, in order to ascertain if $\vec{\xi}$ has all the necessary properties to apply Theorem 4 we need to check if

$$P^{\mu\nu}(\pounds_{\vec{\xi}} g|_S)_{\mu\nu}\Big|_q \geq 0$$

at any point $q \in S \cap \{r \geq 2m\} \cap J^+(EH)$ where S has a local extreme of τ. The Lie derivative can be easily computed

$$(\pounds_{\vec{\xi}} g)_{\mu\nu} = e^\beta \frac{2}{r}\frac{\partial m}{\partial v}\ell_\mu \ell_\nu - \frac{\partial \beta}{\partial r}\left(\delta^r_\mu \xi_\nu + \delta^r_\nu \xi_\mu\right).$$

Then, given that $\bar{\xi}_A|_q = 0$ we obtain

$$P^{\mu\nu}(\pounds_{\vec{\xi}} g|_S)_{\mu\nu}\Big|_q = e^\beta \frac{2}{r}\frac{\partial m}{\partial v}\Big|_S \bar{\ell}_A \bar{\ell}^A \geq 0$$

as required.

8.2. *A past barrier for closed f-trapped surfaces*

The hypersurfaces $\tau = \tau_c =$ const. are spacelike everywhere (and approaching i^0) if $\tau_c < \tau_\Sigma$, while they are partly spacelike and partly timelike, becoming null at AH_1, if $\tau_c > \tau_\Sigma$, where

$$\tau_\Sigma \equiv \inf_{x \in AH_1} \tau|_x.$$

Observe that τ_Σ is also the least upper bound of τ on EH.

Define the hypersurface Σ as

$$\Sigma \equiv \{\tau = \tau_\Sigma\}.$$

Σ is the *last* hypersurface orthogonal to $\vec{\xi}$ which is non-timelike everywhere. See Fig. 11 for a representation of all these facts. It turns out that Σ is a past limit for closed f-trapped surfaces, and this is a direct consequence of the properties of the Kodama vector field and Theorem 4 (or Proposition 3). Thus[10]

Theorem 5. *No closed f-trapped surface can enter the region* $\tau \leq \tau_\Sigma$.

The location of Σ acquires therefore an unexpected importance, and this depends in particular on whether $8\dot{m}_0 > (1 - 2m_0')^2$ or not. Here $\dot{m}_0$ and m_0' are the limits of $\frac{\partial m}{\partial v}$ and $\frac{\partial m}{\partial r}$ when approaching $r = 0$, respectively. In the former case, Σ does penetrate the flat region. It may not be so in the other cases. See Ref. 10 for details.

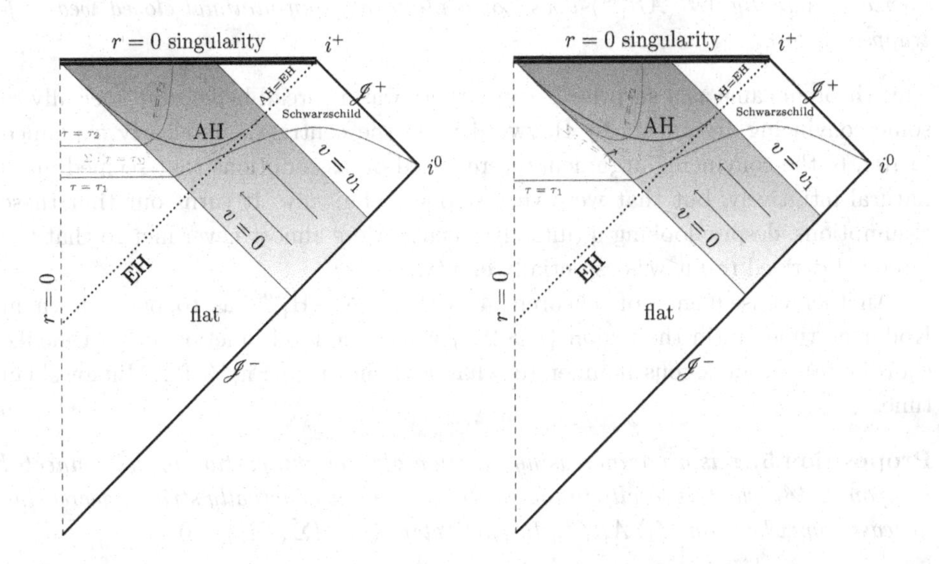

Figure 11. The past barrier Σ represented for some particular cases of the Vaidya spacetime (5.8). The hypersurfaces $\tau = \tau_1$ are spacelike for $\tau_1 < \tau_\Sigma$, while they are partly spacelike, partly timelike for values of $\tau_2 > \tau_\Sigma$, as shown in the figure. The limit case defines $\Sigma : \tau = \tau_\Sigma$, which happens to be a past barrier for f-trapped surfaces due to the convexity properties of the Kodama vector field. This past barrier Σ can enter the flat region or not, and this depends on the properties of the mass function close to the upper left corner with $r = 0$. Here, two possibilities are depicted: the case where Σ never enters the flat region of the spacetime (right) and the opposite case where it does penetrate the flat region (left).

9. Some Properties of $\mathscr{B}$, and about its Location

Some elaborations using the previous results on the Kodama level function τ allows one to derive further properties of the boundary $\mathscr{B}$, and to put severe restrictions on its location. To that end, set $\tau_{\mathscr{B}} \equiv \inf_{x \in \mathscr{B}} \tau|_x$ where $\tau =$ const. are the level hypersurfaces of $\vec{\xi}$. Then the following set of results were obtained in Ref. 10, where the reader may consult the proofs.

Proposition 4. *The connected component $\mathscr{B}_1$ cannot have a positive minimum value of r, and furthermore*

$$\tau_{\mathscr{B}} = \inf_{x\in\mathscr{B}_1} \tau|_x = \tau_\Sigma \,.$$

Corollary 2. *$\mathscr{B}_1 \subset (\mathscr{R}_1 \cup AH_1) \cap \{\tau \geq \tau_\Sigma\}$ and $\mathscr{B}_1$ must merge with, or approach asymptotically, Σ, AH_1 and EH in such a way that $\mathscr{B}_1 \cap (AH_1 \backslash AH_1^{iso}) = \emptyset$ if $G_{\mu\nu}k^\mu k^\nu|_{AH_1\backslash EH} > 0$.*

Furthermore, $\mathscr{B}_1$ cannot be non-spacelike everywhere.

Theorem 6. *$(\mathscr{B}\backslash AH_1^{iso})$ cannot be a marginally trapped tube, let alone a dynamical horizon. Actually, $(\mathscr{B}\backslash AH_1^{iso})$ does not contain any non-minimal closed weakly f-trapped surface.*

This theorem came as a surprise because there was a spread belief, due specially to some convincing arguments by Hayward,[23] on the contrary. As clearly explained in Ref. 6, the convincing arguments were based on assumptions that seemed quite natural intuitively, but that were very strong technically. It turns out that these assumptions, despite looking intuitively natural, were almost never met so that the intended derived result was essentially empty.

Another consequence of Theorem 4 is that $\mathscr{B}_1\backslash \mathrm{AH}_1^{iso}$ has to bend down in Kodama "time" τ (in the region $\{r > 2m\}$ where the level function τ is a timelike coordinate), in analogous manner to what was shown in Fig. 3 for Minkowskian time.

Proposition 5. *τ is a nonincreasing function of r on any portion of the connected component $\mathscr{B}_1$ which is locally to the past of $\mathscr{T}_1$, and it is actually strictly decreasing at least somewhere on $\mathscr{B}_1\backslash AH_1^{iso}$. In particular, $\mathscr{B}_1 \cap (\Sigma\backslash EH) = \emptyset$.*

This result infers that $\mathscr{B}_1$ is to the future of Σ, and we already know that is has to be to the past of AH_1. This also implies that $\mathscr{B}_1$ is squeezed below by Σ and above by AH_1 close to their merging with the event horizon, and thus $\mathscr{B}_1$ must be spacelike close to its merging with Σ, AH_1 and EH.[10] An illuminating pictorial explanation of these results is represented, for a particular case, in Fig. 12.

There remains, as an interesting puzzle, the question of where is *exactly* the boundary $\mathscr{B}$. This is an open question. There are some known restrictions on the 2nd fundamental form (extrinsic curvature) of $\mathscr{B}$,[10] but they are not sufficient to completely determine the position of $\mathscr{B}$ in generic spherically symmetric spacetimes.

The more restrictive known property on $\mathscr{B}$ is given by the following result[10]

Proposition 6. *Any spacelike portion of the connected component $\mathscr{B}_1$ which is locally to the past of $\mathscr{T}_1$ has a second fundamental form with a non-positive (and strictly negative whenever $\mathscr{B}_1$ is not tangent to a τ=const hypersurface) double eigenvalue. In particular, it cannot have a positive semidefinite second fundamental form there.*

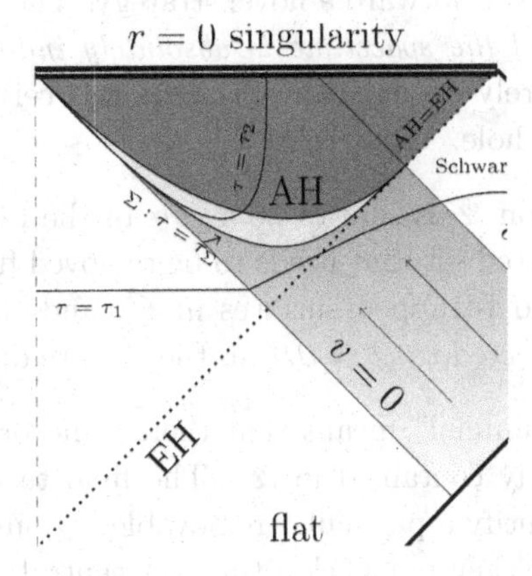

Figure 12. (Color online) This is an enlargement of the conformal diagram on the right in Fig. 11, showing the allowed region for the location of the boundary $\mathscr{B}$. The part shown in red is known to be part of the f-trapped region $\mathscr{T}$. Theorem 1 tells us, however, that $\mathscr{T}$ extends beyond AH, and actually includes it. On the other hand, Proposition 5 implies that $\mathscr{T}$ can never touch Σ (outside EH). Therefore, the zone marked in yellow is the allowed region for the boundary $\mathscr{B}$, keeping in mind that it must be strictly inside the yellow zone: it can never touch Σ, nor AH, there.

Actually, one can find stronger restrictions on the extrinsic properties of $\mathscr{B}$, but they are out of the scope of this chapter.

10. Black Holes' Cores

Let me summarize some of the main conclusions derived so far concerning trapped surfaces, MTTs, EH, $\mathscr{B}$ and the trapped region $\mathscr{T}$.

The clairvoyance property of trapped surfaces is inherited by everything based on them, such as marginally trapped tubes including dynamical horizons. It implies that EH is teleological and that $\mathscr{B}$ is also non-local, even penetrating sometimes into flat portions of the spacetime. In conjunction with the non-uniqueness of dynamical horizons, this poses a fundamental puzzle for the physics of black holes. Four possible solutions have been put forward[6,10,11,19,34,31]

(1) one can rely on the old and well-defined event horizon, and try to put up with its teleological properties.
(2) one can accept the non-uniqueness of MTTs and treat all possible MTTs and dynamical/trapping horizons on equal footing.
(3) one can also use the boundary $\mathscr{B}$ as defined above, despite its non-local properties.
(4) or one can try to define a preferred MTT. Hitherto, the only proposal I am aware of was presented in Ref. 19, based on an evolution principle for the area (entropy) of the marginally trapped surfaces foliating the MTT.

In Ref. 10 we have put forward a novel strategy. The idea is based on the simple question: *what part of the spacetime is absolutely indispensable for the existence of the black hole?* Surely enough, any flat region is certainly not essential for the existence of the black hole. What is?

Definition 3. A region $\mathscr{Z}$ is said to be a *core* of the f-trapped region $\mathscr{T}$ if it is a minimal closed connected set that needs to be removed from the spacetime in order to get rid of all closed f-trapped surfaces in $\mathscr{T}$, and such that any point on the boundary $\partial\mathscr{Z}$ is connected to $\mathscr{B} = \partial\mathscr{T}$ in the closure of the remainder.

Remark 2. Here, "minimal" means that there is no other set $\mathscr{Z}'$ with the same properties and properly contained in $\mathscr{Z}$. The final technical condition is needed because one could identify a particular removable region to eliminate the f-trapped surfaces, excise it, but then put back a tiny but central isolated portion to make it smaller. However, this is not what one wants to cover with the definition.

Obviously, $\mathscr{Z} \subset \mathscr{T}$, but in general $\mathscr{Z}$ is substantially smaller than the corresponding trapped region $\mathscr{T}$. An example of a core is given by the dust Robertson–Walker model of Fig. 3, where the region shown in purple is a core of the larger red region $\mathscr{T}$. If the purple region is removed from the spacetime, then no closed f-trapped surface remains. This example demonstrates that cores are not unique: one can choose any other region $\mathscr{Z}$ equivalent to the chosen one by moving all its points by the group of symmetries on each homogeneous spatial slice of the Robertson–Walker metric. This kind of non-uniqueness is somehow irrelevant, being due to the existence of a high degree of symmetry. Nevertheless, even in less symmetric cases the uniqueness of the cores cannot be assumed beforehand, and one can actually prove that it does not hold in general, see Proposition 8 below.

Theorem 7. *In spherically symmetric spacetimes (with $AH^{iso}\backslash EH = \emptyset$) the region*

$$\mathscr{Z} \equiv \{r \leq 2m(v,r)\}$$

is a core of the f-trapped region.

This follows firstly from the fact that removing $\mathscr{Z}$ from the spacetime short-circuits all possible closed f-trapped surfaces, as they cannot be fully contained in the region where the Kodama vector field is timelike (the complement of $\mathscr{Z}$) as a consequence of Lemma 1. And secondly and more importantly, from Theorem 3, which implies that one cannot remove a smaller region achieving the same goal. (This is where we need the condition of not having isolated-horizon portions in AH, but this is probably technical and the result will hold in general).

The identified cores happen to be unique with spherical symmetry.

Proposition 7. *In spherically symmetric spacetimes (with $AH^{iso}\backslash EH = \emptyset$) $\mathscr{Z} = \{r \leq 2m\}$ are the only spherically symmetric cores of $\mathscr{T}$. Therefore, $\partial\mathscr{Z} = AH$ are the only spherically symmetric boundaries of a core.*

The two previous results are surprising and may have a deep meaning, because the trapped regions and their cores are global concepts, and in that sense they share the teleological and/or clairvoyant properties of EH and of trapped surfaces. However, we have identified at least one boundary of a core which happens to be a MTT — and a very good one in spherical symmetry: the unique one respecting the symmetry — and MTTs are quasi-local objects, they do not need to know future causes or to be aware of things that happen elsewhere. A full interpretation of this surprising result may lead to a better understanding of BHs and of their boundaries.

It arises as an important problem the question of the uniqueness of cores, and their boundaries. As mentioned before, cores are not unique and one can prove the existence of non-spherically symmetric cores in spherically symmetric spacetimes.[10]

Proposition 8. *There exist non-spherically symmetric cores of the f-trapped region in spherically symmetric spacetimes.*

The proof of this result is essentially based on a theorem[7] analogous to Theorem 2 but valid for general DHs, because then one derives that there must be a subset of the future of any dynamical horizon which, when removed from the spacetime, gets rid of all closed f-trapped surfaces. However, as we do not have an analogous Theorem 3 for general DHs, it is unkown whether the core is a proper subset of, or the whole, future of the DH. Therefore, three possibilities arise:

(1) For a generic MTT H, its causal future $J^+(H)$ is a core. This will amount to saying that MTTs are generically boundaries of cores for BHs. This looks like a good result, and an explanation of how a quasi-locally defined object (the MTT) is always the boundary of a globally defined one (the core) should be sought.
(2) Any MTT H other than AH is such that its causal future $J^+(H)$ is *not* a core — the core being a proper subset of $J^+(H)$. Hence, the identified core $\mathscr{Z} = \{r \leq 2m\}$ is special in the sense that its boundary $\partial\mathscr{Z} = \text{AH}$ is a marginally trapped tube. Thereby, AH : $\{r = 2m(v, r)\}$ would be selected as the unique MTT which is the boundary of a core of the f-trapped region $\mathscr{T}$. This would be a very good result, and it might provide ideas to try and find the same type of result in the general, non-spherically-symmetric, situation.
(3) Neither of the previous two, in other words, some MTTs turn out to be boundaries of cores, some are not. This would be a puzzling situation.

Which one of these three possibilities holds remains as a very interesting open question.

Acknowledgements

Some parts of this chapter are based on a fruitful collaboration with I. Bengtsson. Supported by grants FIS2010-15492 (MICINN), GIU06/37 (UPV/EHU) and P09-FQM-4496 (J. Andalucía—FEDER).

References

1. J.E. Åman, I. Bengtsson and J.M.M. Senovilla, Where are the trapped surfaces?, *J. Phys.: Conf. Ser.* **229** 012004 (2010); arXiv:0912.3691
2. L. Andersson, M. Mars and W. Simon, Local existence of dynamical and trapping horizons *Phys. Rev. Lett.* **95**, 111102 (2005)
3. L. Andersson, M. Mars and W. Simon, Stability of marginally outer trapped surfaces and existence of marginally outer trapped tubes, *Adv. Theor. Math. Phys.* **12** 853 (2008)
4. L. Andersson and J. Metzger, The area of horizons and the trapped region, *Commun. Math. Phys.* **290** 941-972 (2009)
5. A. Ashtekar and B. Krishnan, Dynamical horizons and their properties, *Phys. Rev. D* **68**, 104030 (2003)
6. A. Ashtekar and B. Krishnan, Isolated and Dynamical Horizons and Their Applications, *Living Rev. Relativity* **7** 10 (2004). URL (cited on 02-07-2011): `http://www.livingreviews.org/lrr-2004-10`
7. A. Ashtekar and G. J. Galloway, Some uniqueness results for dynamical horizons, *Adv. Theor. Math. Phys.* **9** 1 (2005)
8. I. Ben-Dov, Outer Trapped Surfaces in Vaidya Spacetimes, *Phys. Rev. D* **75** 064007 (2007)
9. I. Bengtsson and J.M.M. Senovilla, Note on trapped surfaces in the Vaidya solution, *Phys. Rev.* **D 79** 024027 (2009)
10. I. Bengtsson and J.M.M. Senovilla, The region with trapped surfaces in spherical symmetry, its core, and their boundaries, *Phys. Rev.* **D 83** 044012 (2011)
11. I. Booth, Black hole boundaries, *Can. J. Phys.* **83** 1073 (2005)
12. H. L. Bray and M. Khuri, P.D.E.'s which imply the Penrose inequality, arXiv:0905.2622 (2009)
13. A. Carrasco and M. Mars, Stability of marginally outer trapped surfaces and symmetries, *Class. Quantum Grav.* **26** 175002 (2009)
14. A. Carrasco and M. Mars, A counterexample to a recent version of the Penrose conjecture, *Class. Quantum Grav.* **27** 062001 (2010)
15. B. Coll, S.R. Hildebrandt and J.M.M. Senovilla, Kerr-Schild symmetries, *Gen. Rel. Grav.* **33**, 649 (2001).
16. M. Dafermos, Spherically symmetric spacetimes with a trapped surface, *Class. Quantum Grav.* **22** 2221-2232 (2005)
17. D.M. Eardley, Black Hole Boundary Conditions and Coordinate Conditions, *Phys. Rev. D* **57**, 2299 (1998)
18. G.J. Galloway and J.M.M. Senovilla, Singularity theorems based on trapped submanifolds of arbitrary co-dimension, *Class. Quantum Grav.* **27** 152002 (2010)
19. E. Gourgoulhon and J.L. Jaramillo, Area evolution, bulk viscosity and entropy principles for dynamical horizons, *Phys. Rev. D* **74**, 087502 (2006)
20. P. Hajicek, Origin of Hawking radiation, *Phys. Rev. D* **36**, 1065 (1987)

21. S.W. Hawking, G.F.R. Ellis, *The large scale structure of space-time*, Cambridge Univ. Press, Cambridge, 1973.
22. S.W. Hawking, R. Penrose, The singularities of gravitational collapse and cosmology, *Proc. Roy. Soc. London* **A314** 529 (1970).
23. S.A. Hayward, General laws of black-hole dynamics *Phys. Rev. D* **49**, 6467 (1994)
24. S.A. Hayward, Spin coefficient form of the new laws of black hole dynamics, *Class. Quantum Grav.* **11**, 3025 (1994)
25. J.L. Jaramillo, J.A. Valiente Kroon and E. Gourgoulhon, From Geometry to Numerics: interdisciplinary aspects in mathematical and numerical relativity, *Class. Quantum Grav.* **25** 093001(2008)
26. M. Khuri, A note on the non-existence of generalized apparent horizons in Minkowski space, *Class. Quantum Grav.* **26** 078001 (2009)
27. H Kodama, Conserved energy flux from the spherically symmetric system and the back reaction problem in the black hole evaporation *Prog. Theor. Phys.* **63** 1217 (1980)
28. M. Kossowski, The total split curvatures of knotted space like 2-spheres in Minkowski 4-space, *Proc. Am. Math. Soc.* **117** 813 (1993)
29. M. Kriele, *Spacetime*, Springer, Berlin, 1999.
30. M. Kriele and S.A. Hayward, Outer trapped surfaces and their apparent horizon, *J. Math. Phys.* **38**, 1593 (1997)
31. B. Krishnan, Fundamental properties and applications of quasi-local black hole horizons, *Class. Quantum Grav.* **25** 114005 (2008)
32. M. Mars, Present status of the Penrose inequality, *Class. Quantum Grav.* **26** 193001 (2009)
33. M. Mars and J.M.M. Senovilla, Trapped surfaces and symmetries, *Class. Quantum Grav.* **20** L293 (2003)
34. A. B. Nielsen, M. Jasiulek, B. Krishnan, and E. Schnetter, The slicing dependence of non-spherically symmetric quasi-local horizons in Vaidya Spacetimes, *Phys. Rev. D* **83** 124022 (2011)
35. B. O'Neill, *Semi-Riemannian Geometry: With Applications to Relativity*, Academic Press, 1983.
36. R. Penrose, Gravitational collapse and space-time singularities, *Phys. Rev. Lett.* **14**, 57 (1965).
37. R. Penrose, *Techniques of Differential Topology in Relativity*, Regional Conference Series in Applied Math. **7** (SIAM, Philadelphia, 1972).
38. R. Penrose, Naked singularities, *Ann. N.Y. Acad. Sci.* **224** 125 (1973)
39. E. Schnetter and B. Krishnan, Nonsymmetric trapped surfaces in the Schwarzschild and Vaidya spacetimes, *Phys. Rev. D* **73**, 021502(R) (2006)
40. J.M.M. Senovilla, Singularity theorems and their consequences, *Gen. Rel. Grav.* **30**, 701 (1998).
41. J.M.M. Senovilla, Trapped surfaces, horizons and exact solutions in higher dimensions, *Class. Quantum Grav.* **19**, L113 (2002).
42. J.M.M. Senovilla, Novel results on trapped surfaces, in "Mathematics of Gravitation II", (Warsaw, September 1-9, A Królak and K Borkowski eds, 2003); `http://www.impan.gov.pl/Gravitation/ConfProc/index.html` (gr-qc/03011005).
43. J.M.M. Senovilla, On the existence of horizons in spacetimes with vanishing curvature invariants, *J. High Energy Physics* **11** 046 (2003)
44. J.M.M. Senovilla, Classification of spacelike surfaces in spacetime, *Class. Quantum Grav.* **24** 3091-3124 (2007)
45. J.M.M. Senovilla, On the boundary of the region containing trapped surfaces, *AIP Conf. Proc.* **1122** 72-87 (2009)

46. H. Stephani, D. Kramer, M.A.H. MacCallum, C. Hoenselaers and E. Herlt, *Exact Solutions to Einstein's Field Equations Second Edition*, Cambridge University Press, Cambridge, 2003
47. J. Thornburg, Event and Apparent Horizon Finders for 3+1 Numerical Relativity, Living Rev. Relativity **10**, 3 (2007). URL (cited on 02-07-2011): `http://www.livingreviews.org/lrr-2007-3`
48. P.C. Vaidya, The gravitational field of a radiating star, *Proc. indian Acad. Soc. A* **33** 264 (1951)
49. Wald, R M *General Relativity*, The University of Chicago Press, Chicago, 1984
50. R.M. Wald, *Living Rev. Relativity* **4** 6 (2001) URL (cited on 02-07-2011): `http://www.livingreviews.org/lrr-2001-6`

Chapter 9

Some Examples of Trapped Surfaces

Ingemar Bengtsson

Fysikum, Stockholms Universitet
106 91 Stockholm, Sweden
ibeng@fysik.su.se

We present some simple pen and paper examples of trapped surfaces in order to help in visualising this key concept of the theory of gravitational collapse. We collect these examples from time-symmetric initial data, 2+1 dimensions, collapsing null shells, and the Vaidya solution.

1. Introduction

Provided an appropriate positivity property holds the existence of a closed trapped surface has dramatic consequences for the future evolution of spacetime. In numerical relativity trapped surfaces provide—assuming that a cosmic censor is active—the practical means by which black holes are recognized, and indeed effective algorithms to spot trapped surfaces in a given initial data set have been developed.[1] There are also theorems in mathematical relativity that tell us much about the existence of marginally trapped surfaces,[2–8] their "evolution" into marginally trapped tubes foliated by such surfaces,[9,10] and their formation from initial data sets that are free of them.[11] A possible application is to quasi-local definitions of black holes:[12–15] the numerical algorithms are not called "horizon finders" for nothing. Other chapters in this book provide the details. The modest aim here is a supplementary one: we will collect some illustrative examples of trapped surfaces that one can obtain with pen and paper only.

After some preliminaries in section 2, we discuss time-symmetric initial data of the Einstein equations in section 3. Early studies initiated by Wheeler and pursued by Brill and Lindquist[16] and many others[17–20] led to the construction of initial data containing an arbitrary number of Einstein–Rosen bridges to new asymptotic regions, and to interesting observations on the minimal surfaces that appear there. Although the term was not used—because the notion of trapped surfaces had not yet been introduced—these surfaces are in fact marginally trapped in the resulting spacetime. In section 4 we restrict ourselves to a 2+1 dimensional toy model of gravity (with a negative cosmological constant),[21–24] where a complete description of all marginally trapped "surfaces" in spacetime can be had.[25] In this toy model we will also see how the trapped surfaces "jump" when we throw lumps of matter into a black hole,[26] and—by increasing the dimension again—what a dynamical horizon can look like in the vacuum case.[27]

In 3+1 dimensions trapped surfaces can be produced in a controlled manner by sending convex shells of incoherent radiation into flat spacetime. This idea was proposed by Penrose[28] as a way to test cosmic censorship; it has been much studied[29–31] and will be discussed in section 5. Following, say, a marginally trapped tube into the future is a more difficult matter, but in section 6 we say more about the spherically symmetric Vaidya solution.[32–37] This is possible because spherical symmetry prevents gravitational radiation from appearing behind the shell. Finally, in section 7 we discuss whether it is possible to observe a trapped surface in its entirety—the answer turns out to depend on which black hole we are in.[38]

For natural reasons there will be a certain emphasis on trapped surfaces I have encountered myself—with apologies to authors who feel, perhaps rightly so, that their own examples are more interesting.

2. Preliminaries

We are interested in the way a submanifold is embedded into some larger space, so for this purpose we decompose the tangent space at a point of the submanifold into a tangential and a normal part. If X and Y are vector fields belonging to the former the Weingarten or shape tensor K will produce a vector belonging to the latter according to its definition

$$K(X,Y) = -(\nabla_X Y)^\perp \ . \tag{2.1}$$

Here ∇_X is the usual covariant derivative along the tangential direction X. For an actual computation we may assume that the submanifold is given in parametric form, $x^a = x^a(u)$, and we find that the Weingarten tensor contracted into a normal vector k^a is

$$K_{ij}(k) = -k_a \frac{\partial^2 x^a}{\partial u^i \partial u^j} - k_a \Gamma_{bc}{}^a \frac{\partial x^b}{\partial u^i}\frac{\partial x^c}{\partial u^j} \ . \tag{2.2}$$

In the examples below such a calculation is typically either straightforward but tedious, or hideously complicated. The details will be relegated to the list of references.

If the submanifold has codimension 1 the normal vector is unique, and the Weingarten tensor is referred to as the second fundamental form—but trapped surfaces have codimension 2 by definition. In any case the first fundamental form γ_{ij} is the metric induced on the submanifold, and the mean curvature vector is defined by

$$H^a = \gamma^{ij} K_{ij}^a \ . \tag{2.3}$$

In Riemannian spaces it can be shown that the surface is minimal—meaning that its area always increases if the surface is subject to variations of sufficiently small compact support—if and only if its mean curvature vector vanishes. For hypersurfaces the mean curvature vector contracted into the unique normal vector is called the mean curvature, denoted K.

As part of its definition a trapped surface is a spacelike surface of codimension 2, in a Lorentzian spacetime. It follows that any normal vector can be expressed as a linear combination of two future directed null vectors, normalised by

$$k_+ \cdot k_- = -2 \ . \tag{2.4}$$

As the notation suggests, one of these vectors (k_+) will be directed "outwards" and the other "inwards". The set of all such vectors will give rise to one outgoing and one ingoing null congruence, and the surface is said to be trapped if the cross sections of both congruences decrease in area as they leave the surface. Whether this is so can be read off from the mean curvature vector, which is

$$H^a = -\theta_+ k_-^a - \theta_- k_+^a \ . \tag{2.5}$$

The surface is trapped if both the null expansions $\theta_\pm$ are negative, meaning that the mean curvature vector is timelike and future directed. The surface is marginally trapped if the outer expansion $\theta_+ = 0$ and the inner expansion $\theta_- \leq 0$.

Actually, in the algorithms that look for marginally trapped surfaces, only the outwards expansion is controlled.[1] Hence they produce marginally outer trapped surfaces, where the meaning of "outer" is provided by some spacelike slice in which the surface is embedded. It turns out that the main body of mathematical relativity theorems concerns marginally outer trapped surfaces, and that such surfaces suffice to prove singularity theorems.[10] But we will concentrate on genuinely trapped and marginally trapped surfaces (with $\theta_- \leq 0$) here, partly because the marginally trapped tubes swept out by a sequence of such surfaces have particularly interesting properties.[13,14]

In 2+1 dimensional Minkowski space, which is helpful for visualisation, a codimension 2 submanifold is simply a curve. Let it be spacelike. It is future trapped if its Frenet normal vector is timelike and future pointing, and untrapped if the normal vector is spacelike. A marginally outer trapped curve is a curve in some null plane, and its inner expansion is negative if and only if that curve is concave. It will be observed that only the untrapped curves can be closed. From now on, when we talk of a trapped surface we often assume that it is closed—since otherwise its existence has no particular consequences. In this sense there are no closed trapped surfaces in Minkowski space. If space has the topology of a cylinder we find that there are closed marginally outer trapped curves in suitable null planes, but still none with $\theta_- < 0$ everywhere.

The Riemannian cousins of marginally trapped surfaces are the minimal surfaces, and many beautiful examples have been published. Based on Plateau's experiments with soap films spanned by thin wires,[39] it was conjectured that one can find a

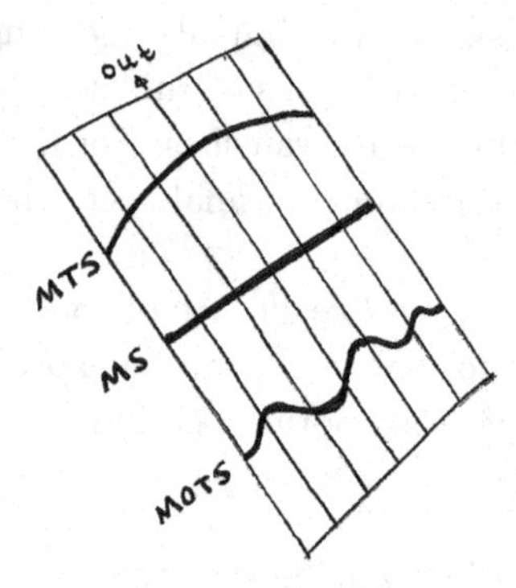

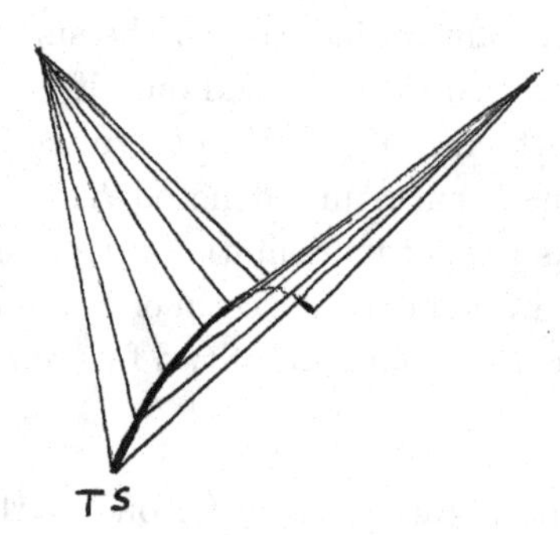

Figure 1. A null plane containing marginally trapped, minimal, and marginally outer trapped curves. Note that any spacelike plane crossing the null plane will contain a curve of the latter type. A locally trapped curve (a piece of a hyperbola bending down in time) is on the right.

minimal surface with any given one dimensional boundary. This was proved with mathematical precision some eighty years later.[40] But the boundary is essential—closed minimal surfaces do not exist in Euclidean space. A way to see this is to observe that the equation obeyed by the functions $x^a(u)$ when the first variation of the area functional vanishes is the Laplace equation.

Thus in flat space a minimal surface needs a boundary, but we can do better if we are inside the 3-sphere. Indeed the solution of the Plateau problem, together with a scheme using reflections, can then be used to produce closed minimal surfaces of any genus.[41] For genus 0 and 1 it is easy to see what these minimal surfaces are. The equator of the 3-sphere is a minimal 2-sphere, and the maximal torus that appears in the Hopf fibration is a minimal 2-torus. Like the geodesics—the great circles on the sphere—they are in a sense "maximal" rather than minimal, because there exists a global deformation that decreases their area. The technical term for such minimal surfaces is "unstable".

Although they are not based on any variational principle, the notion of stability can be generalised so as to apply to marginally outer trapped surfaces.[3,10] Moreover a Plateau problem can be formulated and solved for them.[6]

A totally geodesic submanifold is a submanifold such that the entire Weingarten tensor vanishes. This means that a geodesic starting out tangential to the submanifold will stay in the surface, where it defines a geodesic with respect to the intrinsic metric as well. Unless their dimension is equal to one—a geodesic—such submanifolds typically do not exist, but if the embedding space has a large amount of symmetry they may. In particular, the fixed point set of an isometry is always totally geodesic. (This is easy to see. Consider a vector tangent to the surface at some point. This defines a unique geodesic in the embedding space. If this geodesic were able to move out of the surface a part of it would be moved by the isometry, while its starting point and its initial tangent vector would not be. This contradicts the uniqueness of the geodesic.) In general relativity a totally geodesic spacelike hypersurface occurs whenever time-symmetric initial data exist, and we can then rely on the useful fact that a minimal surface in such an initial data hypersurface is a marginally trapped surface in the resulting spacetime.

3. Time-Symmetric Initial Data

The initial data for Einstein's equations are given by the first and second fundamental forms (two symmetric tensor fields on some space, one of them being its metric), subject to suitable constraints. If we take the second fundamental form to vanish—so that we will have a totally geodesic hypersurface in the spacetime to be constructed—the constraint on the first fundamental form takes the simple form

$$R^{(3)} = 0 \, . \tag{3.1}$$

Its curvature scalar vanishes. This assumes vacuum, with vanishing cosmological constant. We may simplify matters further by assuming that the intrinsic metric takes the form

$$ds^2 = \omega^4(dx^2 + dy^2 + dz^2) \, . \tag{3.2}$$

This was referred to as "geometrostatics" in the early references.[16,17] The reason for the unusual exponent on the conformal factor is that eq. (3.1) now takes the form

$$\Delta\omega = 0 \, , \tag{3.3}$$

where Δ is the flat space Laplacian. From electrostatics we know how to solve this. A suitable solution is

$$\omega = 1 + \sum_{i=1}^{N} \frac{e_i}{r_i} \, , \tag{3.4}$$

where the source points can be placed at N arbitrary positions, and r_i is the Euclidean distance to the ith source point.

If there is only one source point, placed at the origin, and if we use spherical polar coordinates (r, θ, ϕ), this is the $t = 0$ slice of the Schwarzschild solution in isotropic coordinates. Its mass $M = 2e$. This geometry admits a discrete isometry under

$$r \to r' = \frac{e^2}{r} \, . \tag{3.5}$$

From this we learn two things. First, the source points in the solution do not represent singular points in the geometry, they represent the spatial infinities in N (out of $N+1$) asymptotic regions. Second, when $N = 1$ the sphere at $r = e = M/2$ is a fixed point set of an isometry, hence it is totally geodesic and by implication a minimal surface. When the geometry evolves it becomes the bifurcation sphere on the event horizon.

The case when there are more than two asymptotic regions (when $N > 1$) is made intuitively clear in the illustrations provided by Brill and Lindquist.[16] See Fig. 2. There will be N throats leading from the original asymptotically flat sheet to an additional N such sheets. In each throat we will find a minimal 2-sphere, but it will no longer be a round sphere—rather it will be distorted by its interaction with the others. If the throats are well separated this is the end of the story, but if two throats come sufficiently close together an additional minimal sphere will appear and surround the two minimal spheres in the throats.

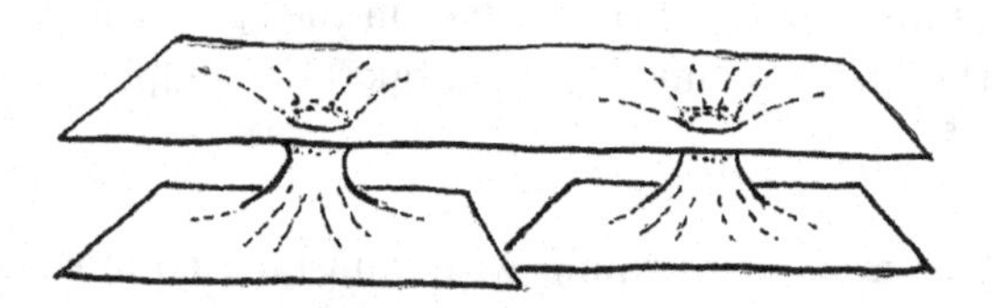

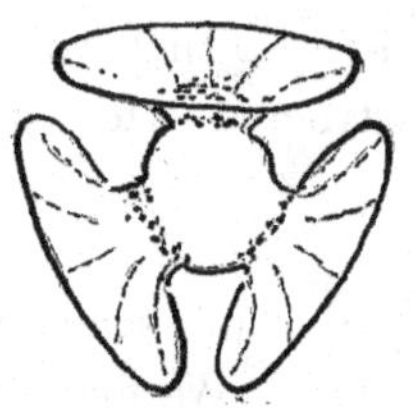

Figure 2. Brill–Lindquist initial data: When the throats are far apart there are only two minimal surfaces. In the symmetric case shown to the right there are obviously three.

To see how this works, let us choose $N = 2$ and $e_1 = e_2$. The conformal factor is

$$\omega + \frac{e}{r_1} + \frac{e}{r_2} = 1 + \frac{e}{\sqrt{\rho^2 + z^2}} + \frac{e}{\sqrt{\rho^2 + (z-a)^2}} , \tag{3.6}$$

where a is the Euclidean distance between the two source points, and cylindrical coordinates were introduced in the second step. It is not difficult to read off the ADM masses in the three external regions, by studying how the metric falls off at infinity (using the coordinate $r_1' = e_1^2/r_1$ as a coordinate in the first region, say). One finds[16]

$$M_1 = M_2 = 2e\left(1 + \frac{e}{a}\right) , \qquad M_3 = 4e . \tag{3.7}$$

It is clear that if $a \gg e$ the two throats will not affect each other much, and there will be only two minimal spheres. But if $a = e$ all three masses will be equal, and by symmetry there will be three throats, and three minimal spheres. Interestingly the area of the outermost minimal sphere—outermost from the point of view of the third external region—is smaller than the sum of the areas of the two minimal spheres it is surrounding. This will be so whenever the distance a is small enough. There must exist a critical value $a_{\rm cr}$ when the third minimal sphere first appears. To calculate this we assume that any minimal surface is axisymmetric and given by $\rho = \rho(\sigma)$, $z = z(\sigma)$, $\phi = \varphi$. The minimal surface equation is then obtained by extremising the area functional

$$A = \int \sqrt{\det \gamma} d\sigma d\varphi = 2\pi \int_{\sigma_0}^{\sigma_1} \rho\omega^4 \sqrt{\dot{\rho}^2 + \dot{z}^2} d\sigma \ . \tag{3.8}$$

We must find solutions corresponding to closed surfaces. Unfortunately—because the stated purpose of this review is to give pen and paper examples—numerical methods must be used to do this. Moreover the numerical calculation is non-trivial, as is clear from the fact that the first attempts to do it did not give quite the correct answer. The conclusion eventually turned out to be that the critical Euclidean distance is $a_{\text{cr}} \approx 1.53e$,[19] and in fact a pair of minimal surfaces are created when we go below this value.[20]

Do take note of the fact that the surrounding minimal sphere is there for a reason. We assume that cosmic censorship holds, and that we have found the area $|S|$ of a—not necessarily connected—trapped surface in our initial data. Then there must be an event horizon intersecting the initial hypersurface, and since it lies outside the trapped surface we expect its area to be larger than $|S|$. In the future we expect the system to settle down into a Kerr black hole, and since the area of the event horizon can only grow the area of the final event horizon is also larger than $|S|$. Some of the initial mass M will be radiated away, so the area of the final event horizon—which is never larger than the area of a Schwarzschild horizon of the same final mass—will be no larger than $16\pi M^2$. Tracing through this string of inequalities we find that

$$16\pi M \geq \sqrt{16\pi|S|} \ . \tag{3.9}$$

The mass M and the area $|S|$ are determined by the initial data. This interesting strengthening of the positive mass theorem is known as the Penrose inequality, and should be a necessary (but not sufficient!) condition for cosmic censorship to hold.[28] But as the two throats in the Brill–Lindquist initial data are moved closer together, the sum of their areas grow.[16] In fact their sum can exceed the bound set by the Penrose inequality. What saves the day is precisely the new minimal surface that appears out of the blue to surround the two, and now counts as outermost.[18]

A possible loophole in the above argument—quite apart from a possible failure of cosmic censorship—appears in its first step, since it is not clear that a surface that surrounds another must have a larger area. However, in the case of time-symmetric data the marginally trapped surfaces are minimal and the reasoning does not need any amendment, provided the trapped surface is taken to be the (not necessarily connected) outermost marginally trapped surface. One can construct simple examples in Oppenheimer–Snyder dust collapse where more care is needed.[42,43]

4. The 2 + 1 Dimensional Toy Model

To bring the multi-black hole problem within the range of pen and paper methods Brill,[23] and Steif,[24] eventually turned to 2 + 1 dimensions where a trapped "surface" is a closed spacelike curve with a future pointing timelike Frenet normal vector. There are considerable simplifications because there is no Weyl tensor and no shear in the Raychaudhuri equation for a null congruence: in 2+1 dimensions

$$\dot{\theta} = -\theta^2 - R_{ab}t^a t^b \ . \tag{4.1}$$

If we impose Einstein's vacuum equation $R_{ab} = \lambda g_{ab}$ the second term vanishes, $\dot{\theta} = -\theta^2$, and hence

$$\theta(0) = 0 \quad \Leftrightarrow \quad \theta(\tau) = 0 \ . \tag{4.2}$$

This has the consequence that a marginally trapped surface must lie on a lightlike plane, or—to use a terminology that is more accurate in the de Sitter and anti-de Sitter cases—on a lightcone with its vertex on $\mathscr{I}$. In turn this means that we can easily get a complete overview of all marginally (outer) trapped curves in a spacetime of this type. Still, even though the vacuum equations imply that spacetime has constant curvature, the model is not completely trivial.

There are no closed trapped curves in Minkowski space, but they do exist in a suitable region of Misner space, which is Minkowski space with points related by some discrete Lorentz boost identified.[2] Since the boost has a line of fixed points there is a kind of singularity to the future of such a trapped curve, as well as a breakdown of the causal structure in the form of closed timelike curves. Misner space does not have a black hole, since it has no sensible notion of future $\mathscr{I}$, but the same construction carried out in anti-de Sitter space has. It gives rise to the BTZ black hole.[21] To see how this works consider the metric of 2 + 1 dimensional anti-de Sitter space,

$$ds^2 = -\left(\frac{1+\rho^2}{1-\rho^2}\right)^2 dt^2 + \frac{4}{(1-\rho^2)^2}(d\rho^2 + \rho^2 d\phi^2) \ . \tag{4.3}$$

At $t = 0$ we have a time-symmetric initial data slice which in these coordinates appears in the guise of the Poincaré disk, a space of constant negative curvature. The group of isometries preserving this slice is isomorphic to the subgroup of Möbius transformations that preserve its conformal boundary (in these coordinates the unit circle). Among them are the hyperbolic Möbius transformations, that have two fixed points on the boundary of the disk and a special flow line which is the unique geodesic connecting the fixed points—we recall that the geodesics on the Poincaré disk are arcs of circles or straight lines meeting the boundary at right angles.

Now pick such a Möbius transformation, obtainable by exponentiating an infinitesimal transformation. We identify all the points one can connect with it. The result is a space of constant curvature and cylindrical topology, having exactly one closed geodesic—a minimal "surface"—corresponding to the unique geodesic flow line. These are the initial data of the BTZ black hole (and occurs as panel a) in Fig. 4). Since anti-de Sitter space is not globally hyperbolic these data do not in fact determine the future evolution, but we can define the spacetime as the quotient of anti-de Sitter space with the discrete isometry group generated by the group element we used to construct our initial data. This spacetime is the BTZ black hole.[21] It has a singular future (since the isometry will have fixed points to the future, and to the past, of the initial data slice), and it will be asymptotically anti-de Sitter in the sense that the first order structure of the quotiented conformal boundary is conformally isomorphic to the 1+1 Einstein universe, the $\mathscr{I}$ of anti-de Sitter space.[22] The event horizon is the boundary of the past of $\mathscr{I}$, as usual.

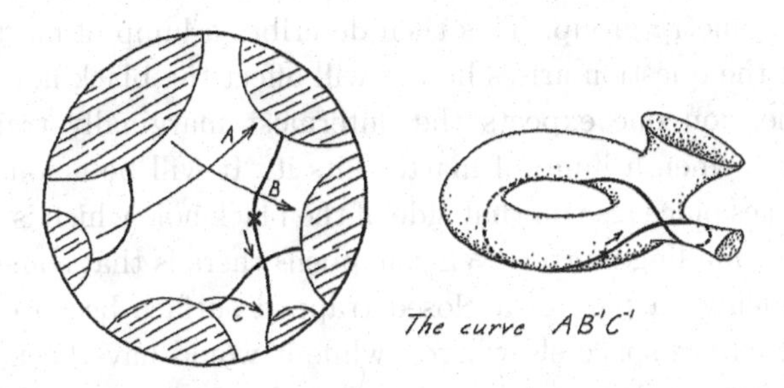

Figure 3. An initial data surface for a black hole spacetime with two asymptotic regions. The white part is a fundamental region. The group Γ is generated by the group elements A, B, C, and a closed curve associated to the group element $AB^{-1}C^{-1}$ is shown. Reproduced with permission.[44]

This idea can be generalised. Given a pair of geodesics on the Poincaré disk at $t = 0$ there is a unique hyperbolic Möbius transformation taking one to the other. If we choose several such pairs of geodesics—in Fig. 3 we choose three pairs—we obtain several group elements. If we make sure that the geodesics do not intersect these group elements generate a free group Γ, and we define our spacetime as the quotient of anti-de Sitter space by Γ. Its initial data surface at $t = 0$ is guaranteed to be a smooth surface of constant negative curvature and one or several asymptotic regions. We can now search for closed geodesics—marginally trapped surfaces—on this initial data surface.

But this is not a hard task. Take any closed curve in the quotient space. It can be associated to a particular element in the discrete group Γ, as shown by example in Fig. 3. This group element is in itself a hyperbolic Möbius transformation, and is therefore associated with a unique geodesic flowline. The latter will become a closed geodesic in the quotient space. In this way we find one closed marginally trapped curve for every topologically distinct class of closed curves. From a spacetime point of view, and from our discussion of the Raychaudhuri equation, we know that a marginally trapped curve must lie on a lightcone with its vertex on $\mathscr{I}$. This vertex can be found by analysing the action of Γ on $\mathscr{I}$, but since this pleasant task has been described elsewhere we do not go into this here.[25]

I should add that the restriction to time symmetric spacetimes can be lifted, at the expense of rather more work.[44]

To get a more lively picture we can try to add some matter to the model. So as to not compromise its basic simplicity this is often done in the form of "point particles", conical singularities with their vertices along timelike or null geodesics.[45] There will be a non-trivial holonomy associated to any spacelike curve surrounding the singular geodesic, and from this one can read off the "mass" of the resulting spacetime. Now suppose we start with a single BTZ black hole, choose a radial null geodesic leaving $\mathscr{I}$ and heading for the event horizon, choose a suitable wedge with that geodesic as its edge, and identify the two boundaries of the wedge using an element of the isometry group. This then describes a lump of matter falling into a black hole, and the question arises how it will affect the black hole when it hits.[26]

In any dimension one expects the outermost marginally trapped surface to "jump" outwards when a lump of matter hits it. It will be a somewhat non-local jump since it takes place also on that side of the black hole which is not yet in causal contact with the infalling matter. What happens there is that some locally trapped surface is "suddenly" a part of a closed trapped surface because of the way the lump of matter curves space elsewhere—while it would have been part of an open surface if no matter had been falling in. But this is not a spherically symmetric situation, and gravitational radiation will prevent any pen and paper calculation from being made—unless we are in 2+1 dimensions, where gravitational radiation does not happen.

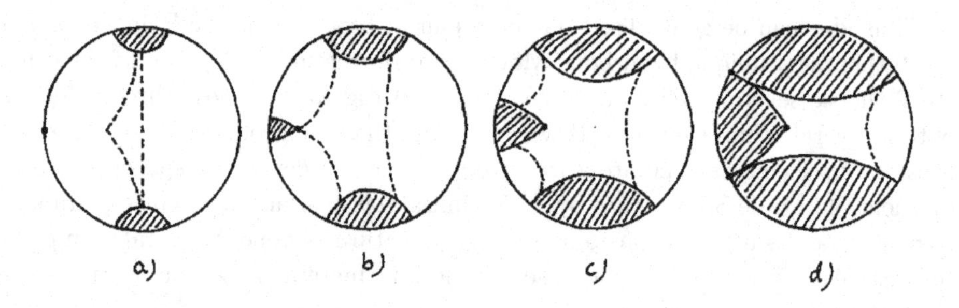

Figure 4. a) The particle comes in from infinity just as a marginally trapped curve is formed in the centre. The event horizon has a kink and does not contain any marginally trapped curves. b) The particle meets the event horizon. c) The event horizon is now smooth and marginally trapped. d) The identification surfaces meet in a singular point at infinity. Reproduced with permission.[26]

Concretely, let a massless particle enter a BTZ black hole spacetime at $t = 0$ (panel a) in Fig. 4). We follow the time development by slicing covering space with a sequence of Poincaré disks at increasing values of t. The particle moves inwards trailing its wedge behind it. Meanwhile the geodesic arcs that bound the fundamental region of the BTZ black hole are moving closer to each other. Eventually they meet the boundary of the wedge; the result is a fixed point on $\mathscr{I}$ that corresponds to a singular point in the quotient space (which will then continue inwards along a spacelike geodesic). The backwards light cone from that singular point is the event horizon of the resulting spacetime.

Following the event horizon backwards in time we observe that it is a smooth surface foliated by closed marginally trapped curves only as long as it passes through the wedge behind the particle (down to panel b) in Fig. 4). At earlier times the event horizon has a kink, and does not contain any smooth and closed spacelike curves. Yet there is a closed marginally trapped curve at $t = 0$, namely the one that would have evolved into the event horizon of the BTZ black hole, had there been no massless particle to complicate matters. This will evolve into an isolated horizon which is destroyed by its encounter with the particle (in panel c), although only the horizon bordering the second asymptotic region is shown in the figure). A spacetime picture of all this is given in Fig. 5.

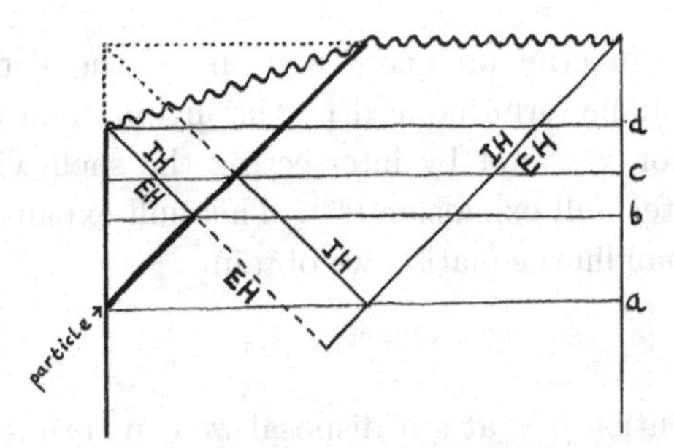

Figure 5. A conformal diagram of a massless particle falling into a BTZ black hole. Four spatial slices corresponding to the panels in Fig. 4 are given, and we also see how the isolated horizon (IH) "jumps".

As a final remark, let us go to 3+1 dimensions by means of the obvious amendment of the metric (4.3), i.e. by adding an extra angular coordinate θ. At $t = 0$ we now have a Poincaré ball, and there are isometries mapping totally geodesic surfaces—spherical caps meeting the boundary at right angles—to each other. Take the quotient of $3 + 1$ dimensional anti-de Sitter space with such an isometry. The result is a genuine black hole spacetime, but a curious one since its event horizon is not a Killing horizon. It grows. One can now look for marginally trapped surfaces in a slicing that respects the symmetries of the quotient space, and one finds that they foliate a spacelike dynamical horizon situated well inside the event horizon.[27] The conformal boundary is connected, and has the topology of a torus times the real line. This pathological feature apart it is an interesting example.

5. Collapsing Null Shells

A rich supply of interesting trapped surfaces can be had by sending a thin shell of incoherent radiation (also known as null dust) into Minkowski space. The surfaces will arise as cross sections of the null shell. We want the interior of the shell to remain flat, meaning that the shell cannot have caustics to the past of the surface. To ensure this we insist that its cross section with some spacelike hyperplane is convex. To the future the generators of the shell cross, and a spacetime singularity results. About the outside of the shell we know very little—unless the shell and the mass distribution are spherically symmetric there will be gravitational radiation there. If a cosmic censor is active a black hole will develop, and eventually settle down to the Kerr solution. Again we have the Penrose inequality (3.9); in fact this was the setting in which the inequality was first proposed.[28] What is remarkable about the shell construction is that both M and $|S|$ can be evaluated without knowledge of the exterior of the shell. They result from calculations in flat space.

The energy–momentum tensor of the shell is

$$T^{ab} = 8\pi\mu k_-^a k_-^b \delta \, , \tag{5.1}$$

where μ is an arbitrary function on a cross section and the delta function is defined with respect to the volume form induced by the ingoing null normal k_-^a. We now choose the cross section we want by intersecting the shell with an outgoing null congruence having outer null expansion θ_+^{int}. This null expansion will jump at the shell. From the Raychaudhuri equation we obtain

$$\theta_+^{\mathrm{ext}} = \theta_+^{\mathrm{int}} - 16\pi\mu \, . \tag{5.2}$$

Since the mass distribution μ is at our disposal we can turn any cross section into a marginally trapped surface in this way. Moreover we can search for trapped surfaces on the shell—in particular for the outermost marginally trapped surface—in an efficient way.

As an example of the kind of insights one can get from this model, consider the case when the shell admits an ellipsoidal cross section. Using a foliation of Minkowski space with the usual $t =$ constant hypersurfaces one then finds that the caustic appears first at the ends, and so does any trapped surface sitting on the shell. The result is that there are no trapped surfaces on any $t =$ cross section before the value of t for which the singularites first appears.[31] This provides some food for thought, given that the numerical algorithms look for marginally trapped surfaces in given spacelike hypersurfaces.[1] (Even in the Schwarzschild solution it is known that there exists a foliation with spacelike hypersurfaces which reaches all the way to the singularity even though its leaves do not contain any trapped surfaces.[38] But that foliation was chosen specifically to make this true, while the foliation used for the collapsing shell is perfectly natural.)

The proof of the Penrose inequality remains partly elusive. It has been proved for the case that the null shell is the backwards light cone of a point.[29] It has also been proved for the case when the cross section is the intersection of the shell with a timelike hyperplane in Minkowski space, in which case it reduces to the famous Brunn–Minkowski inequality.

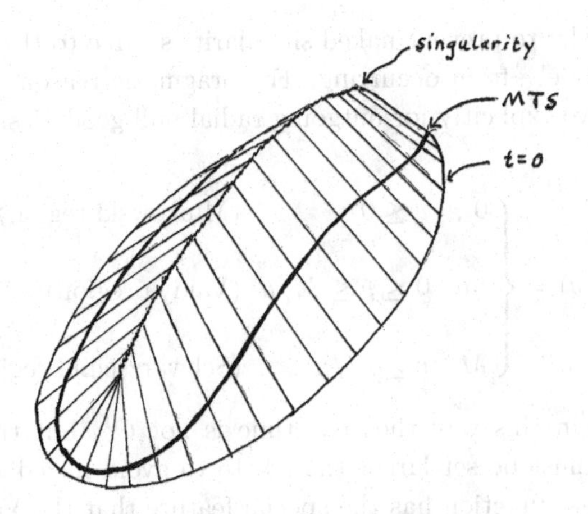

Figure 6. A collapsing null shell considered by Pelath, Tod and Wald:[31] It has an ellisoidal cross section at $t = 0$. Both the singularity and the first marginally trapped surface "bend" in time in such a way that the singularity appears at an earlier value of t than does the last piece of the marginally trapped surface.

$$\oint KdS \geq \sqrt{16\pi|S|}\ , \tag{5.3}$$

which holds for any convex surface in Euclidean space (and K is its mean curvature). Gibbons, who was the first to draw this conclusion, therefore refers to the Penrose inequality as an isoperimetric inequality for black holes.[30] The freedom in choosing the trapped surface can be used to test other conjectures; notably the hoop conjecture has been studied in this way.[31] But this is not the place to pursue these matters. Instead we would like to take a peek behind the shell.

6. The Vaidya Solution

To have a manageable exterior we make the collapsing shell, and its mass distribution, spherically symmetric. We also give the shell a finite width. In other words we will look at the Vaidya solution. The matter is still an infalling null dust. The metric, using Eddington–Finkelstein coordinates, is

$$ds^2 = -\left(1 - \frac{2m(v)}{r}\right) dv^2 + 2dvdr + r^2 d\theta^2 + r^2 \sin^2\theta d\phi^2\ . \tag{6.1}$$

Ingoing null geodesics are given by $v =$ constant. The mass function $m(v)$ is specified, as a monotone function of the advanced time v, on $\mathscr{I}^-$. We will assume that it vanishes for $v < 0$, and becomes constant again at some later value of v, which means that we have a shell of finite thickness, entering Minkowski space and leaving a Schwarzschild black hole behind. There are some restrictions on $m(v)$ that must be imposed in order to prevent naked singularities—due to the unphysical features of the matter model—from occurring. For pragmatic reasons we choose one that enables us to solve explicitly for outgoing radial null geodesics, namely

$$m(v) = \begin{cases} 0\ , & v \leq 0 \qquad\qquad\quad \text{(Minkowski region)} \\ \mu v\ , & 0 \leq v \leq M/\mu \quad \text{(Vaidya region)} \\ M\ , & v \geq M/\mu \qquad\quad\ \text{(Schwarzschild region)} \end{cases} \tag{6.2}$$

When presented in this way the spacetime is not C^1, but this can be mended. The constant μ must be set larger than $1/16$ to avoid naked singularities.[32] The linearly rising mass function has the special feature that the Vaidya region admits a homothetic Killing vector

$$\eta = v\partial_v + r\partial_r\ , \tag{6.3}$$

whose flow lines lie in hypersurfaces with constant $x = v/r$. This is a simplifying feature since a trapped surface remains trapped if moved by a homothety.[8]

Now where are the trapped surfaces? The first observation is that the hypersurface

$$r = 2m(v) \tag{6.4}$$

is foliated by round and marginally trapped spheres. In the Schwarzschild region it is the event horizon, but in the Vaidya region it is a spacelike dynamical horizon lying well inside the event horizon. We will refer to it as the "apparent 3-horizon", because it needs a name. Cosmic censorship requires that all trapped surfaces are confined to the interior of the event horizon.[2] In the Schwarzschild case this is easily proved directly, because a trapped surface cannot have a minimum in t, where t is the parameter along a hypersurface forming timelike Killing vector field.[5] In the Vaidya region there is a variation of this argument employing the Kodama vector field—which gives the direction in which the area of the round spheres is unchanged—leading to the conclusion that any trapped surface must lie at least partly inside the apparent 3-horizon. It may extend partly outside, but there will be a special spacelike hypersurface (of constant "Kodama time") in its exterior which the trapped surfaces cannot reach.[33] See Senovilla's chapter in this book.

The introduction done with, let us look for some examples. We begin by looking for "tongues" sticking out of the apparent 3-horizon. A naive way to do so is to begin by visualising the solution in the (v, r, θ)-coordinates, suppressing one coordinate that we will not use. The apparent 3-horizon then forms a cone with its vertex at the origin, which eventually joins the cylinder representing the Schwarzschild part of the event horizon. We introduce another cone meeting the apparent 3-horizon in a marginally trapped round sphere (a circle in the picture), and look at 2-surfaces that are cross sections of that second cone. In equations

$$v = \frac{k}{2\mu}r - v_0 \ , \qquad v = \frac{1}{2\mu}r + a(\theta) \ . \tag{6.5}$$

When $a = 0$ we are on the apparent 3-horizon, and wherever θ is such that $a(\theta) < 0$ the tongue is sticking out if it. There are some restrictions that must be imposed if it is to do so, for instance that $k > 0$.[9,33] Simple choices are

$$k = 3 \ , \qquad a(\theta) = a_0 + a_1 \cos\theta \ . \tag{6.6}$$

To first order in a perturbation expansion one finds that the tongues are trapped surfaces if and only if $a_0 > 0$. By choosing a_1 suitably we can clearly arrange that they stick partly outside the apparent 3-horizon. But how far out can they go?

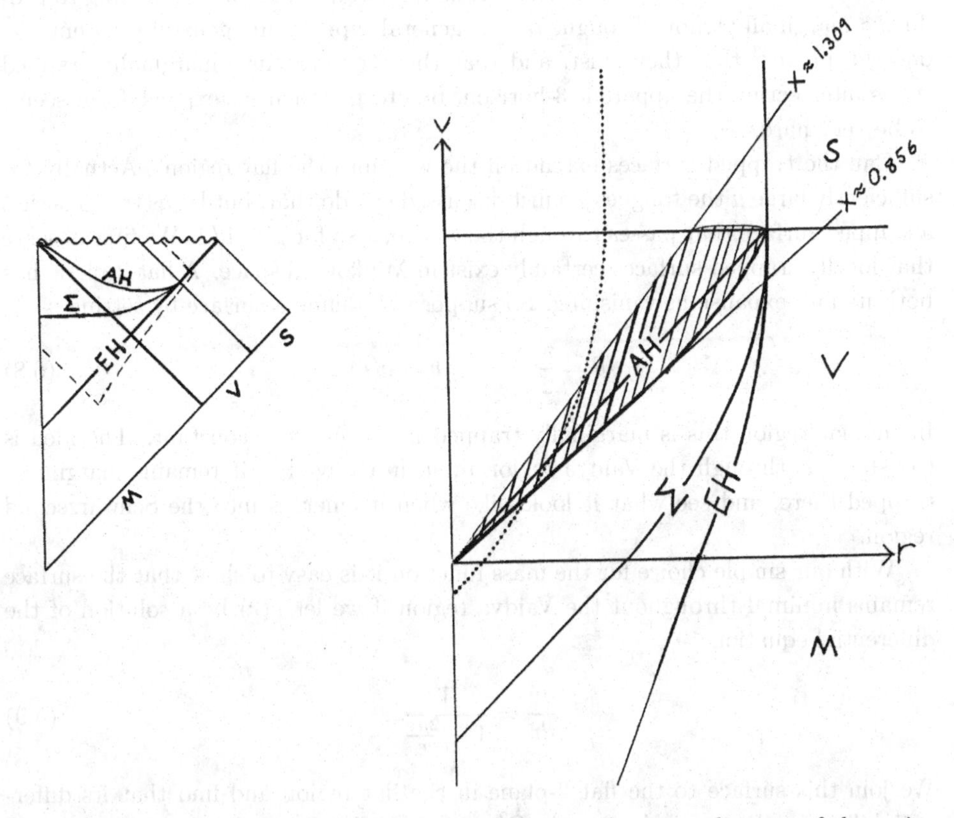

Figure 7. A Penrose diagram of a Vaidya spacetime, and a $v - r$ diagram of the region inside the dashed curve. It includes a part of the event horizon (EH), the spacelike part of the apparent 3-horizon (AH), and a hypersurface Σ below which no trapped surfaces can extend.[33] The tongues discussed in the text are confined between two lines of constant v/r. There are also trapped surfaces to the left of the dotted curve.

The inequality that guarantees that the tongues are trapped is an unwieldy one, even in this simple case, and we had to rely partly on Mathematica in order to analyse it. For the case $\mu = 1/2$ the answer is that the tongues remain trapped provided that

$$0.856 < x = \frac{v}{r} < 1.309 . \quad (6.7)$$

This is how far a maximally extended trapped tongue, of this particular kind, extends—provided it stays in the Vaidya region, where the result in fact reflects the behaviour of trapped surfaces under homotheties.[8] If a tongue hits the Schwarzschild region there are further restrictions; see Fig. 7 for the final result, and Åman et al.[35] for other choices of μ and $a(\theta)$. The Gaussian curvature of this tongue is positive, and maximal at its "tips". Its area grows as it extends more, at least to second order in a perturbation around a round sphere.

Of course this is a very naive construction. It would be more interesting to produce a marginally trapped tongue of this general type. Using general theorems[10,7] one can prove[33] that they exist, and that they "evolve" into marginally trapped tubes intersecting the apparent 3-horizon, but to get them in explicit form seems to be very hard.

Can the trapped surfaces extend all the way into the flat region? Actually for sufficiently large μ the tongues we just discussed can do that, but let us try to design a trapped surface that passes through the centre, also for $\mu = 1/2$. We first observe that locally trapped surfaces certainly exist in Minkowski space. A flat 2-plane has both its null expansions vanishing. So suppose we define a surface by setting

$$\theta = \frac{\pi}{2} , \qquad v = v(r) . \quad (6.8)$$

In the flat region this is marginally trapped if $t = v - r =$ constant. The idea is to "steer" it through the Vaidya region in such a way that it remains marginally trapped there, and see what it looks like when it emerges into the Schwarzschild region.

With our simple choice for the mass function it is easy to show that the surface remains minimal throughout the Vaidya region if we let $v(r)$ be a solution of the differential equation

$$\frac{dv}{dr} = \frac{1}{1 - \frac{2\mu v}{r}} . \quad (6.9)$$

We join this surface to the flat 2-plane in the flat region and find that its differentiability, in these coordinates, is C^{2-}. On the boundary to the Schwarzschild region it forms a round circle. Now it is interesting to observe that the cylindrical Schwarzschild surface

$$\theta = \frac{\pi}{2} , \qquad r = M , \quad (6.10)$$

has both its null expansions vanishing, and it can be joined to one of the surfaces that we have steered through the Vaidya region by a suitable choice of integration constants. The result is a surface which is minimal throughout the entire spacetime—but its topology is that of a sphere with a point removed "at" the singularity, so this is not a closed marginally trapped surface.

But we can wiggle the surface a little bit, so that it becomes locally trapped, and so that it arrives to the Schwarzschild region with negative slope in the $v - r$ diagram. Continuing into Schwarzschild with fixed slope the null expansions become increasingly negative, and once they are sufficiently negative we can close off the surface with a spherical cap—we have constructed a closed trapped surface that extends into the flat part of the Vaidya spacetime. (Full details, for almost the same construction, can be found elsewhere.[34])

On the one hand this looks odd—one might have thought that no closed trapped surfaces could enter the flat region. But indeed in order to see that they are closed one has to collect information from a much larger region of spacetime. From another point of view it is somehow satisfactory, because it means that one cannot reach the singularity without crossing a trapped surface. Actually one can prove[33] that it is impossible for a trapped surface to enter the flat region if $\mu \leq 1/8$. But the smaller one makes μ the closer one gets to a nakedly singular spacetime, so that is presumably at the unphysical end.

A more systematic search for marginally trapped surfaces in the Vaidya solution, using a different mass function, an axisymmetric slicing, and a horizon finder, has been made by Nielsen et al.[36] A more systematic study of spherically symmetric spacetimes in general will be found in Senovilla's chapter in this book.

If we are content with outer trapped surfaces a different picture emerges. Then a trapped surface in the deep interior of the Schwarzschild region can develop a "tendril" reaching down through the Vaidya region and indeed through any point in the interior of the black hole, also in the flat diamond.[37] Once it emerges into the flat region the tendril lies in a tubular neighbourhood of a (null) generator of the event horizon. Casual inspection in the flat region would suggest that the tendril is part of an untrapped surface with its inner expansion negative, but in fact when the global structure of the spacelike slice is examined one sees that it is outer trapped. This behaviour was suggested by Eardley, who in fact conjectured that the boundary of the region through which outer trapped surfaces pass is always the event horizon, in any black hole.[46]

What prevents the tendril from following its generator even further, out of the event horizon? The answer is that should it do so, it would no longer be lying in a suitable spacelike hypersurface, and would not count as "outer" any more. Indeed we assume it to be embedded in a spacelike hypersurfaces containing the trapped surface from which the tendril emerged in the first place. See Fig. 8. The lesson is that we can decide whether a surface is trapped by inspection of the surface itself, while this is not possible for outer trapped surfaces.

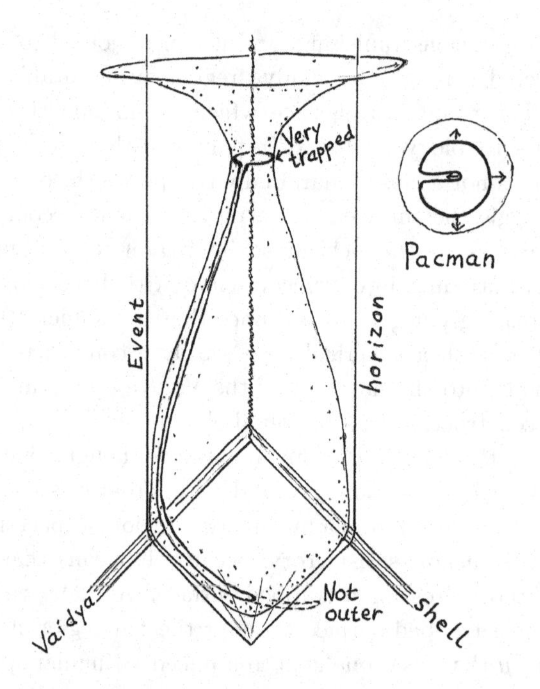

Figure 8. An impressionistic picture — v–r– coordinates would not be so useful here — of Ben-Dov's construction.[37] A spacelike hypersurface sits inside the Vaidya event horizon, and it contains a very trapped surface around its "neck", close to the singularity. This surface has developed a "tendril" reaching down to Minkowski space. It remains outer trapped, as one can see by flattening the spacelike hypersurface (at right). As far as the intrinsic properties of the tendril are concerned, they remain the same if the tendril extends outside the event horizon (shown dashed), but then no spacelike hypersurface can contain both the tip of the tendril and the very trapped surface from which it came.

7. What the astronaut saw

If the cosmic censorship hypothesis holds the reader must enter a black hole in order to "see" a trapped surface. But even if the reader does so, there is no guarantee that she will be able to collect the information needed to verify observationally that a surface is trapped. Indeed, as Wald and Iyer point out, in the Schwarzschild spacetime it is impossible to do so.[38]

To see this note that everything that an observer can ever see is contained in a TIP, the terminally indecomposable past of a timelike curve.[47] The first observation is that none of the round trapped spheres in the Schwarzschild solution is fully contained in a TIP. To see this, we may—without loss of generality—let the TIP end at the "point " $r = t = \theta = 0$. Thus the observer ends up at the North Pole of a round sphere, and we ask whether she has time to see its South Pole. A null geodesic at constant latitude ϕ obeys

$$\frac{d\theta}{dr} = 1/\sqrt{E^2r^4/L^2 + 2Mr - r^2} \; ; \qquad \left(1 - \frac{2M}{r}\right)\dot{t} = E \, , \quad r^2\dot{\theta} = L \, . \tag{7.1}$$

Thus for a given value of the "time coordinate" r the observable part of the round sphere at $t = 0$ has its longitude θ bounded by

$$\theta \le \int_0^r \frac{dr}{\sqrt{2Mr - r^2}} = 2 \arcsin \sqrt{r/2M} \equiv \theta_0(r) \, . \tag{7.2}$$

The observer cannot see the South Pole of the marginally trapped bifurcation sphere before she arrives at the singularity, and certainly none of the trapped round spheres in their entirety.

Are there any other fully visible trapped surfaces? Let us first consider trapped surfaces in a spatial slice defined by settting r constant. Topologically such a space is a cylinder $\mathbf{S}^2 \times \mathbf{R}$, and we have already established that any visible trapped surface there (one that belongs to the interior of our chosen TIP) must be homotopically trivial. Any visible sphere of this kind assumes a maximum longitude $\theta = \theta_{\rm max} < \theta_0(r)$. Let the surface be given by $r =$ constant and $\theta = \theta(t, \phi)$. Then a simple calculation starting from eq. (2.2) verifies that, at a point where the longitude θ assumes a maximum $\theta_{\rm max}$, the outer null expansion θ_+ obeys

$$2\theta_+ = \frac{r - M}{r\sqrt{2Mr - r^2}} + \frac{\cot \theta_{\rm max}}{r} - \frac{r^2 \theta_{,tt}}{2M - r} - \frac{\theta_{,\phi\phi}}{r \sin^2 \theta} \, . \tag{7.3}$$

The last two terms are non-negative because the longitude is at a maximum, and $\cot \theta > \cot \theta_0$ because we assume that we are within the TIP, so it follows that

$$2\theta_+ > \frac{r - M}{r\sqrt{2Mr - r^2}} + \frac{\cot \theta_0}{r} = 0 \, . \tag{7.4}$$

The final equality follows when we make use of eq. (7.2). Hence a surface at constant r cannot be outer trapped if it is inside the TIP.

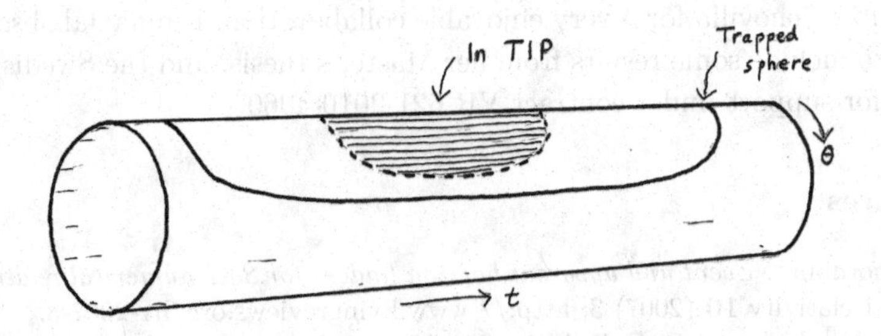

Figure 9. The part of space at constant r within the TIP of an observer is shown dashed. If r is small enough there are homotopically trivial trapped spheres in this space, but they are too large to fit into the TIP.

So much for trapped surfaces at constant r. To show that no trapped surface whatsoever is contained within the TIP requires some thinking.[38] First we consider the function $g = \theta + \theta_0(r)$. Consulting eq. (7.2) we see that $||dg||^2 = 0$, so its gradient defines a null vector tangent to a null geodesic, and in fact a null geodesic with $E = 0$ since the t-component of the vector vanishes. Now we restrict the function to some closed surface T, and we let $p \in T$ be a point where g assumes a maximum on T. At this point the null vector defined by dg is orthogonal to T. Now define a constant r hypersurface Σ that passes just to the future of p, and let the special null geodesic hit Σ at a point q. We arrange Σ so that q lies within the TIP. The orthogonal null congruence emanating from T will intersect Σ to form, at least locally, a surface T_Σ lying within Σ and containing q. Since $E = 0$ for the null geodesic passing through q the longitude θ on T_Σ must have a maximum at q. But then we already know that T_Σ cannot be trapped at q. Raychaudhuri's equation tells us that the null expansion can only increase as we move into the past along the congruence, so we conclude that the surface T cannot be trapped at p, and we are done. No trapped surfaces can exist within the TIP.

To avoid any misunderstanding, in the physically much more realistic Oppenheimer-Snyder model of a collapsing star it is easy to find TIPs that contain trapped surfaces. But the reader eager to establish observationally that a closed trapped surface exists in Nature must jump into the right black hole.

8. Envoi

We hope that the reader enjoyed these examples. If not, the list of references will provide more satisfying reading. Meanwhile the examples may perhaps serve to drive home the point that being trapped or outer trapped is a spatially non-local property of a surface—and also that the condition for being outer trapped is substantially easier to fulfill.

9. Acknowledgments

I thank José Senovilla for a very enjoyable collaboration, Emma Jakobsson for permission to include some results from her Master's thesis, and the Swedish Research Council for support under contract VR 621-2010-4060.

References

1. J. Thornburg, *Event and apparent horizon finders for 3+1 numerical relativity*, Living Rev. Relativity **10** (2007) 3; http:// www.livingreviews.org/lrr-2007-3.
2. S. W. Hawking and G. F. R. Ellis: *The large scale structure of space-time*, Cambridge UP, 1973.
3. R. P. A. C. Newman, *Topology and stability of marginal 2-surfaces*, Class. Quant. Grav. **4** (1987) 277.

4. M. Kriele and S. A. Hayward, *Outer trapped surfaces and their apparent horizon*, J. Math. Phys. **38** (1997) 1593.
5. M. Mars and J. M. M. Senovilla, *Trapped surfaces and symmetries*, Class. Quant. Grav. **20** (2003) L293.
6. M. Eichmair, *The Plateau problem for marginally trapped surfaces*, eprint arXiv:0711.4139, J. Diff. Geom, to appear.
7. L. Andersson and J. Metzger, *The area of horizons and the trapped region*, Commun. Math. Phys. **290** (2009) 941.
8. A. Carrasco and M. Mars, *Stability in marginally outer trapped surfaces in spacetimes with symmetries*, Class. Quant. Grav. **26** (2009) 175002.
9. A. Ashtekar and G. J. Galloway, *Some uniqueness results for dynamical horizons*, Theor. Math. Phys. **9** (2005) 1.
10. L. Andersson, M. Mars, and W. Simon, *The time evolution of marginally trapped surfaces*, Class. Quant. Grav. **26** (2009) 085018.
11. D. Christodoulou: *The formation of black holes in general relativity*, Monographs in Mathematics, EMS 2009.
12. F. J. Tipler, *Black holes in closed universes*, Nature **270** (1977) 500.
13. S. A. Hayward, *General laws of black-hole dynamics*, Phys. Rev. **D49** (1994) 6467.
14. A. Ashtekar and B. Krishnan, *Isolated and dynamical horizons and their applications*, Living Rev. Relativity **7** (2004) 10; http://www.livingreviews.org/lrr-2004-10.
15. I. Booth, *Black hole boundaries*, Can. J. Phys. **83** (2005) 1073.
16. D. R. Brill and R. W. Lindquist, *Interaction energy in geometrostatics*, Phys. Rev. **131** (1963) 471.
17. C. W. Misner, *The method of images in geometrostatics*, Ann. Phys. **24** (1963) 102.
18. G. W. Gibbons, *The time symmetric initial value problem for black holes*, Commun. Math. Phys. **27** (1972) 87.
19. A Čadež, *Apparent horizons in the two-black-hole problem*, Ann. Phys. **83** (1974) 449.
20. N. T. Bishop, *The closed trapped region and the apparent horizon of two Schwarzschild black holes*, Gen. Rel. Grav. **14** (1982) 717.
21. M. Bañados, C. Teitelboim, and J. Zanelli, *The black hole in three-dimensional spacetime*, Phys. Rev. Lett. **69** (1992) 1849.
22. S. Åminneborg and I. Bengtsson, *Anti-de Sitter quotients: When are they black holes?*, Class. Quant. Grav. **25** (2008) 095019.
23. D. R. Brill, *Multi-black-hole geometries in (2+1)-dimensional gravity*, Phys. Rev. **D53** (1996) 4133.
24. A. Steif, *Time-symmetric initial data for multi-body solutions in three dimensions*, Phys. Rev. **D53** (1996) 5527.
25. S. Åminneborg, I. Bengtsson, D. R. Brill, S. Holst, and P. Peldán, *Black holes and wormholes in 2 + 1 dimensions*, Class. Quant. Grav. **15** (1998) 627.
26. E. Jakobsson, *Trapped surfaces in 2 + 1 dimensions*, Master's Thesis, Stockholm Univ., 2011.
27. S. Holst and P. Peldán, *Black holes and causal structure in anti-de Sitter isometric spacetimes*, Class. Quant. Grav. **14** (1997) 3433.
28. R. Penrose, *Naked singularities*, Ann. N. Y. Acad. Sci. **224** (1973) 125.
29. K. P. Tod, *Penrose's quasi-local mass and the isoperimetric inequality for static black holes*, Class. Quant. Grav. **2** (1985) L65.
30. G. W. Gibbons, *The isoperimetric and Bogomolny inequalities for black holes*, in T. J. Willmore and N. J. Hitchin (eds.): Global Riemannian Geometry, Ellis Horwood, Chichester 1984.

31. M. A. Pelath, K. P. Tod, and R. M. Wald, *Trapped surfaces in prolate collapse in the Gibbons-Penrose construction*, Class. Quant. Grav. **15** (1998) 3917.
32. A. Papapetrou, *Formation of a singularity and causality*, in N. Dadhich et al. (eds): A Random Walk in Relativity and Cosmology, Wiley 1985.
33. I. Bengtsson and J. M. M. Senovilla, *The region with trapped surfaces in spherical symmetry, its core, and its boundary*, Phys. Rev. **D83** (2011) 044012.
34. I. Bengtsson and J. M. M. Senovilla, *A note on trapped surfaces in the Vaidya solution*, Phys. Rev. **D79** (2009) 024027.
35. J. E. Åman, I. Bengtsson, and J. M. M. Senovilla, *Where are the trapped surfaces?*, J. Phys.: Conf. Ser. **229** (2010) 012004.
36. A. B. Nielsen, M. Jasiulek, B. Khrishnan, and E. Schnetter, *The slicing dependence of non-spherically symmetric quasi-local horizons in Vaidya spacetimes*, Phys. Rev. **D83** (2011) 124022.
37. I. Ben-Dov, *Outer trapped surfaces in Vaidya spacetimes*, Phys. Rev. **D75** (2007) 064007.
38. R. M. Wald and V. Iyer, *Trapped surfaces in the Schwarzschild geometry and cosmic censorship*, Phys. Rev. **D44** (1991) R3719.
39. J. Plateau, *Sur les figures d'equilibre d'une masse liquide sans pésanteur*, Mém. Acad. Roy. Belgique Vol **23** (1849).
40. H. B. Lawson: *Lectures on Minimal Submanifolds Vol I*, Publish or Perish, Berkeley 1980.
41. H. B. Lawson, *Compact minimal surfaces in S^3*, in *Proc. Symp. Pure Mathematics* Vol. XV, AMS, Providence 1970.
42. G. T. Horowitz, *The positive energy theorem and its extensions*, in E. J. Flaherty (ed.): Asymptotic Behavior of Mass and Spacetime Geometry, Springer, Berlin 1984.
43. I. Ben-Dov, *Penrose inequality and apparent horizons*, Phys. Rev. **D70** (2004) 124031.
44. S. Holst, *Horizons and time machines*, PhD thesis, Stockholm Univ., 2000.
45. S. Deser, R. Jackiw, and G. 'tHooft, *Three-dimensional Einstein gravity: Dynamics of flat space*, Ann. Phys. **152** (1984) 220.
46. D. M. Eardley, *Black hole boundary conditions and coordinate conditions*, Phys. Rev. **D54** (1996) 4862.
47. R. Geroch, E. H. Kronheimer, and R. Penrose, *Ideal points in space-time*, Proc. R. Soc. Lond. **A327** (1972) 545.